International Conference on Road and Airfield Pavement (ICPT 2019)

Steering a Global Pavement Sustainability through Engineering Marvels

IOP Conference Series: Materials Science and Engineering
Volume 1075

Kuala Lumpur, Malaysia
10 - 12 July 2019

Editors:

Ratnasamy Muniandy
Salihudin Hassim
Fauzan Mohd Jakarni
Mohd Shahrizal Ab Razak

ISBN: 978-1-7138-3346-8
ISSN: 1757-8981

Printed from e-media with permission by:

Curran Associates, Inc.
57 Morehouse Lane
Red Hook, NY 12571

Institute of Physics
Dirac House, Temple Back
Bristol BS1 6BE UK

Phone: 44 1 17 929 7481
Fax: 44 1 17 920 0979

techtracking@iop.org

Additional copies of this publication are available from:

Curran Associates, Inc.
57 Morehouse Lane
Red Hook, NY 12571 USA
Phone: 845-758-0400
Fax: 845-758-2633
Email: curran@proceedings.com
Web: www.proceedings.com

TABLE OF CONTENTS

IOP Conf. Series: Materials Science and Engineering 1075 (2021) 011001 doi:10.1088/1757-899X/1075/1/011001

11th International Conference on Road and Airfield Pavement Technology (11th ICPT 2019)
"Steering A Global Pavement Sustainability Through Engineering Marvels"
10-12 July 2019 Kuala Lumpur, Malaysia

Preface

Universiti Putra Malaysia is privileged to host the **11th International Conference on Road and Airfield Pavement Technology (11th ICPT 2019)** at Seri Pacific Hotel, Kuala Lumpur, Malaysia from 10th to 12th July 2019 under the International Committee on Pavement Technology (ICPT). The ICPT is a non-profit governmental organization that was established in 1992 by the National University of Singapore. The conference is a longstanding event that brings together road and airport pavement researchers, engineers, and government officials to promote up-to-date, highly relevant and practical pavement technological advancement worldwide.

Revolve around the main theme of **Steering a Global Pavement Sustainability through Engineering Marvels**, four keynote addresses and 81 technical papers from local and international participants were presented, concerning new indicators and directions relating to the research, development, cutting-edge technologies, construction, management and maintenance of road and airfield pavement engineering. The conference has benefited those in the industry, be they road concessionaires, expressway regulators, related government authorities, consultants, contractors, project managers and engineers from public and corporate sector organisations.

The main objective of ICPT is to promote pavement technological advancement worldwide, with special emphasis in pavement technological development in developing countries by encouraging exchange and transfer of pavement technology among professionals and researchers of different countries and regions. This event received highly encouraging response from several countries such as Indonesia, Malaysia, Thailand, Philippines, Australia, Korea, China, India, Iraq, Japan, Europe, North America, Sri Lanka and Singapore.

Technical papers have undergone a rigorous peer-review process, based on initial Editors' screening, anonymous refereeing by independent expert referees, and consequent revision by the article author(s) when required. Editors would like to thank the author(s) who have contributed to the Conference Proceedings, and also to all delegates for their support towards the successful organization of the conference. Thanks to the outstanding stakeholders such as Public Work Departments of Malaysia (JKR), Construction Industry and Development Board (CIDB) and Malaysian Highway Authority (LLM) for their endless support. Our special thanks also go to the sponsors and booth exhibitors for the financial assistances and continuous support to ensure that this event comes into reality. Also, the Editors would like to extend gratitude to the Program Committee for their devoted assistance and expertise towards the preparation of the conference proceedings as well as all the individual whom are directly and indirectly involved to make the conference possible.

The Editors:

Professor Ir Dr Ratnasamy Muniandy
Associate Professor Ir Salihudin Hassim
Associate Professor Dr Fauzan Mohd Jakarni
Dr Mohd Shahrizal Ab Razak

Department of Civil Engineering, Faculty of Engineering
Universiti Putra Malaysia (UPM)

IOP Conf. Series: Materials Science and Engineering 1075 (2021) 011001 doi:10.1088/1757-899X/1075/1/011001

11th International Conference on Road and Airfield Pavement Technology (11th ICPT 2019) Program Committee

International Steering Committee:

Chairman : **Professor T. F. Fwa**, *National University of Singapore*
Permanent Secretary General : **Dr Low Boon Hwee**, *Highway International Pte Ltd*

Advisors

Professor Dato' Ir Dr Mohd Saleh Jaafar, *Universiti Putra Malaysia*
Associate Professor Ir Dr Raizal Saifulnaz Muhammad Rashid, *Universiti Putra Malaysia*

Organising Committee

Co-Chairs I : **Professor Ir Dr Ratnasamy Muniandy**, *Universiti Putra Malaysia*
Co-Chairs II : **Associate Professor Ir Salihudin Hassim**, *Universiti Putra Malaysia*

Technical Committee

Associate Professor Dr Hussain Hamid, *Universiti Putra Malaysia*
Associate Professor Dr Law Teik Hua, *Universiti Putra Malaysia*
Associate Professor Dr Fauzan Mohd Jakarni, *Universiti Putra Malaysia*
Dr Mohd Shahrizal Ab Razak, *Universiti Putra Malaysia*
Dr Aidi Hizami Alias, *Universiti Putra Malaysia*

Secretariat

Nurul Suhada Mohammad Basir, *Universiti Putra Malaysia*
Norizzaida Ahmad, *Universiti Putra Malaysia*
YM Raja Muhammad Raja Muzamir Raja Mahadi, *Universiti Putra Malaysia*
Norliyana Abd. Karim, *Universiti Putra Malaysia*

IOP Conf. Series: Materials Science and Engineering 1075 (2021) 011002 doi:10.1088/1757-899X/1075/1/011002

Peer Review Declaration

All papers published in this volume of *IOP Conference Series: Materials Science and Engineering* have been peer reviewed through processes administered by the Editors. Reviews were conducted by expert referees to the professional and scientific standards expected of a proceedings journal published by IOP Publishing.

- **Type of peer review:** Single-Blind.

All accepted papers have been peer reviewed through processes administered by the proceedings Editors. Reviews were conducted by expert referees to the professional and scientific standards.

All submitted papers were reviewed based on single-blind review. Authors' identities were known to the reviewers; however, reviewers' identities were hidden from authors.

Two (2) peer reviewers were responsible to review and evaluate one (1) manuscript based on pre-determined criteria and aspects as listed in the evaluation form (as attached). The form needs to be submitted via online platform (**zoho.com**) using the given submission link.

The criteria include as follows.

i. **Title and Overall content** (importance and relevance of the title and overall content to the conference/seminar theme, abstract and keywords contains sufficient overview to represent content of the paper)
ii. **Organization and Format of the Paper** (organization of the paper in terms of sections or subsections and list of figures and tables, the paper in terms of quality of standardized style, page layout, ease of reading and punctuation, paper is free from spelling, grammar and syntax error and enjoyable to read),
iii. **Contents of the Paper** (introduction, literature review, methodology, results and discussions, conclusions and recommendations, references)
iv. **Nomination for the Best Paper Award**.

A Likert scale with a quality range from 1 (Poor) to 5 (Excellent) was used. Also, the reviewers were requested to recommend for **ACCEPTANCE** or **REJECTION** for respective paper to be presented at the conference.

Only one (1) paper was found not suitable (**REJECTED**) for presentation at the conference. However, the author(s) has been given a chance to improve the paper based on the comments and suggestions from the reviewers thus was accepted after revision.

Author(s) of the accepted papers were required to prepare a camera-ready version for inclusion in the Conference/Seminar Proceedings based on the given guidelines of the conference proceedings. Technicality check was then done by the technical conference committee.

- **Conference Submission Management System:** zoho.com and google form

IOP Conf. Series: Materials Science and Engineering 1075 (2021) 011002 doi:10.1088/1757-899X/1075/1/011002

Two online platforms for papers submission and reviewers' responses submission were used; (i) **zoho.com** and (ii) **Google Form**.

Reviewer's responses were recorded in zoho.com and accepted papers were submitted through Google Form.

- **Number of submissions received:** 40 papers
- **Number of submissions sent for review:** 40 papers
- **Number of submissions accepted:** 40 papers
- **Acceptance Rate (Number of Submissions Accepted / Number of Submissions Received X 100):** 100%
- **Average number of reviews per paper:** 2 reviewers
- **Total number of reviewers involved:** 20
- **Any additional info on review process:** None
- **Contact person for queries: ratnas@upm.edu.my; fauzan.mj@upm.edu.my; hsalih@upm.edu.my; ar_shahrizal@upm.edu.my**

IOP Conf. Series: Materials Science and Engineering 1075 (2021) 012001 doi:10.1088/1757-899X/1075/1/012001

High Performance Asphalt for Airfield Pavement

Xavier Guyot*

TIPCO ASPHALT, 118/1 Rama 6 Road Bangkok, Thailand
* Corresponding author: guyot@tipcoasphalt.com

ABSTRACT: Due to the expected increase in air traffic movements around the world and its impact on the environment, the requirement to build more sustainable and low maintenance airfield infrastructures has become very important. One solution consists in using high performance asphalt materials in order to provide a better sustainable whole life cost solution for these pavements.Performances of binder play a key role in the performances of the asphalt pavement as the properties of both are linked. Using Polymer Modified Bitumen Asphalt materials for runway and taxiway represents an alternative solution to traditional Marshall Asphalt (MA). This type of products for wearing course and binder can improve considerably the airfield pavement performances regarding the Rutting resistance, thermal susceptibility or even Fuel resistance.Due to new generation of aircraft more and more aggressive, use of rut resistance asphalt base course is also mandatory. Asphalt Concrete manufactured with Hard Grade bitumen like "High Modulus Asphalt" with improved mechanical characteristics can lead to longer service life, by preserving resources and bringing also cost savings as maintenance is reduced. This type of Asphalt Concrete showed on major airport projects his potential as a viable alternative to the traditional airfield pavement materials for base course such as Crushed Rocks or Dense Bitumen Macadam.

1. Introduction

Most of the world's paved roads are surfaced with asphalt, which gives good performance and durability under the most heavily trafficked conditions. Asphalt mixes are also widely used in the construction of hard standing and parking areas for both light and heavy vehicles. They are therefore eminently suitable for use in the construction and surfacing of access roads, perimeter roads and vehicle parking areas on airfields.

Asphalts give good performance and durability under the wide range of climatic and traffic conditions. Asphalt is also widely used in airport construction, although this usage is not always recognized. A recent survey shows that a majority of airports are in fact constructed with asphalt pavement. For example, runways require good skid-resistance and surface water drainage for good braking while avoiding aquaplaning, an even surface regularity to ensure passenger comfort and minimum risk of damage to delicate electronic components and adequate strength to support the high wheel-loadings of modern aircraft. For aircraft parking areas the main requirement is adequate stability under high wheel-loadings; the principal considerations are adequate stability under wheel-loads and heavy point loads from maintenance machinery as well as good resistance to oil and fuel spillage. Asphalt mixes can be very suitable for some of these applications but less suitable for others and premature failures can occurred if inadequate products and pavement design method are used.

This paper focusses on the presentation of characteristics and performances of asphalt materials using specific type of bitumen and their impact on asphalt pavements durability illustrated by some case studies.

2. Key Parameters For Airfield Pavement

2.1 Adequate Pavement Design Method

(a) French method – STAC Guide

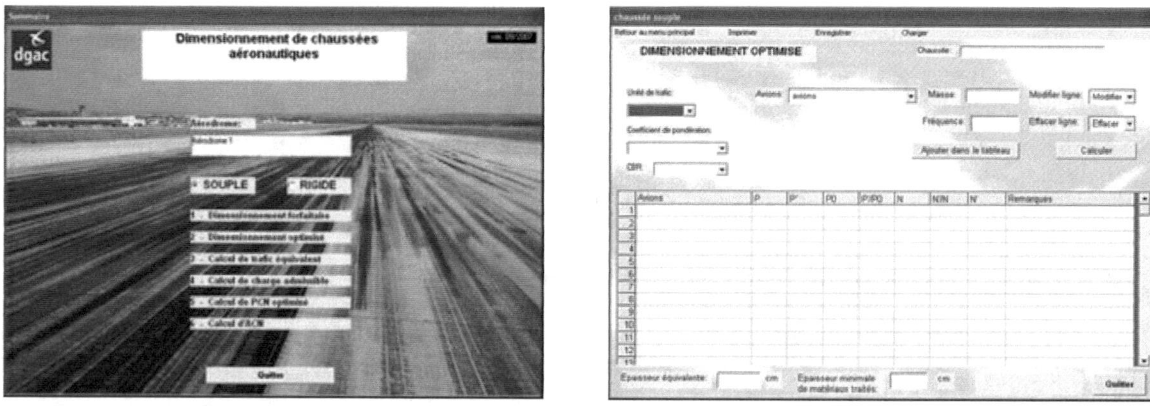

Figure 1. Representation of the DCA software

The official French method is established by the STAC (Civil Aviation Technical Service). The flexible structure is designed by the CBR method in which an equivalent theoretical thickness and a linked thickness of the materials are determined to protect the support soil under the repeated passage of the traffic.

These two minimum allowable thicknesses are determined with calculation charts (programmed since 2007 in the DCA software). Aeronautic structure design is purely empirical and the materials are distinguished by coefficients of equivalence.

The structure to be implemented was established by weighting the thicknesses of the layers by the corresponding coefficients of equivalence so that these two real thicknesses were less than the allowable thicknesses.

The coefficients of equivalence were determined for each type of material based on the American AASHO tests carried out in the 1960s in the United States. The mechanical qualities of the materials evolved, particularly with the introduction of high modulus materials. The STAC therefore carried out life-size experiments at the end of the 1990s in order to take these new materials into account in the pavement design method.

The new instructions thus define coefficient of equivalence for High Performances Asphalt Concrete allowing for the optimization of the pavement structures:

(b)American method - FAARFIELD

The method developed by the FAA includes structural design calculation software called FAARFIELD. This method is not just used in the United States; it has been exported to many other countries.

The thickness to be used is determined by the finite element method in which the structure design criterion is that the stress created under the most constraining landing gear is less than an allowable value which is dependent on the resistance to bending.

Each layer is characterized here by its modulus and its Poisson ratio. The pavement thickness is set so that the CDF (cumulative damage factor), i.e. the sum of the damage caused by all of the aircraft, is less than 1. For each aircraft, the surface deformation of the platform is calculated with a linear elastic model. From an empirical fatigue relationship, we can derive an allowable aircraft traffic level. The elementary damage associated with this aircraft is the ratio between the real traffic and the allowable traffic.

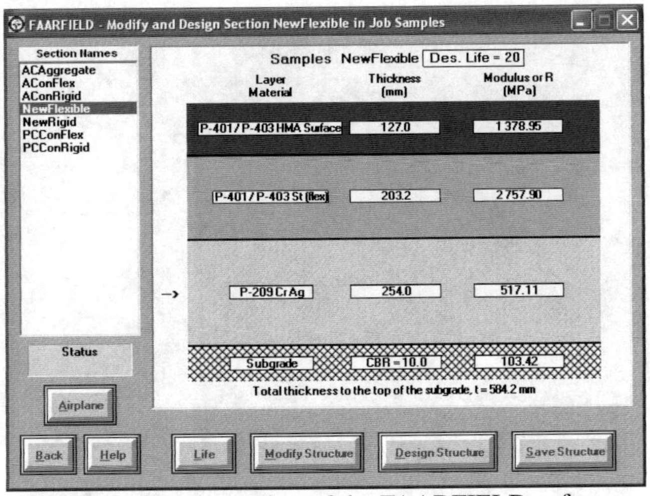

Figure 2. Representation of the FAARFIELD software

This approach is valuable firstly because the feedback from experience is substantial and also because it is based on rational considerations, even though empirical aspects remain. It makes it possible to take into account the real values of the moduli of the materials (as long as the temperature and frequency conditions to be taken into account are set) and to thus make use of High Performances Asphalt using Polymer Modified Bitumen or Hard Grade Binder.

2.2 *Performances based specifications of Asphalt Concrete*

Up to now specifications of Hot Mixes in most of the countries worldwide are still based on the Marshall Method and empirical recipes. Even if this approach has demonstrated its suitability until a recent past it finds its own limits especially in cases of ultimate conditions of traffic In order to meet highest requirement asphalt pavement layers must

i. Resist to rutting at high service temperature under high axle load (Bitumen stiffness)
ii. Resist to fatigue cracking (use of PMB)
iii. Be workable for easy laying and compaction
iv. Be waterproof for preserving underneath structure
v. Be sustainable and resist to the abrasion and effects of water
vi. Be cost effective

Meeting the above requirements imply to re-consider the method of designing and providing predictive testing methods based on performances specifications. In Europe, the formulation method for asphalt mixes used is defined by standards. It is characterized by an approach based as much as possible on the performances of the mixture. For materials that play a structural role, it can be classified in the "fundamental" approach. Asphalt Concrete follows the requirements of the following European standards:

a. NF EN 13108-1 concerning the specifications for asphalt mixes
b. NF EN 13108-20 concerning the formulation method

The mix design method requires specifications on the components and particularly the aggregate, but it involves 5 tests which define the level of the study to perform on the asphalt mix:

a. The Gyrator Compaction test and the water resistance test -> level 1
b. Resistance to rutting -> level 2
c. The stiffness modulus -> level 3
d. Resistance to fatigue -> level 4

Figure 3: The Gyratory Compaction Test EN 12697-31 and Rutting Test Device EN 12697-22

Superpave Mix Design method in the United States introduce also this type of "performances" testing for Asphalt Concrete, not anymore only based on Marshall method.

3. High Performance Asphalt

3.1 High Modulus Asphalt for Base Course

The worldwide trend tends to substitute soft binders by harder grades more adapted to stringent and severe conditions. Bitumen grades can be specified by their penetration, viscosity or Performance Grade. Most of local specifications now have extended their range to harder grades up to 20/30 or 40/50 (penetration grade) or higher Performance Grade (like PG 70-22 in the USA) and even 10/20 or 20/30 in Europe.

Use of harder penetration bitumen grade to strengthen the asphalt layers and thus reducing the risk of rutting is now the trend. It applies to all asphalt layers. Authorities have gradually switched from 80/100 pen bitumen to 35/50 within 25 years. This grade represents now about 70% of asphalt pavement to date in France on national roads.

Harder grades (10/20-20/30) are now commonly used for base course in specific situations such as roundabouts, heavy trafficked lanes, trucks parking, toll gate and industrial platform where standards grades of bitumen can't withstand such high stresses and especially to design High Modulus Asphalt, called "Enrobé à Module Elevé" in France (EME).

The formulation of "Enrobé à Module Elevé" EME is closed to traditional Bituminous Base Asphalt. The aggregate mix is usually continuous graded with approximately 60-65% chippings and the filler content stands around 7%. The current chipping sizes are 10, 14 or 20mm but the most commonly one is 14mm.

Designing Hot Mix Asphalt according to the European standardization is based on mechanical performances. Freedom is given to select the appropriate aggregate mix, bitumen grade, source and dosage to achieve the following requirements given in the table 1 below.

IOP Conf. Series: Materials Science and Engineering 1075 (2021) 012001 doi:10.1088/1757-899X/1075/1/012001

Table 1. Comparative performances GB (Bituminous Base Asphalt) vs EME - NF EN 13108-1

		GB2	GB3	GB4	EME2
Water resistance Duriez ratio r/R		> 0.65	> 0.70	> 0.70	> 0.75
Rutting Test @ 60oC - after 10,000 cycles - after 30,000 cycles	 % %	 < 10	 < 10	 < 10	 < 7.5%
Complex Modulus at 15oC/10Hz	Mpa	> 9,000	> 9,000	> 11,000	> 14,000

To achieve a good workability at the application the mixing temperature of EME should be around 180oC corresponding to the viscosity that provides good coating (0.2 Pa.s). Any type of mixing plant is suitable provided that the burner is set for this temperature.

Regarding airfield pavement, use of "Enrobé à Module Elevé" as base course offer the following benefits:

1. Better strength and durability leading to at least 15% reduction in layer thickness,
2. Better rutting resistance than traditional asphalt,
3. Durable and long life material, therefore maintenance and repair may be limited only to surface course only at longer intervals,
4. Ease of laying with consequence increased productivity, reduction in construction period and construction cost together with less disruption to airport operations

3.2 Polymer Modified Asphalt for Wearing Course

Typical airfield flexible pavement construction can consist of asphalt layers using Dense Marshall Asphalt surface course, which generally provides poor ride quality and skid resistance.

Low skidding resistance at the runway pavement surface represents a major hazard for the aircraft traffic operations in wet weather condition. One of the most important criteria during routine maintenance is the requirement for longitudinal wet friction coefficient. Traditional Marshall Asphalt cannot provide sufficient skid resistance unless they are grooved.

However new standard asphalt material used in Europe are able to provide high performances for wearing course for airfield pavement, especially for runway by using Polymer Modified Bitumen.

The most widely polymer used is the SBS (Styrene–Butadiene–Styrene) because of higher performances provided. It represents more than 80% of the PMB market. Performances are not proportional to the amount of polymer as many parameters have an influence on the final quality (type of polymer, type and origin of the bitumen).

Table 2. Properties of pure binder vs Polymer Modified Binder

PARAMETERS	**INDICATOR ***	**PURE BITUMEN 30/50**	**SBS-MODIFIED BINDER 1.5 %**	**SBS-MODIFIED BINDER 4%**
Adhesion	% of covered surface	50%	90%	90%
Complex modulus	T(°C) for G*/sinδ = 1kPa (DSR SHRP)	71.5°C	72.1°C	76.5°C
Flexibility - Cracking at low temperature	T(°C) for breaking elongation 1 % direct tension (SHRP)	-16.1°C	-19.5°C	-26.4°C
Fatigue Resistance	Elastic recovery at 10°C	Fragile (breaks upon	59.8%	89.0%

Auto recovery		elongation)		
Resistance to Ageing	Residual penetration at 25°C after RTFOT + PAV	38%	not measured	57%
Cohesion	Conventional energy at breaking (5°C, 100mm/min)	< 10 J/cm2	16 J/cm2	33 J/cm2

Table 2 clearly demonstrates the added value of SBS elastomer for all parameters. It is important to mention that SBS action is significant at both high and low temperatures.

The official French Civil Aviation Authorities edit a guide with recommendations regarding bitumen mixtures to use for wearing course for asphalt pavement. The main conclusion is that Polymer-modified bitumen should be preferred for surface courses when the requirement is for:

1. an improvement in mechanical performance, in terms of - elasticity and flexibility
2. greater resistance to rutting
3. reduced sensitivity to surface ripping and shearing efforts,
4. greater resistance to hydrocarbons,
5. enhanced durability (due to the thicker coating of binder), with properties unchanged (resistance to rutting, macrotexture maintained, etc.),
6. greater surface cohesion
7. reduced sensitivity to high and low temperatures.

4. Some Case Studies Worldwide

4.1 SSR Airport projects – Mauritius Island

The Runway Overlay and Parallel Taxiway Projects were funded by Airports of Mauritius Co Ltd (AML) in 2012.

The Runway Overlay project (value 25 Million Euros – duration 12 months) included overlaying the existing runway over a length of 3,370m, widening of existing shoulders by 7.5m on each side and reconstruction and widening of existing turn pads. The Parallel Taxiway project (value 40 Million Euros – duration 18 months) included the new construction of a Code F compliant Taxiway, 2,300m long x 60m wide and associated exit links

Figure 4. Aerial view of SSR Airport

Airfield pavement design was based on the use of innovative Hot Mix Asphalt as Enrobé à Module

IOP Conf. Series: Materials Science and Engineering 1075 (2021) 012001 doi:10.1088/1757-899X/1075/1/012001

Elevé "EME" for the Base Course instead of using Pavement Quality Concrete or traditional Bituminous Bound macadam base. Pavement design was made according to the US Federal Aviation Administration (FAA) method, using the FAARFIELD software, following the Advisory Circular 150/5320-6E Airport Pavement Design and Evaluation. Structures were designed so that the pavements could bear the same aircraft movements expected for the next 20 years. For the alternative design, due to the high temperatures in Mauritius and the low frequency of solicitation, the modulus values were reduced to 6000 MPa for the base course and 1400 MPa for the wearing course. Structure.

Table 3 . Pavement structure – New Taxiway

Course	New taxiway - Subgrade CBR > 15
Wearing	50 mm Marshall Asphalt
Binder	120 mm EME
Base	150 mm Cement Bound
Sub Base	100 mm Granular

Structures using EME layers provide the following advantages:
1. Improved mechanical characteristics leading to pavement thickness reduction and resource preservation (crucial issue for small areas like Mauritius Island)
2. Easy installation and lower maintenance requirements
3. Financial savings (8 % of the cost of pavement structure)
4. Earlier completion time (3 months for each project)

4.2 SIA – Bangkok - Thailand

Thai Slurry Seal, construction branch the building sector subsidiary of Tipco Asphalt, has carried out the partial renovation of the surface of Runway 19L of Suvarnabhumi International Airport in Bangkok, at a cost of 4 million euros in 2017.

The work concerned a section of 1 km, a quarter of the total runway length, with replacement of the pavement structure using hot-mix asphalt over a width of 60 meters, the depth being 55 cm for the 30 meters in the centre and only 10 cm for a 15-meter strip either side.

Every year Thai Slurry Seal performs regular maintenance work for the airport, but this project had certain specific characteristics:
1. Logistic constraints due to working in an airport environment
2. Tight timetable (2 months) to plane mill the 16,000 m3 of existing pavement and lay down 42,000 tonnes of hot-mix asphalt over a substantial depth
3. Use of Modified hot-mix asphalt to meet high performance criteria fixed by the client in terms of rutting resistance (less than 2.5% at 100,000 cycles for the base course) and rigidity stiffness modulus over a broad wide temperature range (from 15°C to 40°C)

Figure 5. SIA Runway renovation project

IOP Conf. Series: Materials Science and Engineering 1075 (2021) 012001 doi:10.1088/1757-899X/1075/1/012001

Highly Modified PMB meting PG 76 criteria has been used for the wearing course and hard grade binder for the Base course, providing exceptional rutting resistance and high value of modulus. Pavement design was made according to the US Federal Aviation Administration (FAA) method, using the FAARFIELD software. The project was delivered a week early in advance with the complete satisfaction of the client.

References

[1] LPC Bituminous Mixtures Design Guide The RST Working Group "Design of bituminous mixtures"
[2] Jean-Luc DELORME, Chantal de la ROCHE, Louisette WENDLING September 2007
[3] European standard EN 13108-1 – Bituminous mixtures – Material Specification
[4] Part 1: Asphalt Concrete
[5] European standard EN 13108-20 – Bituminous mixtures – Material Specification
[6] Part 20: Type testing
[7] US Federal Aviation Administration (FAA) method, Advisory Circular 150/5320-6E
[8] Airport Pavement Design and Evaluation
[9] Bituminous mixtures and surface dressings for airport pavements - Guide to the application of standards
[10] STAC - French civil aviation technical center – 2nd edition – April 2009

IOP Conf. Series: Materials Science and Engineering 1075 (2021) 012002 doi:10.1088/1757-899X/1075/1/012002

Evaluation of controlled low strength materials containing red brick construction waste based on mechanical experiments and BP neural network

J G Wang[1,2] and J X Zhang[1*]

[1] Beijing Key Laboratory of Transportation Engineering, Beijing University of Technology, Beijing 100124, China
[2] Solid waste R & D Center, Beijing Urban Green Source Environmental Protection Technology Co., Ltd, Beijing 102601, China
*Corresponding author: zhangjinxi@bjut.edu.cn

Abstract. In this study, two kinds of controlled low-strength materials (common and rapid-hardening) were designed and prepared by using red brick construction waste fines. Fourteen mixture specimens were prepared with the control parameters of binder to construction waste fine aggregate ratio (α), accelerator to binder ratio (β), and water to solid ratio (γ). The flowability of fresh mixtures, compressive strength of hardened materials were studied. Test results show that it is applicable to use the red brick construction waste fine aggregate to prepare controlled low-strength materials. Increasing α and γ can increase the flowability, however with different the influence mechanisms, where γ has a more significant effect. The 2 hour-strength of rapid-hardening CLSM (controlled low strength material) based on sprayed concrete accelerator can be up to 0.41~0.79MPa, which meet the needs of emergency backfill engineering. Using on the BP neural network, the long-term strength prediction model of CLSM is established, which can help the civil engineers in the material mix-design.

1. Introduction

The construction of large-scale new buildings and the demolition of abandoned buildings have consumed a large amount of natural stone and generated huge construction waste. It is estimated that by 2020, the amount of construction waste in China will reach 30 billion m^3 [1]. Over-exploitation of sand and gravel materials will cause serious damage to cultivated land and vegetation, also the gravel material is a kind of non-renewable natural resources. To solve such problems, the treatment of construction wastes become necessary and urgent. In the recycled aggregates, better-quality recycled coarse aggregate (greater than 5 mm in diameter) has been used for the production of recycled cement concrete [2, 3], recycled blocks [4], etc., but the using of recycled fine material (particle size less than 5mm) with a large amount of red bricks, mortar, and dregs, etc. for recycling is very limited.

According to American Concrete Institute (ACI) Committee 229 [5], controlled low strength material (CLSM) is a self-compacting, self-leveling cement-based material that can replace the traditional backfill materials. CLSM typically consists of cement, fly ash, fines, etc., in which the cement can be partly replaced by by-products (as cementitious materials) such as slag [6], stainless steel slag [7], etc. Fly ash is a mineral admixture to improve workability, and enhance strength. With the largest amount of fine aggregates and higher freedom of material selection, it can make full use of local rich waste, such as surplus clay [8], solid wastes from paper mills [9], waste water treatment sludge [6], etc. In addition, it is convenient for pouring, without vibration or compaction, reducing the use of labor and construction equipment, etc., where these advantages make it widely used in pavement base, bridge abutments fill and trench backfill [5, 9].

In addition, to reduce the disturbance of construction on the daily life of urban residents, it is necessary to carry out rapid construction. To meet the needs of different backfilling projects for hardening time and early strength, two types of CLSM (common and rapid-hardening) specimens are prepared using red brick construction waste fines, in which the early strength properties of rapid-hardening type are realized on the basis of spray concrete accelerator. Through testing the flowability of fresh CLSM and the compressive strength of hardened materials, the engineering properties are investigated and the feasibility of application to different backfill projects is discussed.

2. Experiments

2.1. Material design

The construction waste fine aggregate comes from a construction waste recycling plant in Shijingshan District, Beijing. Its distinctive feature is rich in red brick components. The screening results of the construction waste are shown in figure 1, in which the upper and lower limits of the screening curve are the grading ranges of the concrete fine materials specified in ASTM C33/C33M-13 [10]. The fine powder content (particle size less than 0.075mm) in construction waste fine aggregate is higher (12.2%), so it is not suitable to prepare recycled concrete, but it is very suitable for improving the cohesiveness and water retention of CLSM [11].

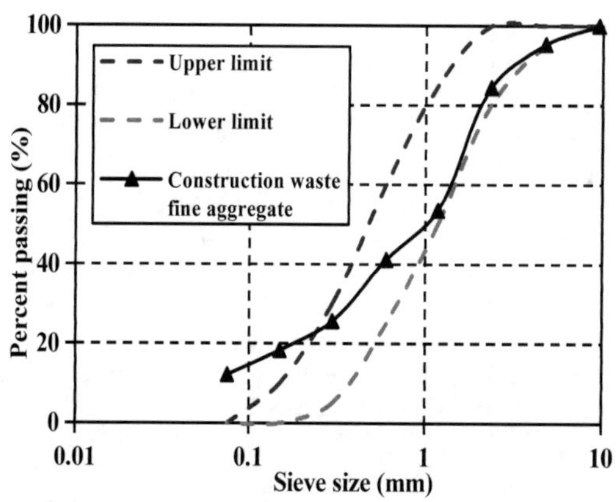

Figure 1. Sieving curve of construction waste fine aggregate

Radioactive substances exist widely in various building materials. Excessive irradiation may cause certain damage to human health [12]. In accordance with GB 6566-2010 [13], the safety inspection of building waste fine materials is executed, and the results are shown in Table 1. From Table 1, it can be seen that the radioactive indexes of construction waste fines meet the specifications, which is mainly due to the sources of construction waste, such as ordinary buildings, transportation infrastructure, etc. These construction waste does not contain special waste like those in the nuclear reactor protection facilities, indicating that the radiation effects of CLSM prepared by construction wastes in this study are safe to the human body.

Water may leach out of construction waste. Leaching harmful substances (hazardous heavy metals and anions, etc.) will migrate and transform, contaminate soil and groundwater, indirectly pollute crops, and threaten the human health. Therefore, according to GB 5085.3-2007 [14], the leaching toxicity detection of construction waste fines was carried out with an inductively coupled plasma atomic emission spectrometer. From table 1, it can be seen that the content of harmful components such as heavy metals in construction waste fines satisfies with the requirements of the standard. Construction waste will not cause secondary pollution to the environment when it is used in backfill projects as the component of CLSM.

Table 1. Detection results of radioactivity and leaching toxicity of construction waste fine materials

Testing items		Test results	Standard requirements [13, 14]
Radioactivity index	Internal exposure index	0.2	≤1.0
	External exposure index	0.4	≤1.0
Concentration limit of hazardous components in leachate（mg/L）	Cuprum, Cu	<0.0001	≤100
	Zinc, Zn	0.023	≤100
	Cadmium, Cd	<0.0001	≤1
	Arsenic, As	<0.0001	≤5
	Hydrargyrum, Hg	<0.0001	≤0.1
	Plumbum, Pb	<0.0001	≤5
	Nickel, Ni	0.0004	≤5
	Chromium, Cr	<0.0001	≤15

In this study, the cement used is 42.5R ordinary Portland cement, the initial setting and final setting time are 140min and 180min respectively, and the 3-day and 28-day strength are 29.2MPa and 53.8MPa respectively. The Class F fly ash with fineness of 35%, water requirement ratio of 110%, is applied in this study. The accelerator used is a spray concrete accelerator produced by a building material company in China. This material is a dry powder formed by grinding a special sulphoaluminate clinker as a base material, which is commonly used in shotcrete. The mixing water uses common tap water.

2.2. Mix proportions, specimen preparation and testing procedures

2.2.1. Mix proportions

Table 2. Mix proportions and fresh properties of CLSM

Type	Binder to CWFA ratio (α)	Accelerator to binder ratio (β)	Water to solid ratio (γ)	Mix proportion (kg/m^3)					Flowability (mm)
				C	A	F	CWFA	W	
Common	0.07	0%	0.28	85	0	182	1210	413	190
	0.10	0%	0.28	118	0	177	1180	413	205
	0.13	0%	0.28	150	0	173	1155	414	213
	0.15	0%	0.28	170	0	170	1135	413	215
Rapid-hardening	0.07	40%	0.28	51	34	182	1210	413	195
	0.10	40%	0.28	71	47	177	1180	413	200
	0.13	40%	0.28	90	60	173	1155	414	215
	0.15	40%	0.28	102	68	170	1135	413	220
	0.13	20%	0.28	120	30	173	1155	414	213
	0.13	30%	0.28	105	45	173	1155	414	209
	0.13	50%	0.28	75	75	173	1155	414	210
	0.13	40%	0.27	90	60	173	1155	399	192
	0.13	40%	0.29	90	60	173	1155	429	229
	0.13	40%	0.30	90	60	173	1155	444	263

Note: C = Cement, A = Accelerator, F = Fly ash, CWFA = Construction waste fine aggregate, W = Water, $\alpha = (m_C+m_A)/m_{CWFA}$, $\beta = m_A/(m_C+m_A)$, $\gamma = m_W/(m_C+m_A+m_F+m_{CWFA})$.

To meet different engineering application requirements, two types of CLSM (common and rapid-hardening) were prepared, in which the rapid-hardening CLSM added sprayed concrete accelerator to realize the early strength. According to the mixing ratio design method in literature [11], 14 mixture specimens were designed and prepared based on the control parameters of binder to CWFA ratio (α), accelerator to binder ratio (β), and water to solid ratio (γ), as shown in table 2. Fly ash can significantly improve the workability of fresh samples, and its amount should be 15% of the amount of construction waste fines [11]. By adjusting the amount of construction waste fines, the total volume of material per unit volume was controlled at 1860-1940 kg/m^3, where this range meet 1840-2230 kg/m^3 for normal CLSM [5].

2.2.2. Specimen preparation and testing procedures

The mixing of CLSM used a mortar mixer. According to the designed mix ratio, the weighed cement, accelerator, fly ash and dried-construction waste fines were added to the mixing pot, and then stirred thoroughly to the homogeneity. Finally, the mixing water was added and then stirred for 120 seconds. The flowability is one of the indispensable control indicators for CLSM engineering applications. Therefore, the flowability of the fresh mixture was tested with reference to ASTM D6103 [15] before pouring into the test mould. The test cylinder size was ø75mm×150mm.

Compressive strength is a typical mechanical indicator for evaluating the bearing capacity of hardened CLSM. According to ASTM D4832 [16], a MQS-2 road material performance tester was used with a loading rate of 1 mm/min. The size of cylinder specimen was ø100mm×200mm. Except for test specimens with early hour strength, other specimens were demoulded after 24 hours of pouring, and immediately moved into a humidity room until test. All test results were obtained by three parallel specimens.

3. Results and Discussion

3.1. Flowability

Flowability is a distinctive feature of CLSM that distinguishes it from traditional backfill materials. The good flowability enables it to be self-compacting without vibration in the pouring process. To ensure that the CLSM can maintain good workability, the minimum acceptable flowability in most applications is 180mm [8, 18]. As can be seen from Table 3, the flowability at different ratios was between 190mm and 263mm, which met the needs of general engineering. The flowability increased significantly with the increase of γ, and its growth tendency was significantly greater than the effect of α on flowability, which is similar to the results in the literature [8, 11]. The reason is that the increase of γ leads to the increase of the free water content in fresh samples, which causes the particles to be in a "suspended" state, and reduces the friction between the construction waste fines and the cohesion between the gelled particles. According to ACI 229R [5], flowability greater than 200mm is a high flowability and can meet the needs of backfill projects such as a narrow operating space or dead corners. Therefore, for projects with high filling performance requirements, γ should not be less than 0.28.

3.2. Compressive strength

3.2.1. Early strength characteristic

Figure 2 shows the early hour strength of CLSM after casting for different mixtures. The strength mainly attributes to the matrix skeleton effect of cement hydration products. The existence of much free water provides conditions for the good flowability of fresh mixtures, while it prolongs the condensation and solidification, and hinders the development of its early strength. Therefore, for the common CLSM (β = 0%), the early strength is 0 for specimens with different α. The accelerator can promote the rapid hydration reaction of the cementitious material, make the fresh mixture solidify and form the early bearing capacity. For the case of β = 40%, the early hour strength develops fastest at α=0.13, which is due to the total amount of cementitious material and the internal free water content. When α is 0.07 and

0.10, the hydration product cannot form a strong skeleton structure; and when α is 0.15, the reduction of construction waste causes the decrease of free water absorbed, resulting in the increase of free water inside the fresh material, which is harmful to the formation of early strength.

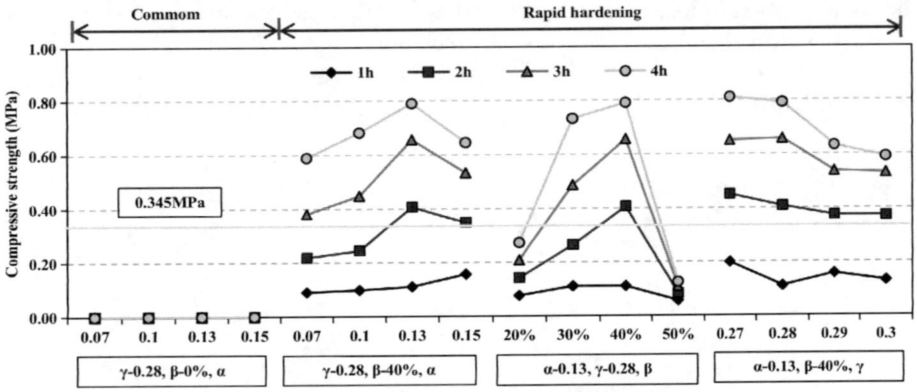

Figure 2. Early hour strength of different mixtures

When α (0.13) and γ (0.28) are constant, the 2-4 hours strength of CLSM at the concentration of 20%, 30%, and 40% of accelerator reaches 0.14-0.27MPa, 0.26-0.73MPa and 0.41-0.79MPa, respectively. However, the 4 hour-strength at 50% is only 0.13MPa, since the excessive accelerator generates insufficient hydration products for the content mismatch of CaO and SiO_2 [19]. When α (0.13) and β (40%) are fixed, the increase of γ (0.27-0.30) leads to the increase of free water inside the fresh mixture, and it takes more time to coagulate and solidify, thereby attenuating its early strength properties.

The rapid-hardening properties of CLSM can accelerate construction progress and save construction costs. Referring to ACI 229R [5], the strength of backfill compacted soil is about 0.3~0.7MPa. Meanwhile, the CLSM minimum strength required in most applications is 0.345MPa to ensure that can support the construction personnel load [8]. Therefore, the subsequent process can be carried out when the CLSM strength reaches 0.345MPa for the road base backfilling. For the accelerator content of 40%, when α is 0.13 or 0.15, the basic bearing capacity can be satisfied after 2 hours of pouring.

3.2.2. Long-term strength

The strength with different ratios at curing age 1d, 3d, 7d, 14d, 28d, 56d and 91d were tested and the results are shown in figure 3. From figure 3(a), it can be seen that the strength of common CLSM increased with the increase of α and curing age, and the strength of 91d can reach 2.16-7.55MPa. The variation of strength with age was faster in the early period and then slowed down, which is consistent with the characteristics of the cement hydration reaction. The excavation strength range of CLSM is 0.3-2.1MPa [5]. So for the pipeline backfilling project, α should not be greater than 0.07.

For the case of β=40%, the change of CLSM strength is similar to that of common type (see figure 3(b)), while it should be noted that the long-term strength with same α is slightly higher than the latter, indicating that the accelerator has a constant promoting effect on the long-term strength development. That is, when the design strength is fixed, the addition of accelerator can properly reduce the total amount of cementitious materials.

From figure 3(c), the strength of CLSM increased first and then decreased with the increase of accelerator content at the same age, and the strength with 40% accelerator was smaller than that of 30%, which is mainly because 40% accelerator promotes the early hydration reaction of the binder, resulting in a relative insufficiency of later hydration products. For the case of β=50%, the strength of 91d was less than 2MPa, which is related to the insufficient content of SiO_2 needed for later hydration reaction due to the excessive amount of accelerator.

According to figure 3(d), there is a negative correlation between the strengths of different ages and γ. This is due to the increase of the free water content in the CLSM, and more capillary pores were formed during the hardening process, which attenuate its strength.

When CLSM is applied to base course and sub-base of flexible pavement, its strength requirement is 2.8-8.3MPa [5]. So for the CLSM designed in this study, α should not be less than 0.10 to ensure sufficient bearing capacity. CLSM with strength of 1-4MPa can be used as a permanent pedestrian plaza [8]. Considering both of mechanical properties and economy, α is preferred as 0.07 or 0.10.

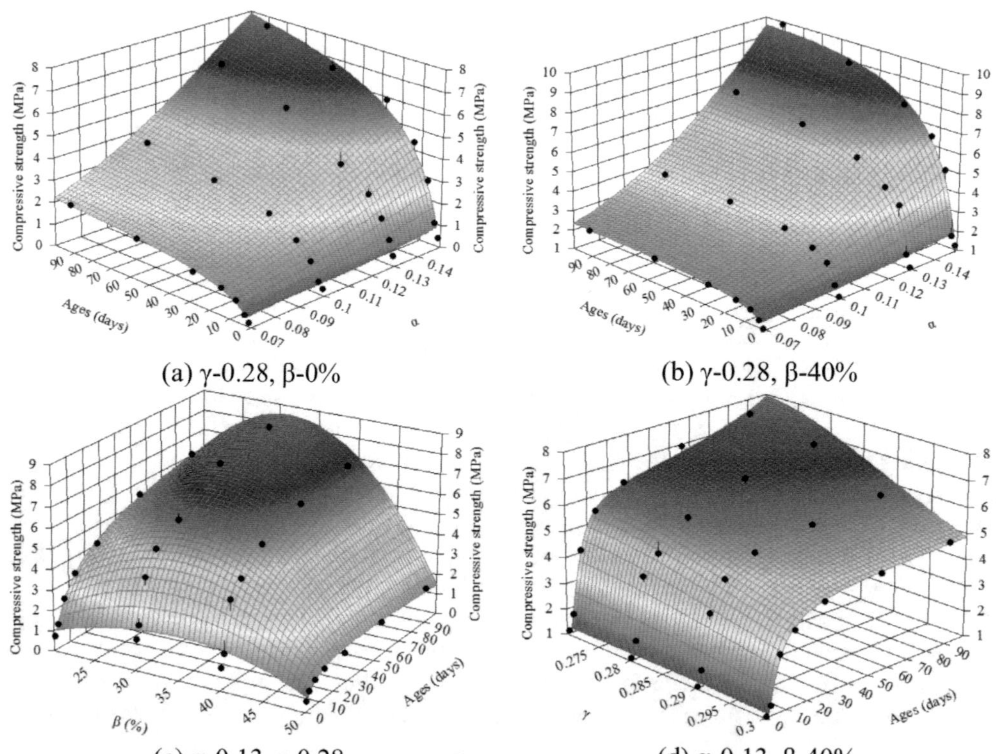

(a) γ-0.28, β-0%
(b) γ-0.28, β-40%
(c) α-0.13, γ-0.28
(d) α-0.13, β-40%

Figure 3. Long-term strength of different mixtures

3.3. Strength prediction based on BP neural network

3.3.1. BP neural network and model building

The long-term strength of CLSM is affected by many factors, which is extremely difficult to establish a model by using the traditional formula fitting method for its complex relationship. However, BP neural network has a good fitting effect especially when dealing with complex nonlinear problems, so it is widely used in many fields [2, 20].

The general steps for establishing a neural network model are as follows: determining the number of network structure layers, selecting the number of hidden layer neurons and determining the network transfer function. According to related research [3], a typical three-layer network structure can complete any mapping from n to m dimensions. Therefore, this study uses a three-layer network structure that includes a hidden layer.

The CLSM strength prediction model is established based on four factors (α, β, γ and t), so the input layer nodes number is 4, and the output layer nodes number is 1. The choice of hidden layer nodes is complicated. In general, the empirical formula (1) is used to determine the number of neurons in the hidden layer.

$$M = \sqrt{n - m} + n \tag{1}$$

where: M — the number of hidden layer nodes; n — the number of input layer nodes; m — the number of output layer nodes. So the number of hidden layer neurons selected in this model is 6.

$$S(x) = \frac{1}{1 + e^{-x}} \tag{2}$$

The Sigmoid function (see formula (2)), which is commonly used in BP neural network, is employed as the transfer function in the network structure in this study. The Sigmoid function has a good effect on the fitting of nonlinear problems, and its threshold is (0, 1), so it is necessary to normalize the inputs. Based on the selected parameters above, a BP neural network structure model is established, as shown in figure 4.

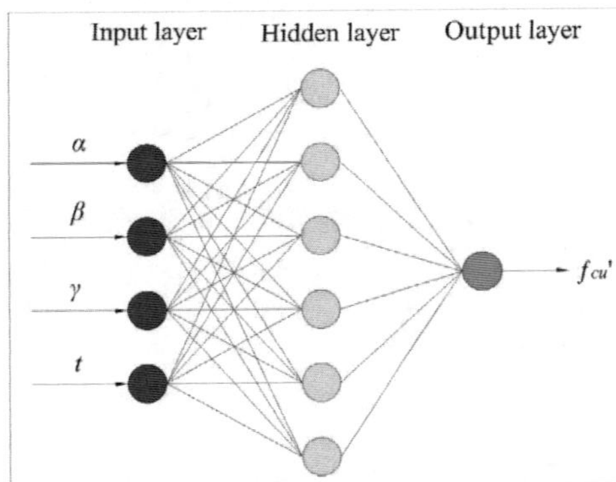

Figure 4. Network structure model

3.3.2. Model training and prediction

This study relies on the BP neural network module in MATLAB software to establish the prediction model. To prevent over-fitting, all sample data are automatically divided into three parts (the training set, the validation set and the test set) before training. During training, the training model data is used to train the model. Each time the training is performed, the system inputs the data in the verification sample data set into the neural network model for verification, thereby obtaining an error. The system then uses this error to correct the weights of the model, so that the model is approached by multiple epochs. The test sample data is used for the final performance test to verify the reliability of the model.

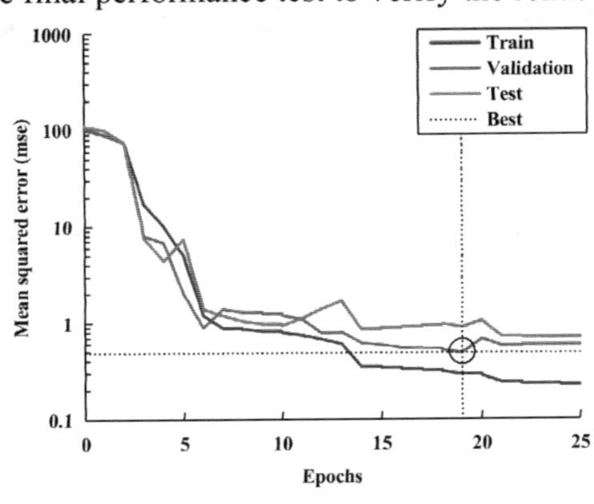

Figure 5. Performance (Best validation performance is 0.48656 at epoch 19)

From figure 5, it can be seen that the training achieves the best effect after 19 epochs, after which the continuous 6 epochs of the mean square error (MSE) no longer fall, so the training stops. At this point, the MSE between the simulated and actual values of the validated sample data is 0.48656. Figure 6(a)(b)(c) demonstrate the regression relationship between the target values of the three data set samples and the model output values, and figure 6(d) shows the relationship of all data representations. Figure 6 signifies that there is a high consistency between the target value of CLSM strength and the output value, and the correlation coefficient is above 0.90. The high prediction accuracy indicating that training and prediction of the strength values based on the BP neural network is feasible; this can provide a good reference for the mix-design of CLSM.

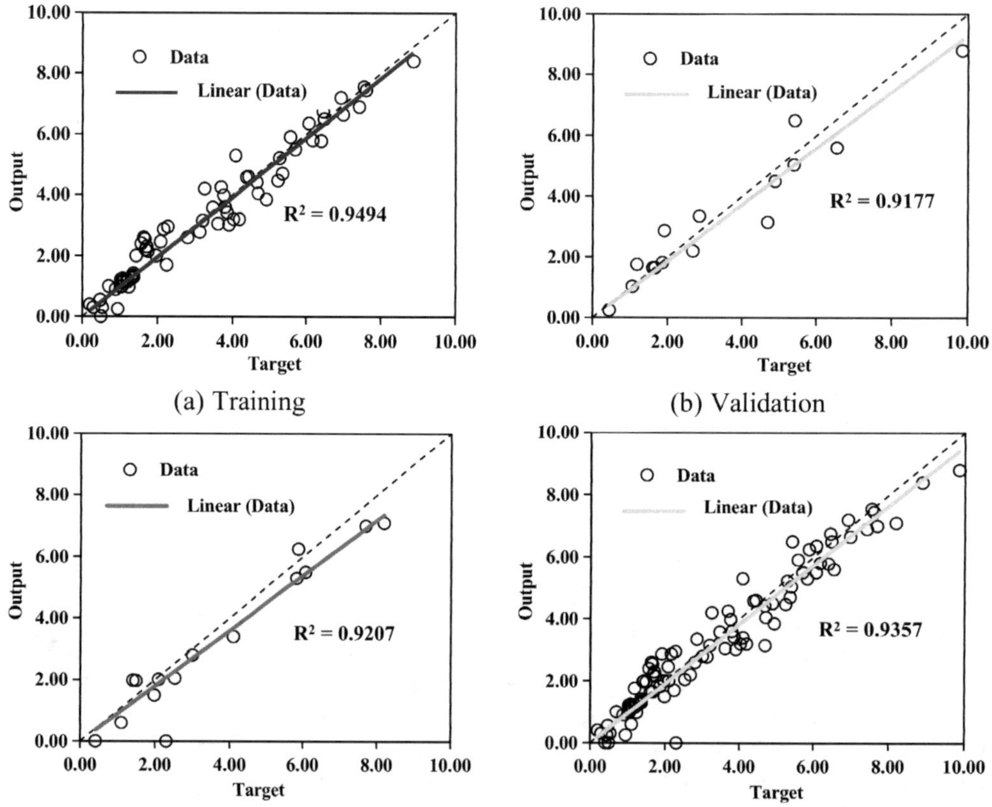

(a) Training (b) Validation

(c) Test (d) All

Figure 6. Regression of target and output strength

4. Conclusions

Construction waste mainly comes from the civil infrastructure construction or building demolition. The internal heavy metal or radioactive material content is extremely low, so the prepared CLSM will not cause secondary pollution to the environment and pose a hazard to the human body in engineering application.

The increase of α and γ promoted the flowability, but the mechanism is different, and the latter is more significant. For narrow backfill projects, γ should not be less than 0.28 to ensure good filling. Rapid-hardening CLSM are prepared based on sprayed concrete accelerator. The strength of 2-4 h can reach 0.41 to 0.79MPa, which meets the requirement of rapid construction. The long-term strength of CLSM was influenced by multiple factors such as α, β, γ and t. The strength prediction model based on the BP neural network is established, which has high precision and good reliability, and can assist the mix-design.

Considering the engineering properties of CLSM in this study, it can be concluded that the preparation of CLSM from red brick construction waste is feasible and conducive to sustainable development. It is recommended that α should not be greater than 0.07 for projects which may be excavated later. For projects requiring strict construction time, 30-40% accelerator may be added. A large amount of mixing water provides excellent flowability for fresh CLSM, while reducing its mechanical properties, so that a suitable mixing ratio needs to be designed for different projects.

Acknowledgments
The authors would like to thank Prof. Yue Hou for his kind assistance in completing this work.

References

[1] Song Y, Wang Y, Liu F and Zhang Y 2017 *Waste Manage.* **59** 350-361
[2] Dantas A, Leite M and Nagahama K 2013 *Constr. Build. Mater.* **38** 717-722
[3] Tanja K, Ivana M and Rafat S 2017 *Constr. Build. Mater.* **148** 757-769
[4] Rafael A, Alexandra M, Jorge M and Ruby M 2016 *Constr. Build. Mater.* **128** 163-169
[5] American Concrete Institute 2005 *Controlled low strength materials (ACI-229R)* (USA: Farmington Hills (MI)) pp 1-17
[6] Hwang C, Chiang C, Huynh T, Vo D, Jhang B and Ngo S 2017 *Constr. Build. Mater.* **135** 459-471
[7] Sheen Y, Huang L, Wang H and Le D 2014 *Constr. Build. Mater.* **54** 1-9
[8] Wu J and Lee M 2011 *Int. J. Pavement Res. Technol.* **4** 293-300
[9] Wu H, Huang B, Shu X and Yin J 2016 *Constr. Build. Mater.* **118** 155-163
[10] American Society for Testing and Materials 2013 *Standard specification for concrete aggregates (ASTM C33/C33M-13)* (USA: Conshohocken) pp 1-11
[11] Ran J, Zhang J, Wang J and Wei J 2016 *Highway* **3** 190-195
[12] Jiang D 2011 *Value Eng.* **30** 254-255
[13] General Administration of Quality Supervision, Inspection and Quarantine of the People's Republic of China 2010 *Limits of radionuclides in building materials (GB 6566-2010)* (China: Beijing) pp 1-3
[14] General Administration of Quality Supervision, Inspection and Quarantine of the People's Republic of China 2007 *Identification standards for hazardous wastes - Identification for extraction toxicity (GB 5085.3-2007)* (China: Beijing) pp 5-7
[15] American Society for Testing and Materials 2004 *Standard Test Method for Flow Consistency of Controlled Low Strength Material (ASTM D6103-04)* (USA: Conshohocken) pp 1-4
[16] American Society for Testing and Materials 2010 *Standard Test Method for Preparation and Testing of Controlled Low Strength Material Test Cylinders (ASTM D4832-10)* (USA: Conshohocken) pp 1-6
[17] Ministry of construction of the People's Republic of China 2009 *Standard for test method of performance on building mortar (JGJ/T 70-2009)* (China: Beijing) pp 1-22
[18] Nataraja M and Nalanda Y 2008 *Waste Manage.* **28** 1168-1181
[19] Nie G, Sun S, Yao X and Tian Y 2014 *Chin. Con. Cem. Prod.* **3** 10-13
[20] Huang Y, Chen H and Song X 2017 *Eng. Constr.* **49** 31-34

Effects of construction time on the performance of cement treated base

A N Win Maung*, Y P K Lee and N Y Ho

Samwoh Innovation Centre Pte Ltd, 25E Sungei Kadut Street 1 Singapore 729333
*Corresponding author: aye.maung@samwoh.com.sg

Abstract. Cement treated base (CTB) has been widely used as a stabilized base layer in pavement for highways, roads, airports, parking areas and some storage areas. CTB comprises mainly three components, namely, cement, water and either natural soil or aggregates. The mixture will become strong and durable after proper compaction and curing. CTB can be produced in central plant, haul to the laying area by dump trucks or can be mixed in-placed using proper materials and equipment. One of the key factors which can affect the performance of CTB is the construction time which consists of mixing to transportation, laying and compaction. This paper describes a study that was conducted to analyse the effects of construction time on the performance properties of CTB. Laboratory tests were conducted for various CTB mixtures. The CTB mixtures were compacted at different time frames and then tested for dry density and unconfined compressive strength. The results were then analysed to study the effects of construction time.

1. Introduction

Using CTB as pavement base layer compared to unbound aggregate base layer is mainly benefited in durability (Gregory et al, 2006). The design, production and construction of cement treated base layers will be based on specific standard procedures and required performance properties. However, the materials characteristics, site conditions, climate and construction procedures can affect the performance properties of CTB. Well-designed CTB is crucial to achieve specified performance and construction of CTB is equally important as well. CTB can be produced in two different ways such as mixed-in-place and mixed at central plant. When comparing the two types of production, CTB produced at central plant gives better mix under close control but highly depends on construction procedure to achieve accomplishment (Gregory, *et al*, 2006). The CTB mix produced at central plant needs to be transported by tipper truck to site area to lay and then compacted to the desired density. In plant mix CTB, construction procedure, time for transportation, laying, compaction after mixing are important factors which can influence the mechanical properties of CTB. As the cement in the CTB starts to interact with water, the hydration process begins. Thus it is crucial to minimize the time between mixing and compaction. Moreover, when moisture in the mix is lost by delay in compaction, the fresh materials become harder and difficult to compact. Furthermore, the bond between the aggregates and cement binder is broken and resulting in the lower strength (Gutherie and Maile, 2010). According to the Portland Cement Association (Gregory et al., 2006), the recommended compaction time after mixing CTB is within 2 hours. Nonetheless, some factors can cause delay in compaction and exceed the recommended time of 2 hours. This study aims to analyse how construction time affects the performance of plant mixed CTB.

2. Background

As previously mentioned, the construction process of CTB layers is as critical as mix properties which can affect the performance of the constructed CTB layers. Practically, the CTB mixing plant is set up within a short distance from the site area to minimize transport time. CTB mixes are transported by tipper trucks with proper cover to minimize moisture loss, lay at the site by highly efficient continuous paving method to minimize the paving time, and compacted by roller (as shown in figure 1) to achieve the specific degree of compaction. However, there are unforeseen conditions which can cause delay in compaction. These include longer distance to transport due to changes in site conditions, larger or longer area of paving, breakdown or error at mixing plant, machinery breakdown, etc. As such, this study aims to investigate how compaction time could affect the performance of CTB.

Figure 1. Laying and Compaction of Cement Treated Base.

3. Methodology

The CTB samples were designed according to the Federal Aviation Administration (FAA) Advisory Circular 150/5370-10G (FAA, 2014). A mixture of coarse and fine granite aggregates were used to prepare the CTB samples and the gradation limits for the combined aggregate gradation is shown in table 1. Ordinary Portland cement was used for the CTB mixing. To evaluate the effects of compaction timings, the CTB samples were compacted at different time intervals after mixing with the cement. A total of six batches of CTB samples were prepared. After mixing, each batch of samples was compacted at hourly intervals up to 4 hours. The dry density and moisture content were determined according to ASTM D558. After compaction, samples were conditioned at moist condition till the age of testing. The unconfined compressive strength for the age of 7 and 28 days were determined according to ASTM D1633 (method A). According to the FAA (2014) requirements, the 7-day unconfined compressive strength shall be 2758 - 5516 kPa, and not exceeding 6895 kPa at 28 days age. Air temperature was also recorded as reference.

Table 1. Specification for Aggregate Gradation (FAA, 2014).

Sieve Size	Allowable Passing Percentage
50mm	100
4.75mm	45 – 100
2.0mm	37 – 80
0.425mm	15 – 50
0.18mm	0 – 25

4. Results review and discussion

4.1. Results summary

Gradation analysis for the sand-granite aggregate mixture was conducted on every batch of CTB samples. The aggregate gradation results are shown in figure 2.

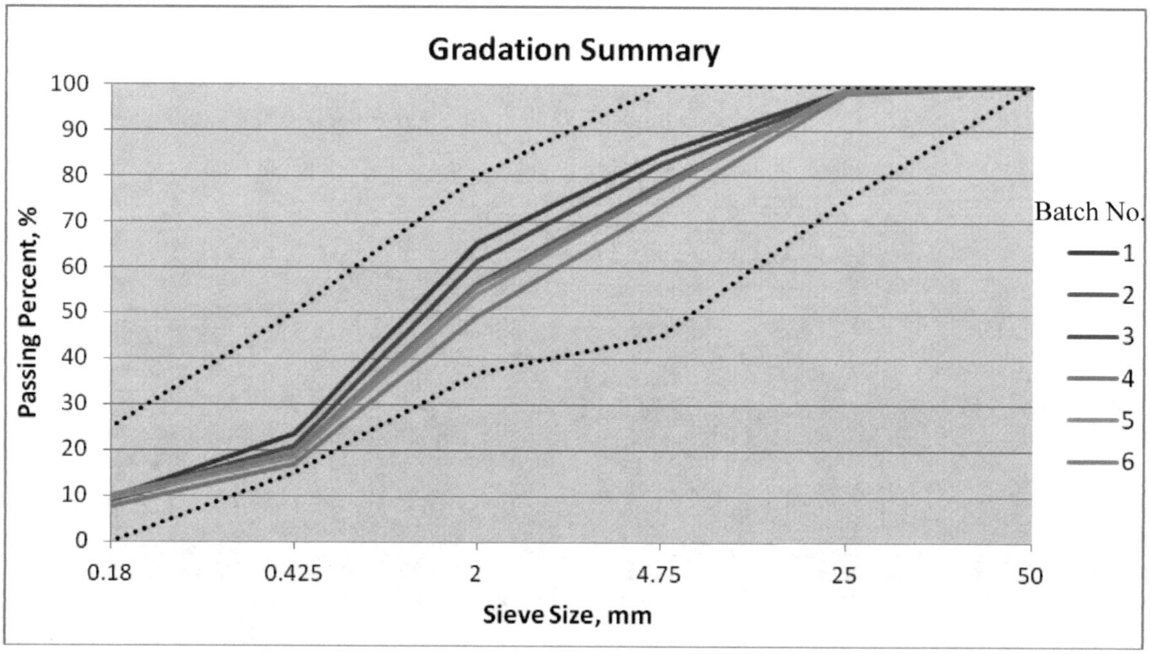

Figure 2. Gradation summary of aggregates.

After mixing the materials, the CTB samples were compacted at various time intervals ranging from 0hr (immediately after mixing) to 4hr at every 1hr interval. At each time interval, the dry density, moisture content and the unconfined compressive strength at 7 days and 28 days were determined. The average results based on 6 batches of samples are shown in table 2.

Table 2. Summary of Results.

Time interval between mixing and compaction	Air Temperature, °C	Average Moisture Content (%)	Average Dry Density (Mg/m^3)	Average 7 day compressive strength (kPa)	Average 28 day compressive strength (kPa)
0hr	28.4	7.04	2.10	5368	5925
1hr	29.8	6.81	2.08	4871	5449
2hr	30.3	6.56	2.06	4394	4984
3hr	31.1	6.26	2.04	3882	4362
4hr	31.2	6.00	2.01	3048	3442

IOP Conf. Series: Materials Science and Engineering 1075 (2021) 012003 doi:10.1088/1757-899X/1075/1/012003

4.2. Discussion of results

It is observed that CTB mixtures become drier as the compaction time (i.e. time between mixing and compaction) increases. As shown in figures 3 and 4, the moisture contents and dry densities of CTB decrease as the compaction time increases. It can be seen that dry density reduced approximately 0.02Mg/m^3 per hour. Based on the test results, the moisture content reduces about 1% and the dry density reduces 0.1Mg/m^3 after 4hrs.

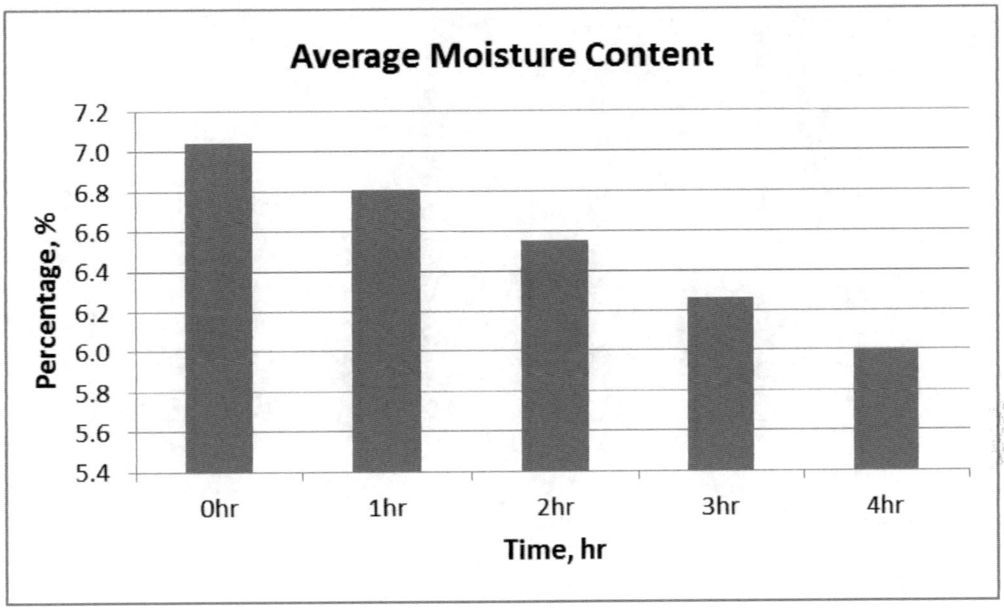

Figure 3. Average Moisture content of cement treated base mixtures.

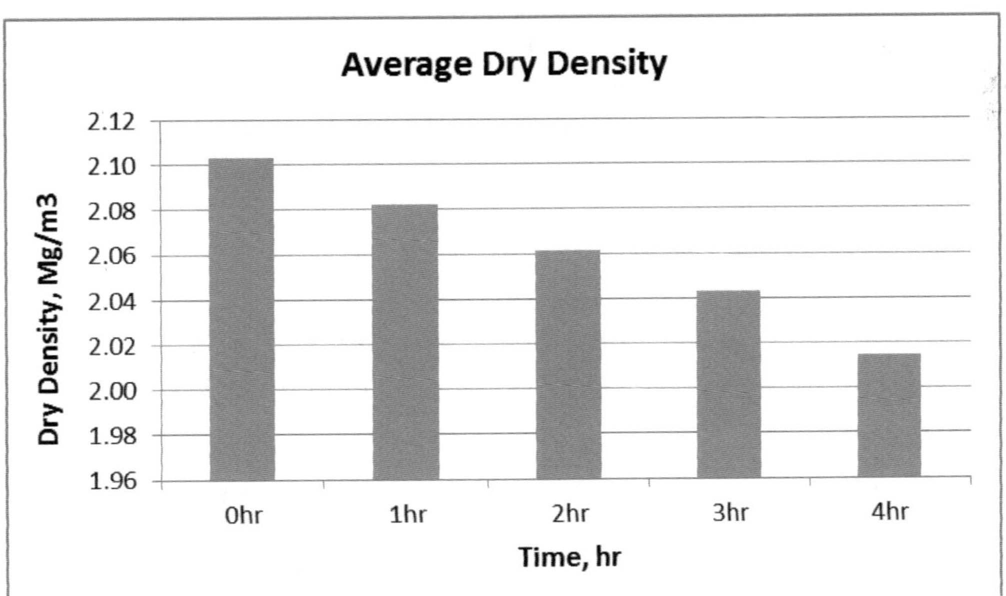

Figure 4. Average Dry Density of cement treated base mixtures.

As shown in figure 5, the 7-day compressive strength reduces as the compaction time increases. At 2 hours after mixing, the strength reduced to 4394 kPa which is still within the strength limit of 2758-5516 kPa.

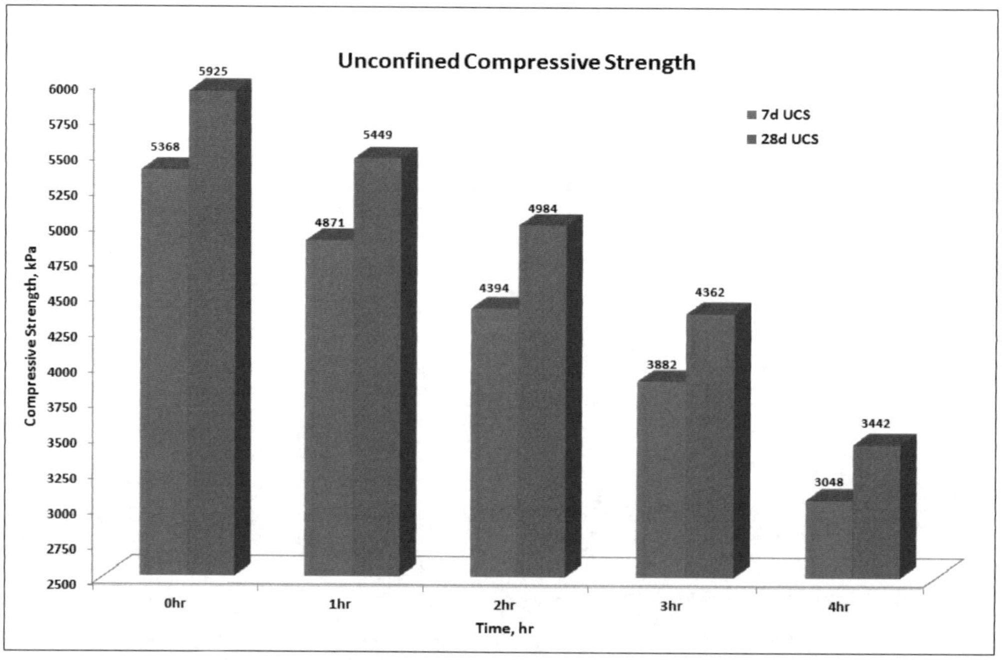

Figure 5. Average 7-day & 28-day strength of cement treated base mixtures.

Similarly, for the case of 28-day compressive strength, the strength also reduces as the compaction time increases. The average 28-day compressive strength reduced to 4984 kPa at 2hrs, 4362 kPa at 3hrs and 3442 kPa at 4hrs.

Table 3. Average Loss of Performance Properties by Time.

Time between mixing and compaction	1 hour	2 hour	3 hour	4 hour
Loss in Moisture content, %	0.23	0.48	0.78	1.04
Loss in Dry density, Mg/m^3	0.02	0.04	0.06	0.09
Loss in 7-day unconfined compressive strength, kPa	497	974	1486	2320
Loss in 28-day unconfined compressive strength, kPa	476	942	1564	2484

Table 3 shows a summary of the average loss in performance properties of CTB mixtures by compaction time. According to the FAA specifications (2014), all placement, compaction, and finishing operations of the CTB shall be completed within two hours from the start of mixing. Based on the test results obtained from this study, the average loss of moisture is about 0.5% which is within the FAA (2014) moisture loss requirement of within ±2%. Nonetheless, even though the moisture loss is within the acceptable range, the dry density reduced by about $0.04 Mg/m^3$ at 2 hours after mixing. In the case of compressive strength, both the 7-day and 28-day strength reduced by about 1000 kPa at 2 hours after mixing.

5. Conclusions

Based on the test results obtained in this study, the following conclusions can be drawn:

a) Longer construction time significantly affects the performance properties of CTB even though all properties are still within the FAA requirements after 2 hours compaction time.
b) The 7-day and 28-day unconfined compressive strength were reduced by about 1000 kPa after 2 hours compaction time which is about 20% reduction.
c) The mix design should not be based on minimum required strength to avoid strength failure.

Further research is being carried out to investigate effects of climate (temperature and humidity) with compaction time on the performance of CTB.

References

[1] Federal Aviation Administration (FAA) Advisory Circular 150/5370-10G. 2014 *Item P-304. Cement-Treated Base Course.*
[2] Guthrie W. Spencer, Sebesta Stephen, Scullion Tom. 2002 Selecting optimum cement contents for stabilizing aggregate base materials, Taxas Transportation Institute, Texas.
[3] Guthrie W. Spencer and Maile A. Rogers. 2010 Variability in construction of cement-treated base layers: material properties and contractor performance, Transportation Research Board 89th Annual Meeting, Washington D.C.
[4] Gregory E. Halsted, David R. Luhr and Wayne S. Adaska. 2006 Guide to cement-treated base (CTB), Portland Cement Association, Illinois, USA.
[5] Hossain M. Shabbir, Nair Harikrishnan, Ozyildirim H. Celik. 2017 Determination of mechanical properties of cement treated aggregate base, Virginia Transportation Research Council, Charlottesville, Virginia.
[6] Kevin J. Gaspard. 2000 Evaluation of cement treated base courses, Louisiana Department of Transportation and Development, Baton Rouge.
[7] Rasha. Abd Al-RedhaGhani. Mohammed A. Al-Jummaily, Ahlam K.R. Al-Zerjawi. 2018 Study of cement treated base aggregate properties for pavement structure, University of Kufa, Iraq.
[8] Yang Sheng Yeo, Peerapong Jitsangiam, Hamid Nikraz. 2011 Mix design of cementitious basecourse, Curtin University, Perth, Australia.

IOP Conf. Series: Materials Science and Engineering 1075 (2021) 012004 doi:10.1088/1757-899X/1075/1/012004

Effect of Anti-Stripping Agent with Bipolarity to Strip-Ravelling Damage Caused by Surface Shear Stress

Hiroki Imai[1*], and Hiromitsu Nakanishi[2]

[1]Engineering Research Laboratory of Taiyu Kensetsu Co., LTD. Nagoya, Japan
[2]Global Business Team of Taiyu Kensetsu Co., LTD. Nagoya, Japan
[*] Corresponding author: imai-hiro@taiyu.co.jp

Abstract. The stripping is one of serious distresses because it causes the decline in the bearing capacity of the asphalt pavement. As the result of evaluation of the bearing capacity of the pavement by FWD, the dissipation amount of work becomes larger at the stripping area in comparison with non-stripping area. An anti-stripping agent is effective for preventing such stripping in advance. The new anti-stripping agent which used in this research forms the strong electrochemical bonding between the asphalt and the aggregate by giving bipolarity to the asphalt. Furthermore, it is clarified that the new anti-stripping agent can give improve and enhance the shear strength of the asphalt as well. The shear strength is closely related to the rutting and the fretting of the asphalt pavement in accordance with past research. Therefore, enhancing the shear strength should contribute to not only the improvement in the rutting and the fretting but also the prevention of the strip-ravelling that the fine aggregates and the asphalt mortar are washed out from the pavement surface and it often takes place in Malaysia and Asian countries. In this research, we will focus on the strip-ravelling of the asphalt pavement, and simulate the strip-ravelling by using an immersed-wearing tester. As a result, this research clarifies the relationship between the shear strength of the asphalt and the strip-ravelling of the asphalt mixture.

1. Introduction

The stripping in asphalt pavement is one of the serious distresses because it occurs at the bottom or in the middle of the asphalt mixture layer and causes the decline in the bearing capacity of the asphalt pavement. As the result of evaluation of the bearing capacity of the pavement by the dissipation amount of work, which is obtained by FWD, the dissipation amount of work becomes larger at the area where the stripping is taking place in comparison with the non-stripping area.

In order to obviate such stripping which declines the bearing capacity of the pavement, it is widely known that the anti-stripping agent is very effective. The newly developed anti-stripping agent used in this research gives bipolarity to asphalt and forms electrochemical bonding between the aggregate and the asphalt. Furthermore, this new agent improves and enhances the shear strength of asphalt in addition to the anti-stripping efficiency.

In the past paper, authors already reported that the shear strength of asphalt was closely related to the rutting and the fretting of the asphalt pavement [1]. Therefore, enhancing the shear strength of asphalt should lead to the prevention of the strip-ravelling, which is caused by the friction between the tire and the surface of the pavement as shown in figure 1, as well as the rutting and the aggregate fretting.

Figure 1. Photos of strip-ravelling at surface of asphalt pavement

In this research, we focus on the strip-ravelling, which is the typical distress of asphalt pavement in Malaysia and other Asian countries, and evaluate the anti-stripping performance of asphalt mixture by the immersed wearing test simulating the strip-ravelling.

In this paper, it is clarified that there is the close relationship between the shear strength of the asphalt and the strip-ravelling, and the strip-ravelling is caused by the decline in the anti-stripping property of the asphalt mortar in the asphalt mixture, which is brought by the decrease in the shear stress of the asphalt.

2. Decline of bearing capacity by stripping

2.1. Dissipation amount of work by FWD

The dissipation amount of work expresses the energy dissipated into the pavement when subjected to the loading of 49 kN by FWD. As shown in figure 2, on the coordinates where the deflection is on the horizontal axis and the load is on the vertical axis, the dissipation amount of work is expressed by the area of hysteresis loop obtained from the chronological value of the load and the deflection when the loading values are positive. The area of the hysteresis loop is expressed as the difference in the loads at loading process and unloading process, and it is calculated by equation (1).

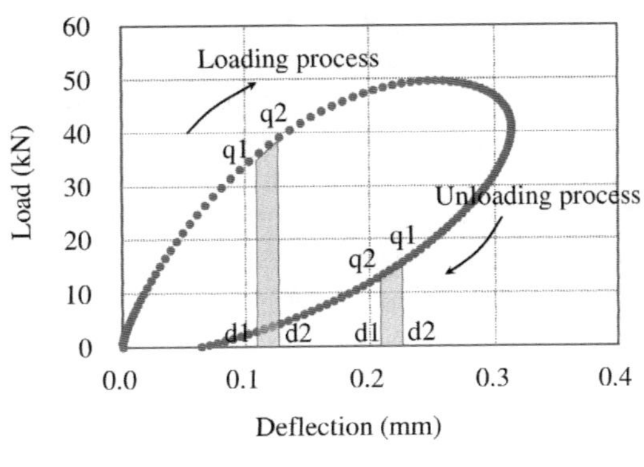

Figure 2. Dissipation amount of work (the relation between deflection and load)

$$Dissipation\ amount\ of\ work(kN \cdot mm) = \sum[(q1 + q2) \times (d2 - d1)/2] \quad (1)$$

2.2. Dissipation amount of work at stripping area

The dissipation amount of work depends on the bearing capacity of the asphalt pavement, and if the pavement is distressed by the fatigue clacks, the dissipation amount of work becomes larger [2]. Therefore, we measured the dissipation amount of work by FWD at the area where stripping occurred in order to grasp the influence of the stripping of asphalt mixture on the bearing capacity of pavement. The stripping area was specified by the observation of the road surface condition and the section of core taken from the pavement.

Figure 3 shows the relationship between the thickness of the asphalt mixture layer and the dissipation amount of work. This figure shows the measurement results of the dissipation amount of work on 10 stripping areas and 10 non-stripping areas each. As shown on the left side of figure 3, the dissipation amount of work at the stripping area is obviously larger than that at the non-stripping area. And, the right side of figure 3 shows the difference in the size of the dissipation amount of work. Therefore, the stripping, as well as the fatigue cracks, can be regarded as one of serious distress which will bring the decrease in the bearing capacity of the pavement. Once such stripping occurs on the surface of the pavement, the distress will expand widely to the middle and the bottom of the asphalt mixture layer and bring the early deterioration of the pavement. Therefore, it is essential to obviate preventively the occurrence of the stripping.

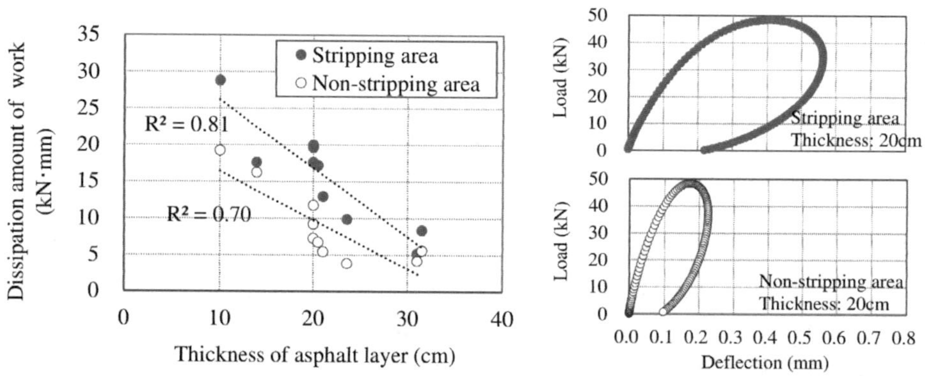

Figure 3. Dissipation amount of work at stripping area

3. Mechanism of stripping and anti-stripping

The aggregate with the main ingredient of SiO2 charges negatively because Oxygen with the negative charge is allocated on the surface of the aggregate. Therefore, if the asphalt mixture using such aggregate is immersed in water, Hydrogen of H2O will adhere to the surface of aggregate instead of asphalt due to the electronegativity of Oxygen. As a result, the asphalt around the aggregate is stripped from the aggregates by the hydrogen bonding, which is stronger than the intermolecular force between aggregate and asphalt, as shown on the left side of figure 4. In contract, if the anti-stripping agent which gives the bipolarity to the asphalt, the electrochemical bonding between the aggregate and the asphalt is formed as shown on the right side of figure 4. Since such electrochemical bonding is stronger than the hydrogen bonding, the asphalt with the anti-stripping agent which provides the asphalt with the electrochemical bipolarity can exert the high anti-stripping performance.

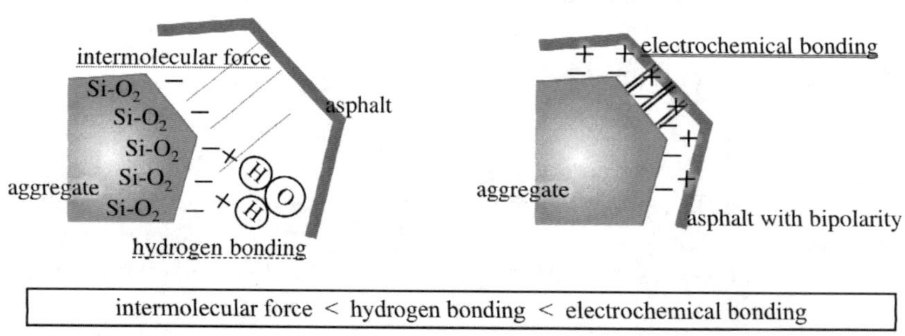

Figure 4. Mechanism of stripping and anti-stripping

4. Materials used and general properties

4.1. Materials used

In general, Granite and Andesite are known as the aggregates which easily arise the stripping. In this research, Andesite is used for the coarse aggregate and the fine aggregate because Granite is hardly used as aggregates for asphalt pavement in Japan. Calcium carbonate is used as a mineral filler.

Type of the asphalt mixture used for this research is dense-graded asphalt mixture (13). The asphalt used is straight asphalt with penetration 60-80 (StAs60-80). And, the anti-stripping agent used is "Tough Fix Hyper (TFH)" which can give bipolarity to asphalt [3]. The mix proportion and the lithology of the aggregates used for this research and the mix gradation are shown in table 1 and table 2, and the asphalt properties after adding TFH into StAs60-80 are shown in table 3.

Table 1. Mix proportion and lithology of aggregates

Material	Crashed stone (13)	Crashed stone (5)	Coarse sand	Screenings	Filler
Lithology	Andesite	Andesite	Andesite	Andesite	$CaCO_3$
Mix proportion (%)	37.0	15.5	29.0	14.5	4.0

Optimum asphalt content: 5.8%

Table 2. Mix gradation of dense graded asphalt mixture (13)

Sieve size (mm)	19.0	13.2	4.75	2.36	0.6	0.3	0.15	0.075
Passing weight percentage (%)	100	98.5	64.1	46.0	23.5	15.2	9.4	5.9
Gradation envelope (envelope, %)	100	95-100	55-70	35-50	18-30	10-21	6-16	4-8

Table 3. Properties of straight asphalt using TFH

Base asphalt		Straight asphalt 60-80			
TFH content	(%)	0.0	0.1	0.15	0.2
Penetration	(1/10mm)	62	57	58	59
Softening point	(°C)	47.0	48.5	48.5	48.0
Ductility	(cm)	>100	>100	>100	>100
Viscosity at 135°C	(mPa·s)	366	366	368	368
Viscosity at 160°C	(mPa·s)	132	128	128	130
$G^*/\sin\delta$ at 64°C	(Pa)	1781.3	1944.2	2318.3	2412.6

As shown in table 3, the elastic component of the complex shear modulus "$G^*/\sin\delta$" becomes higher along with the increase in TFH. It is considered that because the shear strength of the asphalt is becoming stronger due to the bipolarity brought from TFH.

4.2. General property for resistance to stripping

4.2.1. Test method of immersed wheel tracking test.

The immersed wheel tracking test is normally adopted to evaluate the anti-stripping property of asphalt mixture in Japan. In this test, the stripping would be caused at the middle and the bottom of the asphalt mixture by moving a wheel longitudinally and transversely in hot water of 60 °C. Table 4 shows the test conditions and two kinds of water level for the tests. The anti-stripping performance is evaluated by measuring the stripping area percentage obtained from the specimen after the test. The stripping area percentage is obtained by checking the stripping parts in the cross-section of the specimen on the sheet with 5 mm mesh. And the stripping area percentage is calculated by equation (2) [4].

Table 4. Condition of immersed wheel tracking test

Item	Condition	Water level	
		bottom of specimen	top of specimen
Wheel weight (N)	686		
Contact pressure (MPa)	0.63		
Curing temperature (°C)	60		
Curing time (hours)	12(air)/ 1(water)		
Test temperature (°C)	60		
Test time (hours)	6.0		

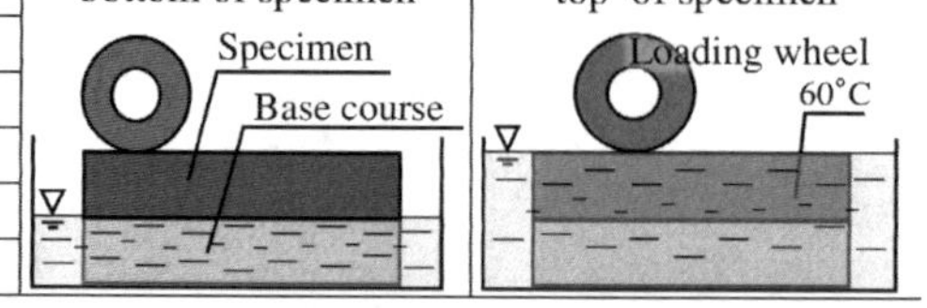

$$Stripping\ area\ percentage\ (\%) = \frac{Stripping\ area}{Sectional\ area} \times 100 \quad (2)$$

4.2.2. Effect of anti-stripping agent "Tough Fix Hyper".
The immersed wheel tracking test is normally adopted to evaluate the anti-stripping property of asphalt mixture in Japan. In this test, the stripping would be caused at the middle and the bottom of the asphalt mixture by moving a wheel longitudinally and transversely in hot water of 60 °C. Table 4 shows the test conditions and two kinds of water level for the tests. The anti-stripping performance is evaluated by measuring the stripping area percentage obtained from the specimen after the test. The stripping area percentage is obtained by checking the stripping parts in the cross-section of the specimen on the sheet with 5 mm mesh. And the stripping area percentage is calculated by equation (2) [4].

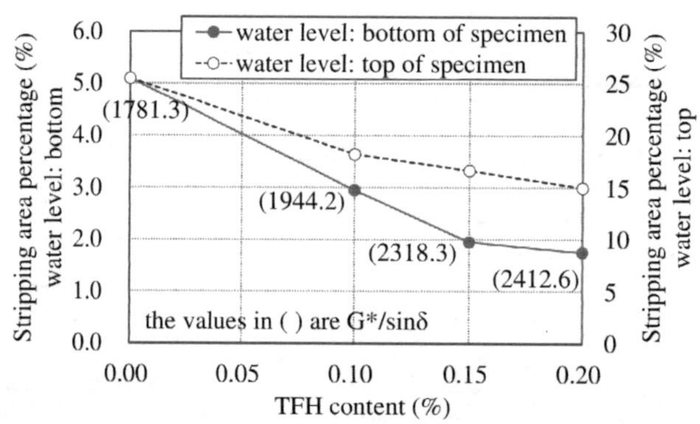

Figure 5. Relationship between TFH content and stripping area percentage

5. Evaluation of strip-ravelling

It is important to obviate and mitigate the risk of the occurrence of the strip-ravelling because the strip-ravelling may cause the serious distress of the asphalt pavement in the future as shown in figure 1. At present, however, the resistance to the strip-ravelling cannot be evaluated properly because the suitable test method is not established. Thus, in the following study, we will establish the test method to be able to simulate the strip-ravelling and will develop the asphalt mixture with the superior resistance to this problem.

5.1. Immersed wearing test

5.1.1. Test equipment and evaluation method.
The strip-ravelling is the distress in which only the fine aggregate and the asphalt mortar are worn away from the surface of the asphalt pavement. And it would become worse by the being subjected to the friction (shear) between the tire and the surface of the pavement. In this research, therefore, using the equipment which is normally adopted for the evaluation of the aggregate fretting of porous asphalt mixture, we evaluate the strip-ravelling as shown in figure 6. Furthermore, the loading weight, the rotational radius, and the rotational speed were in accordance with the conventional method of the

porous asphalt mixture [5]. The test situation and the specimen after the immersed wearing test are shown in figure 7.

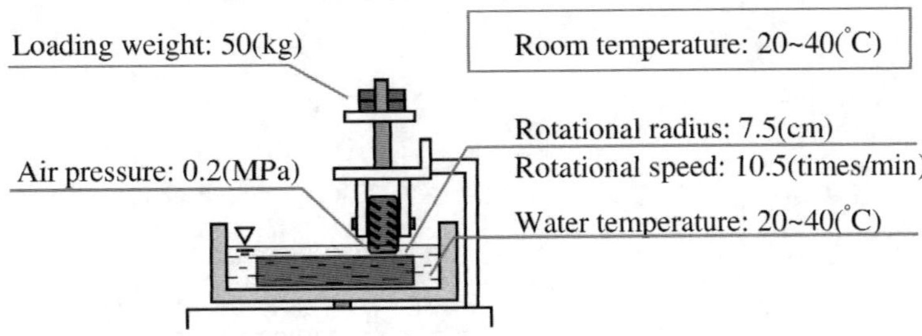

Figure 6. Equipment of immersed wearing test

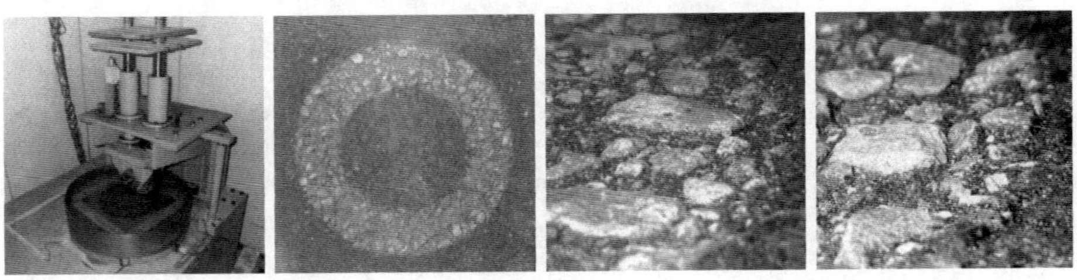

Figure 7. Test situation and specimen of immersed wearing test

The strip-ravelling is evaluated by the wearing depth obtained by comparing the surface shapes of the specimen before and after the test. Such wearing depths are measured on the same line before and after the test as shown on the left side of figure 8. And, the average of the wearing depths of all points measured on the horizontal axis at the intervals of 5 mm are adopted as the index value of the strip-ravelling (mm) as shown on the right side of figure 8 in this research.

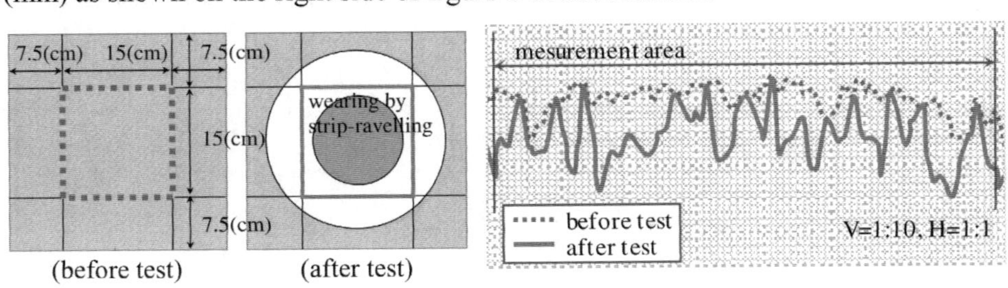

Figure 8. Evaluation for strip-ravelling

5.1.2. Curing temperature and curing time. The curing temperature and the curing time of specimens are very important because they would influence significantly on the anti-stripping performance of the asphalt mixture. Hence, in order to determine these two conditions for the curing of the specimens, referring the curing conditions for ASTM D 1559 Residual Marshall Stability, we have carried out several types of immersed wearing test in advance based on the conditions of the curing temperature of 60 °C and the curing time of 48 hours.

The relationship among the curing temperature, the curing time, and the strip-ravelling is shown in Figure 9. In this case, the test temperature is provisionally set at 20 °C and the test time is set to 1 hour, the validity of these conditions will be re-examined in the next section. As shown in this figure, the strip-ravelling shows a tendency to converge to a certain value by setting the curing temperature to 60 °C and the curing time to 48 hours or more. Therefore, the curing temperature shall be set to 60 °C and the curing time to 48 hours in this research.

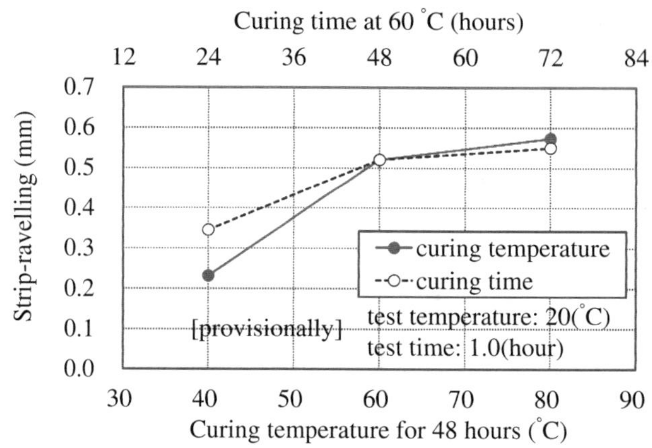

Figure 9. Relationship among curing temperature, curing time, and strip-ravelling

5.1.3. Test temperature and test time. It is also important in order to evaluate the strip-ravelling to decide the test conditions like the test temperature and test time as same as to have decided the curing conditions as mentioned above. Especially, the test temperature should be examined carefully because this research is aimed to evaluate only wearing of fine aggregate and asphalt mortar. Specifically, it is necessary to set the temperature which would not cause the rutting and the aggregate fretting of the asphalt mixture.

The relationship among the test temperature, the test time, and the strip-ravelling is shown in figure 10. The specimens used in this test have been cured at 60 °C for 48 hours. As shown in this figure, the strip-ravelling increases with the rise of the test temperature. However, the wearing depths of more than 5 mm, which shows the fretting of the coarse aggregates, were observed in several points and the plastic deformation like the rutting was also occurring at the test temperature of 40 °C. Therefore, the test temperature was set to 30 °C because this temperature did not provoke such distresses on the specimens. And, the test time was set to 1.0 hour because there was no significant difference between 1.0 hour and 2.0 hours.

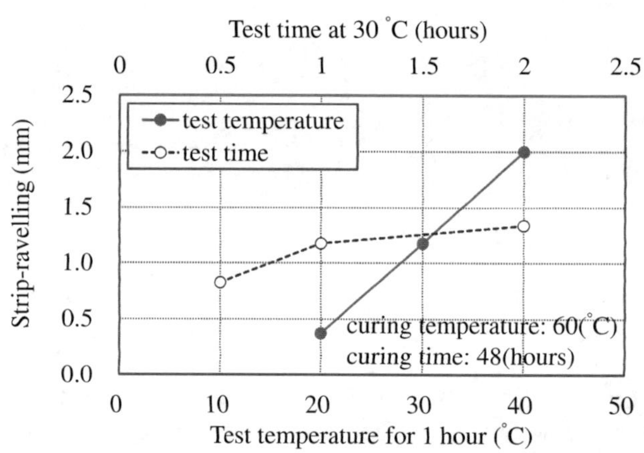

Figure 10. Relationship among test temperature, time, and strip-ravelling

5.2. Effect of anti-stripping agent "TFH" on strip-ravelling

The strip-ravelling when adding TFH of 0.1 to 0.25 % into StAs60-80 is shown in figure 11. As shown in this figure, the strip-ravelling is greatly improved by the addition of TFH of 0.15 % or more, and the strip-ravelling is able to be reduced by 1/2 compared with the case of no addition.

This result shows that the shear strength is improved at the same time as the anti-stripping performance, such improvement in the shear strength will be proved by the increase in the elastic component of the complex shear modulus, G*/sinδ as having shown in table 3.

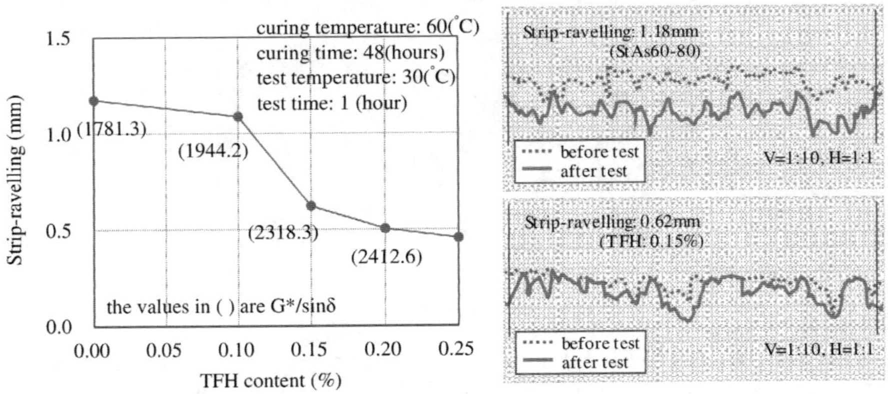

Figure 11. Relationship between TFH content and strip-ravelling

6. Influence of asphalt mortar on strip-ravelling

The immersed wearing test was able to simulate the wearing of the fine aggregate and the asphalt mortar of the asphalt mixture as having shown in figure 1, and it was clarified that the relationship between the strip-ravelling and the shear strength of asphalt. Such an improvement in the shear strength of asphalt, which is provided by the addition of TFH, is brought from the improvement in the anti-stripping performance by forming the electrochemical bonding between the aggregate and the asphalt as having shown in figure 4. Therefore, the stripping test of the asphalt mortar was carried out in order to verify how TFH would result in the effects against the fine aggregates and the asphalt mortar, which would be worn by the strip-ravelling.

6.1. Residual Marshall Stability test using asphalt mortar

Dense-graded asphalt mixture (13) using the coarse aggregate and the fine aggregate of Andesite as shown in table 1 or 2 has been evaluated in this research so far. In this section, the specimens using only the coarse sand in the aggregates constituting the dense graded asphalt mixture (13) were fabricated in order to evaluate the anti-stripping performance of the asphalt mortar consisted of only coarse sand of Andesite. And such asphalt mortar consisting of only the coarse sand is evaluated by Residual Stability after the curing times of 48 and 96 hours referred to ASTM D1559. Furthermore, the optimum asphalt content of the asphalt mortar consisting of only the coarse sand is set to 6.6 % so that the thickness of asphalt film is 9.5 μm as the same as that of dense graded asphalt mixture (13) by the calculation of the gradation of the coarse sand.

6.2. Effect of anti-stripping agent "TFH" on Residual StabilityDense-graded

The Residual Stability of the asphalt mortar consisting of only the coarse sand when adding TFH of 0.15 % into StAs60-80 is as shown in figure 12. And in this test, the Residual Stabilities of the dense graded asphalt mixture (13) shown in table 1 or 2 are also evaluated as the comparison as well as the asphalt mortar.

As shown in this figure, the Residual Stability tends to decrease with the increase in the curing time from 48 to 96 hours. And the Residual Stability of asphalt mortar consisting of only coarse sand shows the bigger decrease than that of the dense graded asphalt mixture (13). Therefore, the strip-ravelling will be caused by the low anti-stripping performance of asphalt mortar like coarse sand. In contract, it is clarified that the resistance to the strip-ravelling is improved by using TFH because TFH added into StAs60-80 would be able to restrain the decrease in the Residual Stability.

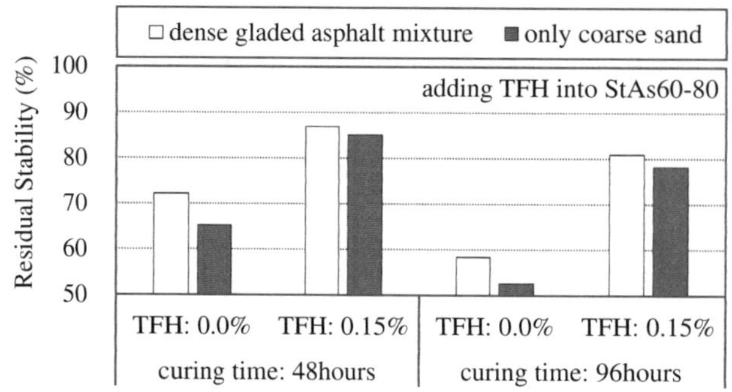

7. Conclusion

The findings in this research are as follows:

a) The stripping is the serious distress which causes the decline in the bearing capacity of the pavement. The increase in the dissipation amount of work by FWD will explain clearly such a fact.

b) The strip-ravelling occurring in Asian countries can be recreated properly by the immersed wearing test carried out on the conditions that the specimens are cured at 60 °C for 48 hours are tested at 30 °C for 1 hour.

c) The strip-ravelling is clearly related to the shear strength of asphalt. And, the shear strength and anti-stripping performance of the asphalt are significantly improved by adding TFH of 0.15 % into StAs60-80.

d) The Residual Stability of the asphalt mortar is easily to decrease by the stripping. And the stripping of the asphalt mortar causes the stripping of the asphalt mixture. Therefore, it is necessary to prevent the stripping by adding TFH into StAs60-80.

e) In this research, we have focused on the striping and the strip-ravelling in case of using StAs60-80. Hereafter, it will be necessary to expand the scope of the research to the modified asphalt and others.

References

[1] H. Imai, H. Nakanishi. Relationship between asphalt viscoelasticity obtained from rotation viscometer and property of asphalt mixture, Journal of Japan Society of Civil Engineers, Ser. E1 (Pavement Technology), Vol.73, No.3, I_131-I_138 (2017).

[2] K. Maruyama, M. Kumagai. **Evaluation of fatigue damage in asphalt pavement using FWD dissipated work**, *Journal of Japan Society of Civil Engineers, Ser. E1 (Pavement Technology)*, Vol.67, No.3, I_27-I_34 (2011).

[3] H. Imai, N. Quang Phuc, H. Nakanishi, T. Ando. **Anti-stripping agent for hot mix asphalt pavement**, *The transport journal (International conference on sustainability in civil engineering)*, pp45-48 (2016)

[4] H. Imai, N. Quang Phuc, H. Nakanishi, A. Kato. **Evaluation of anti-stripping property of modified asphalt under severer conditions**, *The transport journal (International conference on sustainability in civil engineering)*, pp198-203 (2018)

[5] Japan Road Association. **Torsional aggregate scattering value**, *Pavement performance evaluation method supplement*, pp38-53 (2008)

IOP Conf. Series: Materials Science and Engineering 1075 (2021) 012005 doi:10.1088/1757-899X/1075/1/012005

Using Crumb of Tires in Hot Asphalt Mixture as a Part of Aggregate

Al-Hasani Mohammed[1], Qun Yang[1], Khalil Al-Bukhaiti[2,*]

[1]College of Transportation Engineering, Tongji University, Shanghai, China.
[1]College of Transportation Engineering, Tongji University, Shanghai, China.
[2]School of Civil Engineering, Southwest Jiaotong University, Chengdu, Sichuan, China.
[*]Corresponding author: khalil2020@my.swjtu.edu.cn

Abstract. Scrap tires are a major part of the global solid waste management problem. In recent years, the problem of waste tires has become very acute that there is an urgent need to find an optimal and efficient way to use scrap tires in asphalt mixtures. The previous studies showed that the utilization of crumb of tires has more effect on asphalt mixture performance, which is represented increasing in the Marshall Stability and increasing crumb of tires causes increasing in Marshall Flow and increasing in air voids content additionally more than original mixes. This paper used crumb of tires in asphalt mixture as a part of aggregate. Three sizes of the grade of crumb tires were used in asphalt mixture No. 4 (4.75 mm), No. 8 (2.36) mm and No. 50 (300μmm). Three percentages of asphalt binder (4, 5 and 6) by weight with three percentages of crumb tires (2, 4 and 8) % by weight also were used with aggregates for preparing asphalt mixture specimens. Asphalt mixture specimens were conducted according to Marshall Methods. Thirty-six specimens were equipped for evaluating Marshall Properties (Marshall Stability and flow, air voids percentage, bulk density, maximum bulk density, and Marshall Stiffness). Indirect Tensile Strength test (ITS) has been applied to obtain the cracking resistance of asphalt mixture using twenty-four specimens, which contains crumb tires. The results showed that the applying of the crumb tires has a significant performance of asphalt mixture by increasing the Marshall stability, flow, air voids, and decreasing bulk density and indirect tensile strength compared to the original mixtures.

1. Introduction

The nation faces a major ecological problem due to an accumulation of automobile waste and truck tires. Environmental regulations prohibit the open burning or burying of tires is a solid waste facility. These tires are accumulating at the rate of about (3/4) billion per year, An immense problem affecting environmental pollution is the increase of waste tyre vehicles. In an attempt to decrease the magnitude of this issue, crumb rubbermodifier obtained fromwaste tyre rubber has gained interest in asphalt reinforcement [1],[2].

There are approximately 270 million are waste tire generated annually in state of these 230 million are passenger car tire and40 million truck tire according to the 800 million scrap tire currently stock piles throughout the country A scrap tire passenger car weight approximately 9 kg and will provide 60% rubber 20% steel and 20% fiber and other waste product [3].

The paving industry uses (1 - 2) million tires per year. Each metric ton of hot mix asphalt containing rubber can utilize (2 - 6) tires [4],[5].The purpose of modification includes: increase the viscosity at high temperature, increase the flexibility and elasticity of binders at low temperature, improve the adhesion to aggregates, and improve high thermostability and aging resistance [6].The aim of this paper is to study the effect of crumb rubber as part of aggregate on the performance of asphalt mixture for different percentage of adding (2%, 4% and 8%) by weight of aggregate. For this purpose the performance changes were evaluated by Marshall tests and Indirect Tensile Strength test at 25°C.

Munder A Bilema and Mohamad Y Aman [7],calculate the optimum bitumen content for each percentage and evaluate the moisture sensitivity of crumb rubber modified asphalt at two different compacting temperatures.On the other hand crumb rubber dissolved in the binder and stiffened to increase it is rutting resistance. Even though only up to (2.5%) coarse rubber particles could be used as compared to (15%) of crumbrubber.Rubberized asphalts resisted rutting better than conventional (unmodified) asphalt as well as polymer modified binder asphalts. Wheel tracking performance was similar for dry and wet mix asphalts, but fatigue life for dry mix was three times as long as that of wet mixes. Katelyn Nicole Love [8] the effects of rubber modified binder on the performance of porous asphalt by evaluating the porosity, permeability, long term draindown, abrasion resistance, fatigue resistance and dynamic modulus. Result study on cores of asphalt rubber mixtures indicated higher wet indirect tensile strength and tensile strength ratios than the control mixture. In addition , the result of friction testing indicates that the rubberized mixture is performing better than the control mixture [9].

Asphalt rubber is defined by the ASTM Standard D6114 as a "mix of grade asphalt cement paving, ground recycled tire (that is, vulcanized) rubber and other additives, as needed, for use in binder construction in pavement. Rubber is mixed and reacted in hot asphalt cement enough to cause swelling in rubber particles before use. "Asphalt and rubber binder is a mixed mixture (in a hot mixing plant) and requires specialized mobile mixing equipment to produce it. crumb rubber modifier (CRM) ranges from 18 to 22 percent.

The crumb rubber modifier used in rubber asphalt is in the 10-16 mesh range [10].Powder rubber or polymers should be added to AC-30 to form the binder used in asphalt bases [11]. The resulting structural coefficient (a-value) for the powder rubber base is (0.45) compared to (0.40) for conventional base coarse using AC-30, addition of the powder rubber increases the cost of the binder (10%) only while increasing its coefficient by (12.5%)[12].Many studies have reported that crumb rubber modifier (CRM) binders can produce asphalt pavements that pavement life, decreased traffic noise, reduced maintenance costs and resistance to rutting and cracking [13],[14] and[15].

The objective of this study was to evaluate how the properties of rubber affect the properties of asphalt bonding material. The study used a rubber crumb of tires products to modify asphalt bonding material. The test of Marshall Stability, and flow, air voids percentage, bulk density, maximum bulk density, and Marshall Stiffness test. The objective was achieved by analyzing the performance characteristics of modified volumes to determine the impact of rubber properties on binder performance.

2. Materials and tests

2.1. Asphalt Cement

All data presented in this study had been collected in Jiangsu road Mstar Technology LtdThe Road Laboratory of Department College of Transportation EngineeringinTongji University. One binder of asphalt cement was tested, from Venezuela heavy oil Refinery with a grade of (40-50) penetration. The physical properties (according to ASTM specification) of this type are illustrated in table 1.

Table 1. Physical Properties of Asphalt Cement

Test	Unit	Venezuela heavy oil (40-50)
Penetration at 25 C°	0.1 mm	45
Ductility at 25 C°	Centimeter	100 +
Softening Point	C°	50
Flash Point	C°	339
Specific Gravity	---	1.033

Kinematic Viscosity	cst.	270

2.2. Aggregate (coarse and fine materials)

Crushed aggregates are used in this work with a fixed dense gradation for all the specimens and from Liyang city quarry. The middle limits of the (19 mm) size dense gradation has been selected as a basic gradation in accordance with (ASTM D-3515) as shown in figure 1.

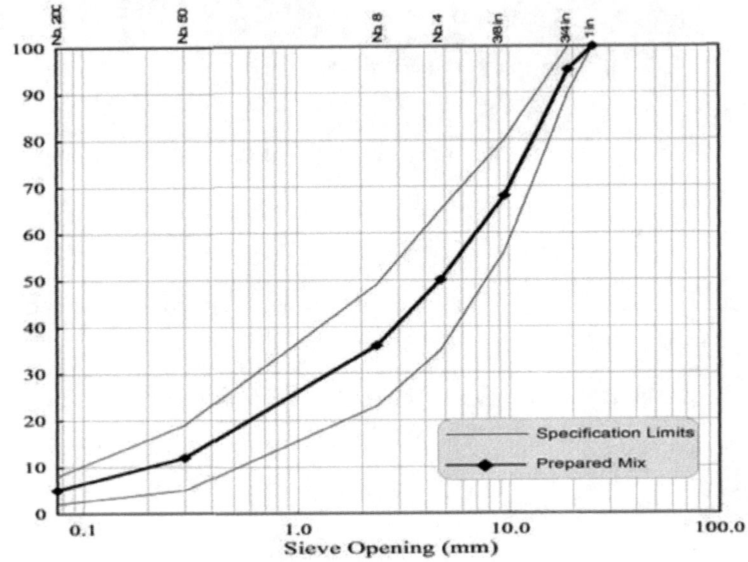

Figure 1. Specification Limits and Selected Gradation of Aggregate Maximum Size (19 mm)

2.3. Crumb of tires

Crumb of tyres was prepared by crushing, and grinding used tyres to small pieces about (3x3) cm, and using Nitrogen containers where the small pieces of tyres are put for (45 seconds) and crushing these freezing particles by using Marshall's hammer to produce crumb of tyres.

The crumb of tyreswas sieved to the sizes required, figure 2 shows crumb of tyres after freezing and crushing. Sieve analysis of crumb of tyresis shown in table 2.

Table 2. Properties of Crumb of Tires

Sieve size (mm)	% Passing
9.5	100
No.4 (4,75)	67
No.8 (2.36)	32
No.50 (0.3)	11
No. 200 (0.075)	3
Specific Gravity	1.237

Figure 2. Crumb of Tires Particles after Free and Crushing.

2.4. *Mineral filler*
One type of filler is used in this work which was Limestone dust obtained from Changjiang river Factory.

3. Test methods
The following tests were used in this work to evaluate the asphalt concrete mixture:
- Resistance to plastic flow (Marshall Stiffness) at (60°C).
- Indirect Tensile Strength (ITS) at (25°C).

3.1. *Preparation of mixtures*
The aggregates were first dried to constant weight at (110°C), separated into the desired sizes and recombined with mineral filler in order to meet the required gradation for each specimen. The aggregates were then heated to a temperature of (155°C) before mixing with asphalt cement. The asphalt cement is heated to a temperature, which produces a kinematic viscosity of (170±20) centistokes up to (163°C) as an upper limit. Then, asphalt cement was weighted to desired amount, added to the heated aggregates, and mixed thoroughly until all aggregate particles were coated with asphalt binder.

3.2. *Preparation of asphalt mixture containing crumb of tyres*
To prepare one sample of aggregates containing crumb of tyres, calculate weight of each grade size tires of (No.4), (No. 8) and (No. 50) were prepared according to added percentage of (2,4, and 8)% by weight. After mixing the three sizes were stored in other container for final heating and mixing with aggregates. The total weight of crumb of tyres was reduced from the weight of aggregate for each selected sizes. The combined aggregates were put in the oven for heating to the temperature of mixing.
Before mixing the aggregate with asphalt, and the combined crumb of tyres were put in oven for (15) minutes and after that added to the combined aggregate then the asphalt binder was added for mixing together.

3.3. *Resistance to plastic flow (Marshall Method)*
This method covers the measurement of the resistance to plastic flow of cylindrical specimens of bituminous paving mixture loaded on the lateral surface by means of the Marshall apparatus according to (ASTM D-1559). The test specimens were compacted using one compact effort, which is (75) blows each surface. The bulk specific gravity density (ASTM D-2726), theoretical (maximum) specific gravity (ASTM D-2041) and percent air voids (ASTM D-3203) are determined for each specimen. Marshall Stability and flow tests were performed on each specimen according to the method described by (ASTM D-1559). The cylindrical specimens (2.5" (62.5 mm) height * 4" (101.6 mm) diameter) were compressed on lateral surface with a constant rate of (50.8 mm/min) until the maximum load was reached. The

maximum load resistance and corresponding flow values were recorded. Three specimens for each combination were prepared and the average results were reported.

3.4. Indirect tensile strength test

The same methods illustrated in items (6.1. and 6.2.) were used for preparing (24) samples for Indirect Tensile Strength test (ITS) according to (ASTM D-4123). The specimens were left to cool at room temperature for 24 hours, immersed in a water bath at 250C for 30 minutes, then tested for ITS at rate of 50.8 mm/min. (2 in/min) in a compression machine until recording the ultimate load resistance.Equation (1) used for calculating the indirect tensile strength after testing the specimens is [16].

$$ITS = 2P/\pi TD \quad (1)$$

Where:
ITS = indirect tensile strength (calculated for three specimens). N/mm^2.
P = Ultimate applied load required to fail specimens (N).
T = Thickness of the specimen, mm.
D = Diameter of the specimen, mm.

4. Results and Discussion

The data of the studyare presented in table 3 and table 4.

Table 3. Marshall Properties of Asphalt Mixture

No	% Crumb of Tires	Stability KN			Flow mm			% Air voids			Theoretical Density			Bulk density gm/cm^3		
		% asphalt			% asphalt			% asphalt			% asphalt			% asphalt		
		4	5	6	4	5	6	4	5	6	4	5	6	4	5	6
1.	0	9.45	7.33	6.63	2.7	3.2	4	3.8	2.4	1.9	2.49	2.46	2.42	2.397	2.406	2.374
2.	2	10.27	9.8	9.52	4.9	5.8	6.7	6.9	3.8	2.8	2.483	2.429	2.393	2.328	2.349	2.34
3.	4	9.03	7.95	7.8	5.6	6.6	7.1	7.8	6	4.3	2.447	2.414	2.384	2.269	2.283	2.298
4.	8	6.95	6.37	5.8	6.8	7.3	8	8.8	6.4	5.6	2.396	2.351	2.335	2.198	2.215	2.19

Table 4. Indirect tensile strength properties of the asphalt Mixture.

No	% Crumb of Tyres	ITSN/mm^2		
		% Asphalt		
		4	5	6
1	0	0.993	0.744	0.718
2	2	0.684	0.813	0.503
3	4	0.727	0.684	0.624
4	8	0.856	0.521	0.607

According to the asphalt-rubber concrete mixes tested at the specifications followed the result obtained are:

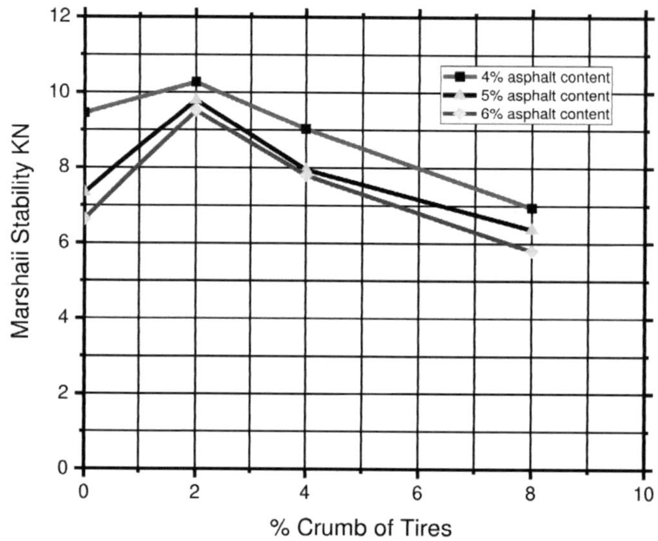

Figure 3. The effect of the crumb of tires on Marshall Stability

Figure 3 shows the effect of the crumb of tires on Marshall Stability at different asphalt contents. Stability increases with increasing Crumb of Tyres content until (2%) after that, there is a reduction in the stability. At (4%) the stability remained near the original value but, at (8%) the stability became less than the original value.

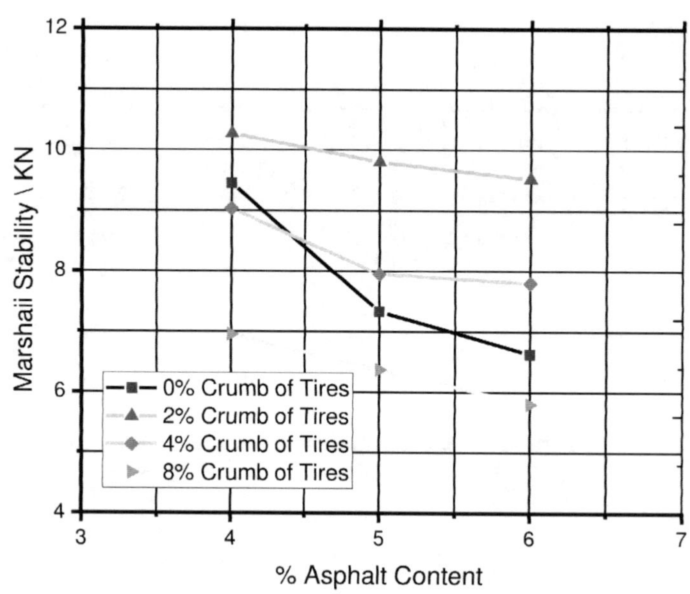

Figure 4. The effect of asphalt content on Marshall Stability

Figure 4 shows the effect of asphalt content on Marshall Stability at the different crumb of tires contents. Stability decreases with increasing asphalt content but, the rate of decreasing with crumb of tires content is lower compared with the original mixture.

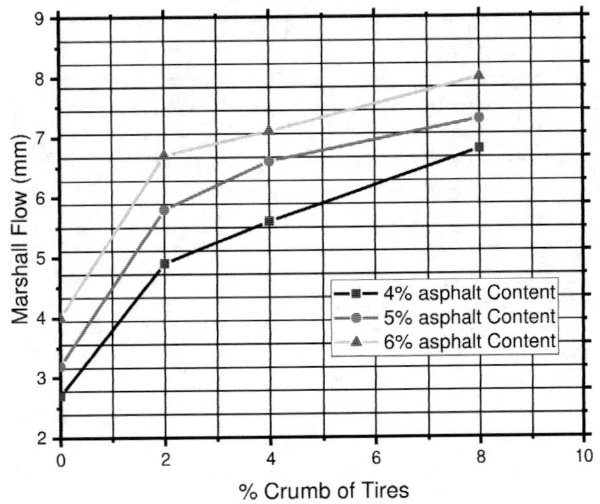

Figure 5. Effect of Crumb of Tires Contents on Marshall Flow of Asphalt Mixture at Different Asphalt Contents.

Figure 5 shows the effect of the crumb of tires on Marshall flow. The flow increases with increasing in crumb of tires contents. The percent of increase at (2%) is about (75%) more than the original value. It increased gradually until it reached more than (100%) as compared to the original value. To explain this behavior, the Marshall flow with a crumb of tires was composed of deformations among the particles of aggregate and the deformations in the crumb of Tyres particles, while in the original mixture deformations were among the particles of aggregateonly.

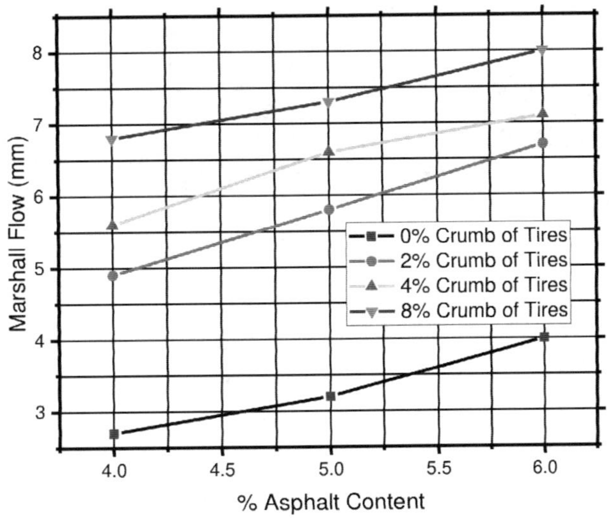

Figure 6. Effect of Asphalt Content on Marshall Flow of Asphalt Mixture at Different Contents of Crumb of Tires.

Figure 6 shows the effect of asphalt contents on Marshall flow at the different crumb of tires contents. The flow increases with increasing the asphalt content for different crumb of tires content. The flow increases with increasing Crumb of Tyres contents.

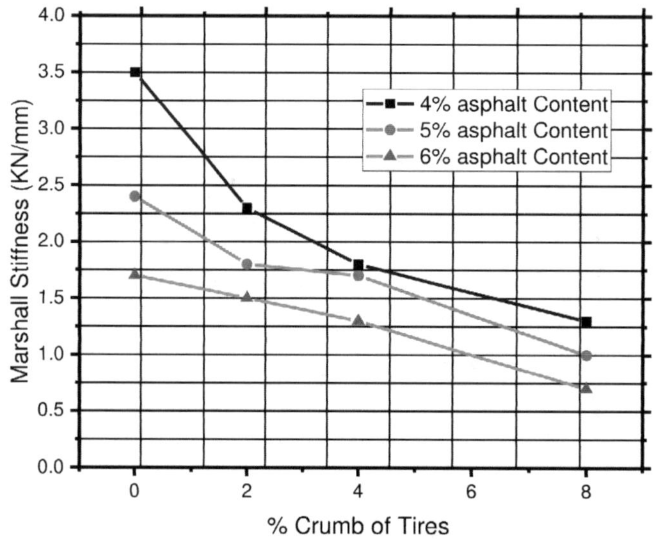

Figure 7. Effect of Crumb of tires Contents on Marshall Stiffness of Asphalt Mixture at different Asphalt Contents

Figure 7 shows the effect of Crumb of Tyres on Marshall stiffness,where:-

$$\text{Marshall Stiffness} = (\text{Marshall Stability} / \text{MarshallFlow})(\text{kN/mm}) \quad (2)$$

Marshall Stiffness decreases with increasing crumb of tires contents. The rate of variance decreased with increasing crumb of tires content. This result is natural because the flow increases with increasing the crumb of tyrecontents.

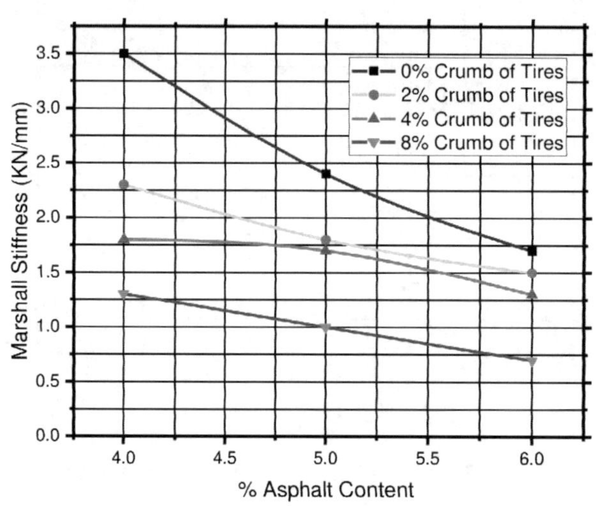

Figure 8. Effect of Asphalt Contents on Marshall Stiffness of Asphalt Mixture at the differentcrumb of tires contents.

Figure (8) shows the effect of asphalt contents on Marshall Stiffness fordifferent crumb of tires contents. The rate of variance in Marshall Stiffness decreases with increasing crumb of tires contents,whereas the rate of change in Marshall Stiffness for the original mixture is higher than mixture content crumb of tires contents.

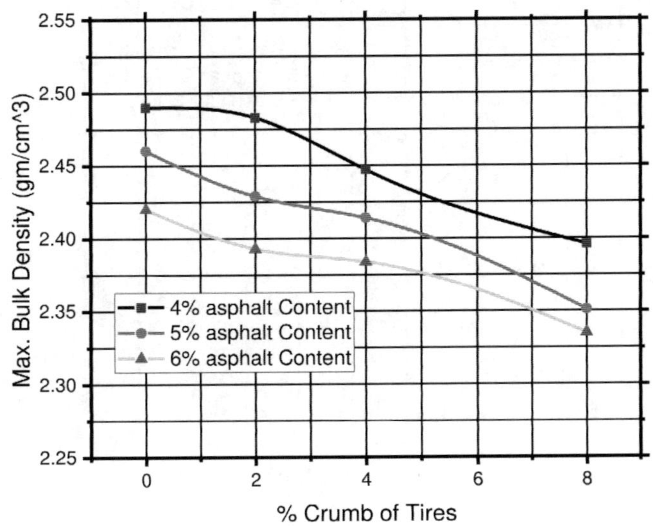

Figure 9. Effect of Crumb of tires Contents on Max Density of Asphalt Mixture at Different of Asphalt Contents.

Figure 9 shows the effect of the crumb of tires contents on maximum density. The crumb of tires content has a significant effect on maximum density because the specific gravity of crumb of tires is less than the specific gravity of aggregate.

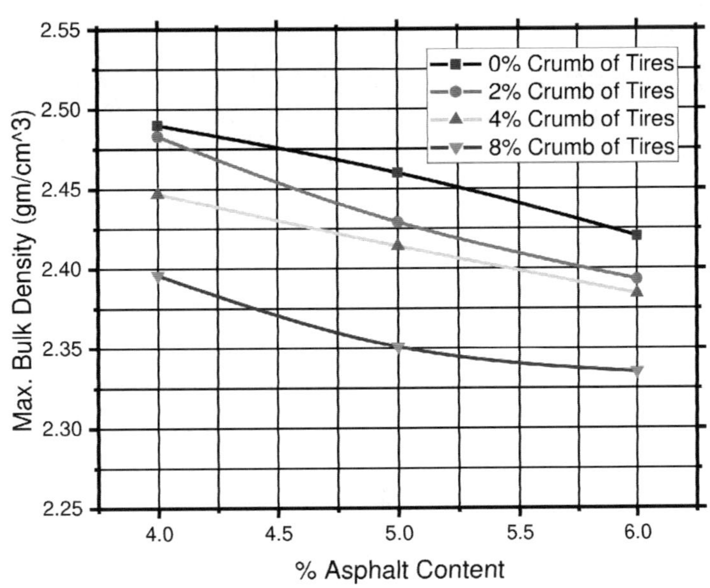

Figure 10. Effect of Asphalt Content on Max Density of Asphalt Mixture at Different Crumb of tires Contents.

Figure 10 shows the effect of asphalt content on maximum density for different crumb of tires contents. There is a decrease in maximum density with increasing asphalt contents.

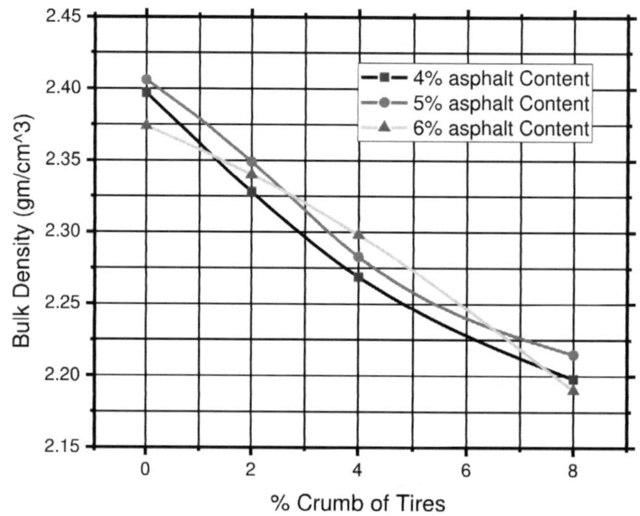

Figure 11. Effect of Crumb of tires Contents on Bulk Densityof Asphalt Mixture at Different Asphalt Contents.

Figure 11 shows the effect of the crumb of tires contents on the bulk density of mixture. The crumb of tires contents has a significant effect on bulk density. By increasing the crumb of tirescontents, the bulk density of mixture decreases because the specific gravity of crumb of tires is less than the specific gravity of aggregate and will operate larger volume as compared to the same weight of aggregate.

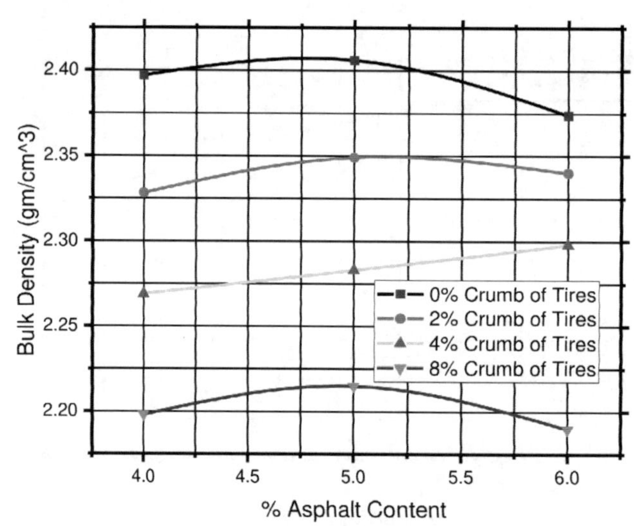

Figure 12. Effect of Asphalt Content on Bulk Density of Asphalt Mixture at Different Content of Crumb of tires.

Figure 12 shows the effect of asphalt contents on the bulk density of mixture for different crumb of tires contents. The variance in asphalt contents hasnota significant effect on the bulk density of mixture for a different crumb of tires contents.

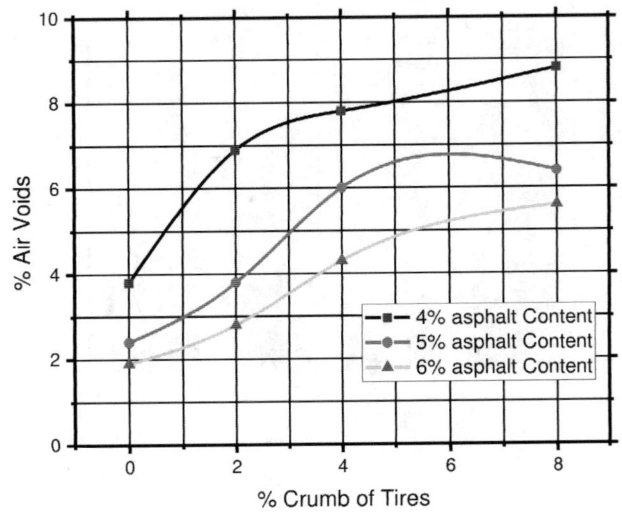

Figure 13. Effect of Crumb of tires Contents on Air Voids of Asphalt Mixture at Different Asphalt Contents

Figure 13 shows the effect of crumb of tires contents on the percentage of air voids in mixture. Air voids increase with increasing of crumb of tires contents because crumb of tireshavespecific gravity less than the specific gravity of the aggregate. On the other hand, the combined aggregate with crumb of tires particles has surface areagreater than pure aggregate Therefore, percent of air voids increase with increasing crumb of tires contents.

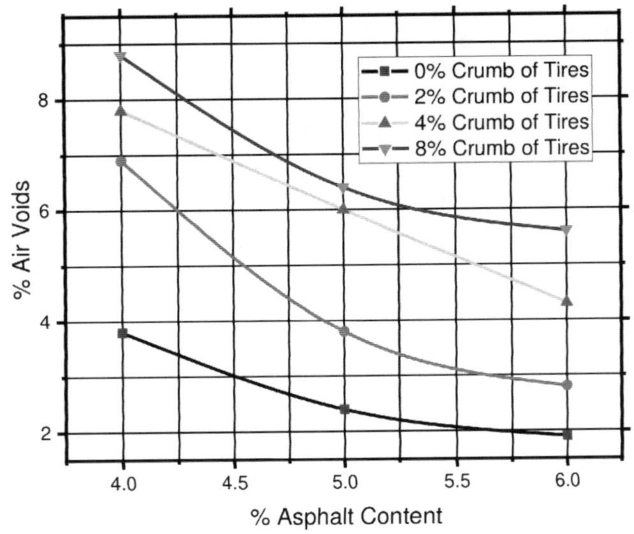

Figure 14. Effect of Asphalt Contents on Air Voids of Asphalt Mixture at Different Contents of Crumb of tires.

Figure 14 shows the effect of asphalt content on the percentage of air voids in mixture for different crumb of tyres content. Air voids decrease with increasing of asphalt content.

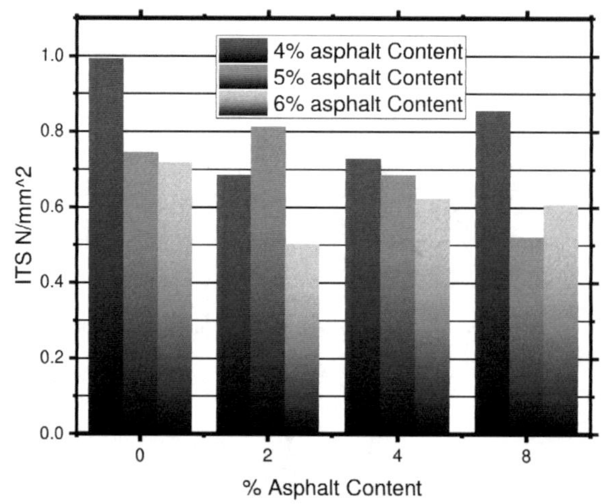

Figure 15. Effect of Crumb of tires Contents on Indirect Tensile Strength of Asphalt Mixture at Different Asphalt Contents

Figure 15 is a Histogram showing the effect of crumb of tires on the Indirect Tensile Strength (ITS) of asphalt mixture at different asphalt contents. The behavior of the variations gradually decreases at (0 and 4)% of crumb of tires additive with increasing asphalt contents. Variations at otherpercent's of added crumb of tires take random behavior.

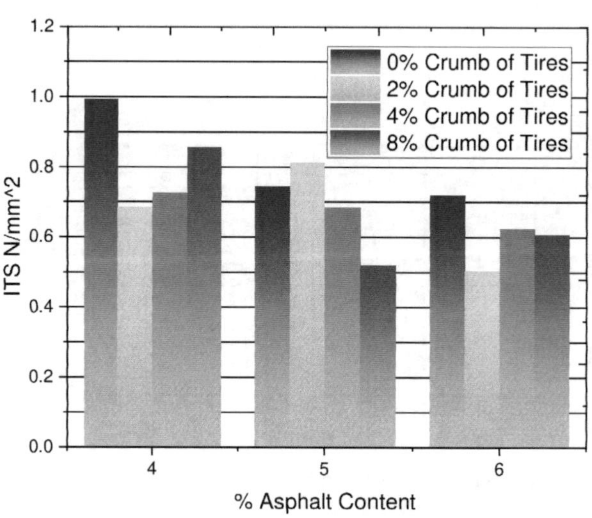

Figure 16. Effect of Asphalt Contents on Indirect Tensile Strength of Asphalt Mixture at Different Crumb of tires Contents

Figure 16 is a Histogram showing the effect of asphalt contents on the Indirect Tensile Strength (ITS) of asphalt mixture at different crumb of tires contents. The behavior of the variations for the (ITS) at (4 and 6)% of asphalt contents increases with increasing the crump of tyres contents but, the percent of reduction in (ITS) as compared to original mixture (0% crump of tyres), is about (10% to 30%). At asphalt content (5%) (ITS) decreases with increasing crumb of tires contents, but the original mixture (0% crump of tires) is less than (2% crump of tires).

5. Conclusions and Recommendations

The study showed that the utilization of the crumb of tires have more effect on asphalt mixture performance as follow:-

a) Increasing the Marshall stability until (2%) Crumb of tirescontent and reduction in Stability after that percentage.
b) Increasing crumb of tires causes increasing in Marshall Flow about (75 - 100)%,in the same time anincrease in the air voids content compared tothe original mixture.
c) The crumb of tires additives caused a decreasing in the bulk density of the asphalt mixture.
d) Increasing crumb of tires causes a reduction in the indirect tensile strength about (10 to 30)% less than original mixtures.
a. Based on the results of this study, the effect of the crumb of tires on asphalt mixture performance at high temperature, need to a furtherstudy, moreover, the effect of aging onthe performance of asphalt mixture containing crumb of tires.

References

[1] Nuha Salim Mashaan "A Review on Using Crumb Rubber in Reinforcement of Asphalt Pavement" Hindawi Publishing Corporation the Scientific World Journal Volume 2014, Article ID 214612, 21 pages

[2] Munder A Bilema "Moisture Sensitivity of Crumb Rubber Modified Modifier Warm Mix Asphalt Additive for Two Different Compaction Temperatures" IOP Conference Series: Earth and Environmental Science.

[3] Sanjay Ghuge "A Study On The Properties Of Bitumen & Aggregate By Replacing Waste Tyre Rubber With Bitumen & Aggregate Construction Of Road Pavement" International Research Journal of Engineering and Technology (IRJET) Volume: 05 Issue: 01 | Jan-2018.

[4] California Department of Transportation "**Cost Differential Analysis Between Asphalt Containing Crumb Rubber and Conventional Asphalt**" Public Resources Code Section 42703, 2015 CRUMB RUBBER REPORT.

[5] Todd Genovese "**Asphalt Binder Modification with Ground Tire Rubber**" CMS Identification Number: 32162 Original Contract Routing Number: PO FAA EHS12032162.

[6] Taher M.A. Al-ani "**Modification of Asphalt Mixture Performance by Rubber-Silicone Additive**" Anbar Journal of Engineering Sciences (AJES-2009).

[7] Munder A Bilema, Mohamad Y Aman "Moisture Sensitivity of Crumb Rubber Modified Modifier Warm Mix Asphalt Additive for Two Different Compaction Temperatures" IOP Conf. Series: Earth and Environmental Science 140 (2018) 012072.

[8] Katelyn Nicole Love "INFLUENCE OF DIFFERENT RUBBER MODIFIED BINDER VARIABLES ON PERFORMANCE PROPERTIES OF POROUS ASPHALT MIXTURES" May 2014, Clemson University.

[9] Serji N. Amrikhanian, Ph.D. " **Utilization of Crumb Rubber in Asphalt Concrete Mixture-South Carolina's Experience**" Department of Civil Engineering Clemson University , January 2001.

[10] https://www.vaasphalt.org/used-tires/

[11] Saad Issa Sarsam "**A Study On Aging and Recycling of Asphalt Concrete Pavement**" University of Sharijah Journal of Pure and Applied Sciences, Volume 4, No. 2, June 2007.

[12] Freddy L. Roberts, T.L. James Louay N. Mohammed, Hanlin Qin, and Baoshan Huang "**Comparative Performance of Rubber Modified Hot Mix Asphalt Under ALF Loading**" LTRC, July 2004. Technical Summary Report 374.

[13] Raden Hendra Ariyapijati "**Contributions Crumb Rubber in Hot Mix Asphalt to the Resilient Modulus**" Green Process, Material, and Energy: A Sustainable Solution for Climate Change

AIP Conference Proceedings 1855, 030005 (2017).

[14] H.T. Tai Nguyen "Effects of crumb rubber content and curing time on the properties of asphalt concrete and stone mastic asphalt using dry process, International Journal of Pavement" Research and Technology 11 (2018) 236–244.

[15] Yuqiao Yang "**Preparation and Performance of Asphalt Compound Modified with Waste Crumb Rubber and Waste Polyethylene**" Hindawi Publishing Corporation Advances in Materials Science and Engineering Volume 2016, Article ID 5803709, 6 pages.

[16] Asal Farid Naom "**Some Investigation Into The Properties Of Polymer Modified Asphalt Cement**" Thesis, Department of Civil Engineering Baghdad University, November 1999, P 31.

[17] Clayton Daniel Plemons"**Evaluation of the Effect of Crumb Rubber Properties on the Performance of Asphalt Binder**" Thesis , Department of Civil Engineering Auburn University, May 5, 2013.

Blending Between Aged and Virgin Asphalt Binders in Recycled Pavements: A Review Study

P Pirzadeh[1, *], S Kapilan[2] and P Kriz[3]

[1]Asphalt Americas BSA Technology Center, Sarnia Technology & Application Research, Imperial Oil Ltd., Sarnia, Canada
[2]ExxonMobil Asia Pacific Private Limited, #06-00 Harbour Front Tower One, Singapore
[3]Asphalt Technology, ExxonMobil Research & Engineering Co., Sarnia, Canada
[*] Corresponding author: Payman.Pirzadeh@esso.ca

Abstract. It is the intent of this work to provide a background on works on diffusion in asphalt binders and review the researches carried out at Imperial Oil and ExxonMobil Technology Centres which have investigated nature of diffusion in asphalt material. These investigations include diffusion between Reclaimed Asphalt Pavement (RAP) and virgin binders, progress of diffusion in laboratory-produced mix samples from rheology perspective, and impact of silo storage on diffusion progress between binders and its manifestation on the performance of plant-produced asphalt mix. RAP is a major component of manufacturing hot mix asphalt (HMA) in North America. RAP utilization promotes sustainability of asphalt industry with economic and environmental incentives for manufacturing process. Recent statistics in North America show that RAP content of HMA has increased from 15 to 20%. This RAP content increase further prompts development and implementation of science-based practices in manufacturing plants to achieve optimum quality of RAP containing mixes. Quality of blending between aged and virgin binders has been shown to significantly impact performance and durability of final mix in the field. It has been demonstrated that mechanical blending helps achieving effective contacts between RAP and virgin binder. However, diffusion is the dominant process to help blending between the binders. It has been demonstrated that binder diffusion follows Fick's Law and the rate increases with temperature. Field verification indicated that around 12 hours is required to complete the diffusion at the typical hot mix production and silo storage temperatures. Mix samples with more-progressed diffusion exhibited improved rutting resistance and higher number of cycles to fatigue failure. Field work also validated that there is an optimum storage time during which diffusion is the dominant process impacting mix properties; but eventually mix hardening dominates due to oxidation, evaporation and absorption.

1. Introduction

Reclaimed Asphalt Pavement (RAP) has been part of asphalt mixture manufacturing process for several decades [1-5]. The economic and environmental incentives have been the main driving force behind increased usage of RAP [6-8]. Recent report from National Asphalt Pavement Association (NAPA) suggests that the US national average RAP content in asphalt mixture production has increased from 15% to 20% between 2009 and 2016 [9]. However, extra caution should be exercised when RAP is utilized in a new mix. This additional caution is to (1) manage the increased stiffness of the aged binder in RAP, (2) achieve the correct final mix gradation when RAP aggregates get incorporated and (3) reach proper cohesion between virgin and RAP binders as well as adhesion to aggregates [5]. Reaching the appropriate cohesion and adhesion is impacted by the degree of blending between the new virgin and aged RAP binders [5, 10, 11]. Blending between virgin binder and aged RAP binder can be modelled to occur in two stages: mechanical mixing and diffusion. Mechanical mixing mainly occurs during manufacturing process, where mixture of virgin and RAP aggregates are combined with virgin binder and are mechanically mixed together. Mechanical mixing brings the RAP and virgin-binder-coated aggregates next to each other and produces a contact surface between the two binders to interact with each other. Plant design, operating temperature, production rate and residence time during mixing can impact the quality of mechanical mixing [7, 12-21]. Diffusion, on the other hand, starts when contact surface between the binders are established and fractions of binders migrate in the direction of concentration gradient, and eventually reach an ideally uniform distribution between the binders. Previous studies mainly on asphalt binders and laboratory mix samples have indicated that diffusion is the major process that governs the blending quality between RAP and virgin binders [12, 22-24]. It is the intent of this work to provide a brief background on diffusion studies in asphalt binders and review the research carried out at Imperial Oil and ExxonMobil Technology Centres on the nature of diffusion between RAP and virgin binders. These studies span a range of investigations starting from rheological examination of diffusion progress in binders [22] up into laboratory-produced mix samples [23].

One of the earliest works on diffusion in asphalt binder was the work by Oliver using radioactive tracers [26]. Rejuvenators/recycling agents were radioactively tagged and their concentration-time profile were utilized to estimate an average diffusion coefficient. Oliver observed an Arrhenius type dependence of average diffusion coefficient on temperature. Carpenter and Wolosick later investigated diffusion of rejuvenators into RAP binders through staged solvent washing and binder recovery of RAP-containing mixes [27]. Their work indicated that the distinction between rejuvenator-rich and aged asphalt-rich recovered layers disappeared as blending time got longer, and more uniform binder was obtained at longer blending times. Other researchers, who also utilized multistage solvent recovery have pointed to the fact that not all the RAP binder might necessarily be participating in diffusion, which results in formation of inhomogeneous binder blends around aggregates [28-31]. Rrejuvenators diffusing into asphalt binders were subject of study through Attenuated Total Reflectance-Fourier Transform Infrared spectrophotometry (ATR-FTIR) [32, 33]. FTIR signals of specific chemical groups in rejuvenator markers were used to construct the concentration-time (either peak height or peak areas) profile. Results indicated that Fick's Law of diffusion could be applied to the data and diffusion coefficient were estimated for the rejuvenators. The ATR-FTIR data also supported the fact that diffusion coefficient followed an Arrhenius-type behavior as a function of temperature. On a different approach, same authors later developed a method to verify diffusion in asphalt binders through dynamic shear rheometers (DSR) [34], which was further refined by Kriz et al. [22, 23]. Their results confirmed the observation from ATR-FTIR study, i.e. an Arrhenius type of behaviour of diffusion coefficient was also noted in examining rheological properties of binders. Recently, there has been developments in the field of fluorescent microscopy to track fluorescence-active components in asphalt binders to examine diffusion between RAP and virgin binders [35, 36]. Results suggested that diffusion coefficients estimated from DSR methods are a bit larger than those estimated from Fluorescence method, however, both are within the same order of magnitude.

Investigations on blending quality and diffusion progress in RAP-containing HMA produced in asphalt plants suggest that a majority of studies carried out have mainly focused on mechanical mixing of the binder-coated aggregates [12-15]. Given the complexities associated with conducting plant trials, controlling the thermal history of samples is not trivial, specifically post-manufacturing before examining the collected samples start. Lack of consistency in thermal treatment of samples typically has resulted in unusual trends in mechanical and performance properties. For example, it has been reported that dynamic modulus of RAP-containing mixes were less than the control virgin mixes. These studies usually did not control the required time to achieve a homogeneous blend between virgin and aged binder. The unclear thermal history of samples makes it difficult to properly track diffusion progress, and draw clear conclusion on the impact RAP on performance of the manufactured asphalt mix.

2. Objective

It is the intention of this work to provide a review specifically on diffusion process and its application to blending between virgin and aged binder in RAP. This review covers the studies where thermal history of samples were tracked, whether in the lab or in the field, to ensure the observed changes in properties of asphalt material, both binder and mix, are due to progress of diffusion [22, 23].

3. Diffusion in asphalt

3.1. Diffusion between virgin and RAP binders

Diffusion is a kinetic phenomenon that is driven by differences in concentration of material. Fick's Law is the differential equation that connects evolution of concentration (ϕ) in time and space with the diffusion coefficient (D) under appropriate boundary conditions:

$$\frac{\partial \phi}{\partial t} = D \frac{\partial^2 \phi}{\partial^2 x} \tag{1}$$

where t and x represent time and distance of diffusion, respectively, and ϕ can describe the asphalt binder concentration. The diffusion coefficient (D) is a function of temperature. Dependence of D on temperature is typically formulated in the form of Arrhenius equation:

$$D = Ae^{\left(-\frac{E_a}{RT}\right)} \tag{2}$$

where A, E_a, R and T are pre-exponent factor, activation energy, universal gas constant and temperature (in Kelvin), respectively. Detailed investigation of diffusion in asphalt binders has not been an easy task due to complexities associated with asphalt binder, e.g. diversity of molecular composition and relatively high viscosity of the binder at ambient temperatures. Also tracking progress of diffusion requires a very accurate and stringent control over sample thermal history. In the work of Kriz et al. [22,23], ϕ was considered for RAP, and its evolution over time was tracked. Utilizing an expanded differential equation solution to Equation 1, and applying separation of variables and appropriate boundary conditions, the following integrated solution to Equation 1 was obtained:

$$\phi_{RAP}(x,t) = (1-\alpha)\left(1-\phi_0^{virgin}\right) - \frac{2\left(1-\phi_0^{virgin}\right)}{\pi}\sum_{n=1}^{\infty}\frac{\sin(n\pi\alpha)}{n}\cos\left(\frac{n\pi x}{L}\right)\exp\left(-\left(\frac{n\pi}{L}\right)^2 Dt\right) \tag{3}$$

where L is the total diffusion distance in DSR experiments, and α represents the fraction of RAP in a blend with virgin binder. $\phi_{RAP}(x,t)$ represents concentration of RAP at a particular location x along the diffusion distance L, i.e. DSR gap, at a time t. Viscosity of the blend at a particular location x is dependent on the relative concentrations of RAP and virgin binders at that location. Utilizing the blending rule [22,24,34], the instantaneous viscosity of the blend at a particular location of diffusion distance x at a time t can be estimated as:

$$\eta(x,t) = e^{\phi_{RAP}(x,t)\ln \eta_{RAP} + [1-\phi_{RAP}(x,t)]\ln \eta_{virgin}} = \frac{\eta_{RAP}^{\phi_{RAP}(x,t)}}{\eta_{virgin}^{(\phi_{RAP}(x,t)-1)}} \tag{4}$$

Integrating Equation 4 over the diffusion distance, i.e. DSR gap, gives the viscosity of the blend as a function of time, which is measured by DSR instrument. The rate of viscosity change is then governed by diffusion coefficient (D). Therefore, fitting the measured viscosity data from the integrated Equation 4 to Equation 3 provides an estimate of diffusion coefficient.

In their study on binder diffusion, Kriz et al. [22] examined the diffusion phenomenon in a 50:50 blend of RAP binders with Superpave grades of PG 94-10, PG 88-22 and PG 82-22 with virgin binders of PG 58-28 and PG 64-22. However, in addition to the more comprehensive mathematical model to measure diffusion coefficients, Kriz et al. [22] also looked into factors that could influence measuring diffusion coefficients, including:

- Impact of experimental setup on data acquisition, i.e. influence of evaporation and oxidation on viscosity profiles, sample thickness, impact of gravity and applied strain level
- Understanding impact of temperature on diffusion progress, i.e. applicability of Fick's Law to RAP-virgin binder diffusion
- Estimating diffusion coefficient based on correlation with liquid viscosity according to free volume theory of diffusion

Diffusion is a kinetic process that temperature plays a significant role in its progress. According to Equation 2, higher temperatures favor diffusion progress. However, higher temperatures also favor other processes such as evaporation and oxidation. Occurrence of the latter processes can have detrimental impact on the quality of material, i.e. binders become stiffer, and can impact progress of diffusion. Therefore, as viscosity of the RAP-virgin binder blend evolves over time towards the ideal blend viscosity, corrections should be applied to distinguish the impact of diffusion from evaporation and oxidation. Such correction was applied by first measuring the complex viscosity (η^*) of an ideal blend of RAP and virgin binder over the same duration of a diffusion experiment. The incremental increase in η^* towards the end of measurement time was considered as evaporation and oxidation influence on η^* (see Figure 1). This increase was then subtracted from raw η^* profile of diffusion process to obtain the sole influence of diffusion. Figure 1 provide a summary of correction procedure applied to measured η^* of the binder blend before fitting Equation 3 to estimate diffusion coefficient (D).

Influence of sample thickness on measured diffusion coefficients was also examined. Binder blends of 100-500μm thickness, i.e. DSR gap size, were assessed. A gradual increase in diffusion coefficient was noted as the sample thickness increased. Impact of binder layer densities on the measured diffusion coefficient was investigated under two scenarios: (i) RAP sample in the top layer, (ii) RAP sample in the bottom layer. The collected data demonstrated an increase and a decrease in the measured diffusion coefficients with increased sample thickness for scenarios (i) and (ii), respectively. This impact of layer configuration on measured diffusion coefficients pointed to the potential impact of barodiffusion, meaning that the difference in density of RAP and virgin binder could have resulted in appearance of a gravitational driving force. This driving force aids diffusion when the denser RAP layer is on the top. Therefore, it was found to set the sample thickness to a minimum of 100μm to minimize dependence of diffusion coefficients to sample configuration while maintaining reasonable repeatability of results. The impact of strain level on the measured diffusion coefficient was also explored. It was demonstrated that higher strain levels in DSR dynamic testing resulted in artificially larger diffusion coefficients due to more intense mechanical mixing. Therefore, it was recommended to conduct the experiment at lower strain levels within the linear regime of viscoelastic properties, e.g. strain of 1%.

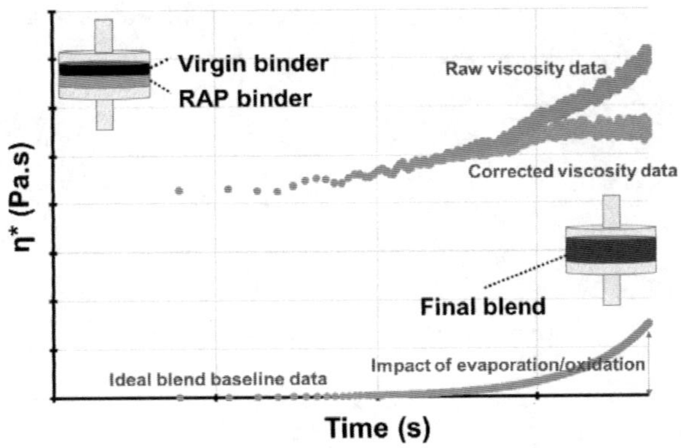

Figure 1. Collection and correction of complex viscosity (η^*) profile of diffusion between virgin and RAP binders

At the beginning, there are two separate binder layers. As diffusion progresses and binders blend, viscosity starts to increase (blue dots). Part of viscosity increase is due to evaporation and possible oxidation during the test. A viscosity profile of the ideal blend of the two components is collected to estimate impact of evaporation and oxidation on viscosity. The constant ideal blend viscosity is subtracted from this function to obtain the increase function due to oxidation & evaporation (orange dots). Next, this viscosity-increase function of blend is subtracted from experimentally measured η^* data (blue dots) to separate the effects of diffusion and aging processes during DSR experiment (gray dots).

The diffusion coefficients between the listed binders were then measured, utilizing the corrected η^* data at different temperatures using the optimized experimental setup. Corrected η^* data were fit to the integrated Fick's Law of diffusion in Equation 3. Fitting was performed under the assumption that the maximum achievable η^* would be that of the ideal 50:50 blend. The collected data, presented in Figure 2, suggested that diffusion coefficients followed the Arrhenius functional form in Equation 2 for the influence of temperature.

Kriz et al. [22] also utilized a simplified version of free volume theory to estimate diffusion coefficients of straight run asphalt binders from their measured viscosity–temperature profiles. Figure 3 depicts an example of viscosity-temperature profile of a PG 52-34, a sample RAP binder and the 50:50 ideal blend of these two binders. The viscosity of a 50:50 ideal blend of the virgin and RAP binders is estimated from their corresponding viscosity-temperature data using the general form (i.e. independent of time and location) of blending rule presented in Equation 4. The slope of the viscosity-temperature data of the blend provides the activation energy of flow ($E_{a\eta}$). The diffusion activation energy (E_a) and the pre-exponential factor (A) in Equation 2 can then be calculated through a set of calibration equations derived from the free volume theory. The latter numbers can then be plugged into Equation 2 and diffusion coefficients of the virgin and RAP binders can be estimated at different temperatures. It was found that the differences between the measured and the estimated diffusion coefficients were well within the uncertainties associated with experimental measurements and approximations applied to physical models.

With the developed models and estimated diffusion coefficients, an attempt was made to explore how long it might take the diffusion process between RAP and virgin binders to approach completion in a real case of hot mix asphalt (HMA). A hypothetical manufacturing and construction temperature profile was simulated for two average thicknesses of a binder blend in a HMA, 20 and 100 µm. Figure 4 summarizes the results of this simulation for the modelled hot mix asphalt (HMA) samples. It was found that thinner binders and higher temperatures facilitate diffusion to reach completion. Based on simulations, about 20% and 70% of the ideal blend viscosity might be achieved at the end of the manufacturing process for a HMA with 100 µm and 20 µm average binder thickness, respectively. Upon storing the HMA in a silo at a typical temperature of 140°C, the viscosity of the HMAs can reach between 80-100% of that of the fully blended binders.

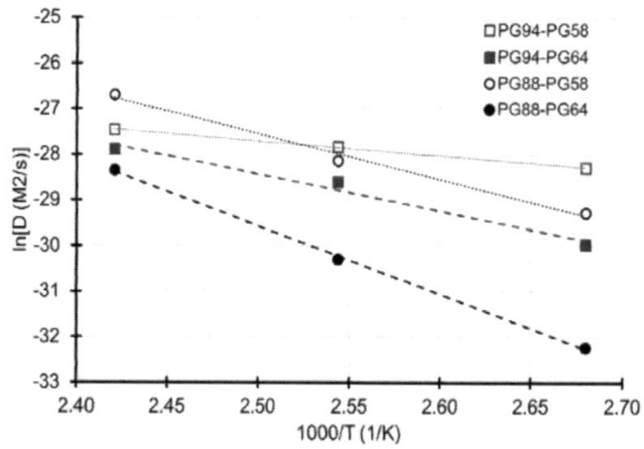

Figure 2. Estimated diffusion coefficients of various RAP and virgin binders estimated from fitting corrected measured η^* data to Equation 3

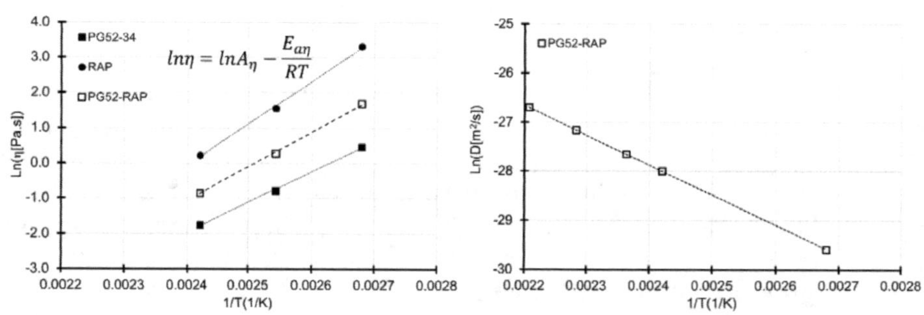

Figure 3. (Left) viscosity-temperature data for a PG 52-34 virgin binder and a recovered RAP binder using blending rule (Right) estimated diffusion coefficient for PG 52-34 and RAP at different temperatures

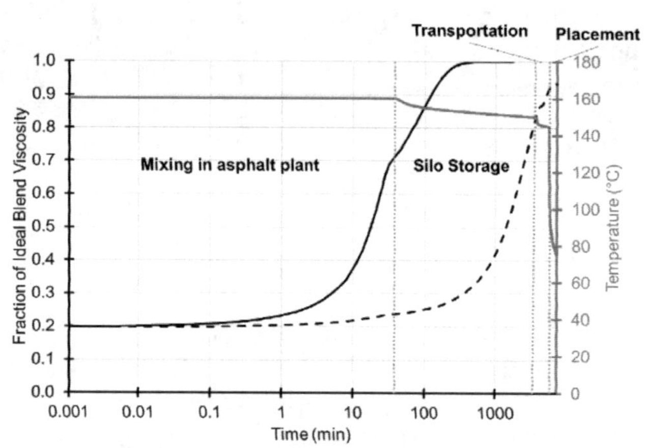

Figure 4. Evolution of blending between RAP and virgin binders

Two different binder thicknesses were considered: 20 μm (solid black) and 100 μm (dashed black). Temperature profile is presented in solid red line. Depending on the average diffusion distance, i.e.

binder thickness, the degree of blending could be significantly different at the end of manufacturing process. However, silo storage can help further progress of blending and better blending can be achieved.

3.2. Impact of diffusion on laboratory-produced RAP-containing asphalt mix

Investigation on quality of blending between virgin and RAP binders have been mainly conducted on laboratory-produced asphalt mixes [10-12, 16, 20, 23, 30, 31]. However, utilizing DSR instrument to monitor rheological properties of an asphalt mix has been a novel approach. Kriz et al. [23] had utilized torsional bars and monitored changes in complex viscosity of RAP-containing asphalt mixes. This approach helped better understanding in situ progress of diffusion between virgin and RAP binders from rheological perspective.

A mix design for a surface course (based on Ontario Canada specifications) was considered with 30% RAP. The utilized virgin and RAP binders had a Superpave grade of PG 58-28 and PG 82-10, respectively, same materials studied in section 3.1. This study included four mix sample sets: three control sample sets and one diffusion set. Two of the control sets were produced to measure η^* of asphalt mix samples with individual virgin and RAP binders. These η^* data were fed into the fitting procedure for Equation 3, the blending rule. An additional control mix was produced to measure the η^* for the fully blended RAP and virgin binder with the correct asphalt mix design ratio to assess the impacts of binder evaporation and absorption into aggregates over the duration of diffusion tests. The results from the latter mix was utilized to correct the η^* of the diffusion sample set for evaporation and absorption (a procedure similar to the protocol applied to binders in Section 3.1). All control samples proportionally included washed RAP aggregates as part of their aggregate content to limit the changes in rheology to diffusion between RAP and virgin binders. The mix samples were prepared in small beams and were conditioned at three temperatures of 90, 120 and 150°C in a pressurized aging vessel under Nitrogen atmosphere for various time periods. The temperature profiles were closely monitored to ensure all samples were treated the same way to minimize variability in η^* measurements. After conditioning, samples were mounted on torsional bars to measure η^* of the mix. Measurements were carried out at 20°C. Figure 5 summarizes the corrected measured η^* values for the diffusion sample set.

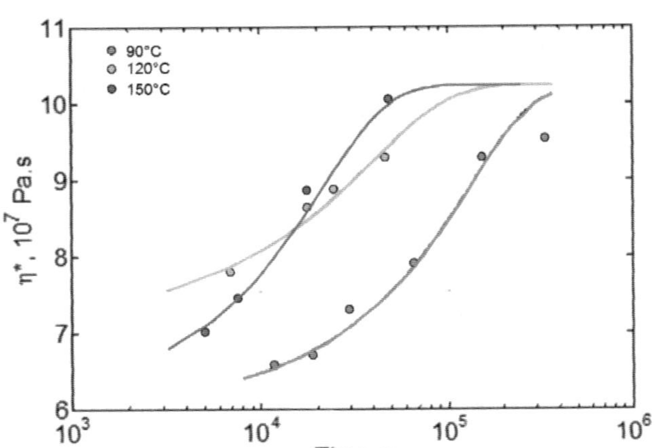

Figure 5. Complex viscosity of asphalt mix samples measured at three different conditioning temperatures

The measured η^* data were corrected for the influence of evaporation and oxidation of each mix at the conditioning temperatures. Measured η^* data were fitted to integrated Fick's Law in Equation 3. The applied diffusion coefficients were from those presented in Figure 2. The best fit to Equation 3 was achieved when the average diffusion distance was set at 800 μm.

Since the binders utilized in the examined mixes were those studied in Section 3.1, the diffusion coefficient between RAP and PG 58-28 binders at 90, 120 and 150°C were calculated from established correlations in 3.1. Similar to a binder diffusion phenomenon, it was observed that η^* of a RAP-containing asphalt mix also evolved faster towards that of the ideal blend at higher temperatures. However, challenges were noted in fitting the measured η^* of asphalt mix to Equation 3, compared to fitting binders' viscosity described in Section 3.1. Heterogeneity in an asphalt mix due to aggregate gradation, air voids, distribution of binders across the sample (voids filled with asphalt), necessitated more flexibility in the range of fitted parameters such as binder fraction (α) and average diffusion distance (L). The best fit of measured mix η^* was achieved when average diffusion distance was set to 800μm (in contrast to 100μm mentioned in Section 3.1). This observation is consistent with a recent study that has measured a relatively wide range for binder thickness in an asphalt mix [38].

Simulations of HMA, conducted on binders in Section 3.1 (presented in Figure 4), were then repeated with 800 μm average diffusion distance. Results indicated that binder blending achieved about 70% of an ideal blend by the time of mix placement on the road. However, diffusion does not stop after pavement cools down to ambient temperature; instead, it will continue to progress with a much slower pace during the course of service life of the pavement. Therefore, the placed mix might experience inhomogeneous distribution of RAP and virgin binders and can become susceptible to local stress accumulation due to variability in stiffness. That potentially exposes the pavement to rutting in places with higher portions of virgin (softer) binder while other areas might be susceptible to fatigue cracking due to higher portions of RAP binder. A second round of simulation with adding silo storage, however, indicated that keeping asphalt mix at typical asphalt silo storage temperature can significantly help diffusion to approach completion. Simulated data suggested that 12 hours of silo storage for HMA at 150°C helped the final complex viscosity to achieve 95% of an ideal blend. This observation can also be noted in Figure 5 for evolution of η^* where ideal blend viscosity was achieved around 12 hours at 150°C. Therefore, this study not only confirmed applicability of diffusion model to asphalt mix and impacts of higher temperature on facilitating progress of diffusion, but also provided an estimated timeframe required for achieving better blended binders in the type of examined asphalt mix, i.e. approximately 12 hours.

3.3. Impact of diffusion on plant-produced RAP-containing asphalt mix

Investigating blending and diffusion in plant-produced material has been a challenging task due to numerous parameters that can impact the properties of final material. As mentioned earlier, plant design and operation as well as thermal history of manufactured samples are among these parameters [4, 5, 13-15].

In both Sections 3.1 and 3.2 the applicability of diffusion model was demonstrated to asphalt binders and mixes. It was also demonstrated that higher temperatures and longer contact time help diffusion of virgin and RAP binder to further progress toward completion (see Figures 3 to 5). It was predicted through simulations that silo storage can have significant impact on progress of diffusion since the manufactured mix is kept at higher temperatures for an extended period of time. Silo storage can potentially be a stage where diffusion between virgin and RAP binders can be promoted. Given the gradual nature of diffusion, it is expected that changes in manufactured asphalt mix properties and performance should be observed over the course of silo storage, as blending between binder approaches completion. It should be noted that since no mechanical mixing occurs in a silo, changes in rheological and performance properties of the asphalt mix could mainly be attributed to the net competing impacts of diffusion, evaporation, oxidation and absorption.

Two asphalt plants (Plants 1 and 2) were asked to manufacture surface (HL-3) and base (HL-8) mix courses according to pavement specifications in the province of Ontario in Canada. These materials spanned a RAP content range from 15% up to 40%. The virgin binders included PG 52-34 and PG 58-28 for base and surface courses, respectively. Samples were collected at 0, 1, 4, 8, 12 and (when possible) 24 hours of silo storage. Mix samples were analysed for volumetric properties, dynamic modulus ($|E^*|$), rutting depths and moisture damage, as well as fatigue cracking performance. Thermal history of samples were tightly controlled to ensure all samples are treated equally, and diffusion progress was mainly limited to silo storage duration.

An example of dynamic moduli data is presented in Figure 6 for the surface mix sample manufactured at Plant 2. It is observed that $|E^*|$ decreases over 12 hours of storage particularly at lower reduced frequencies, corresponding to higher temperatures. This is a counter-intuitive observation since it has been historically observed that silo storage usually results in an increase in mix stiffness [39-44]. This observation points to the fact that diffusion is the dominant process among all other concurrently occurring phenomena such as absorption, evaporation and oxidation up to about 12 hours. The master curve in Figure 6 at 24 hours, however, exhibited an increase in $|E^*|$ values, resembling the historical trends. This behaviour points to the fact that aging processes start to dominate after 12 hours of storage. Measured mix volumetric data also pointed to a dynamic nature of the RAP-containing mix. At a constant air void content, voids in mineral aggregates (VMA) and voids filled with asphalt (VFA) evolved over duration of silo storage. While VMA decreased over time, VFA had an increasing trend. Both properties started to plateau after 12 hours of storage. The direction of changes in VMA and VFA are indicative of diffusion.

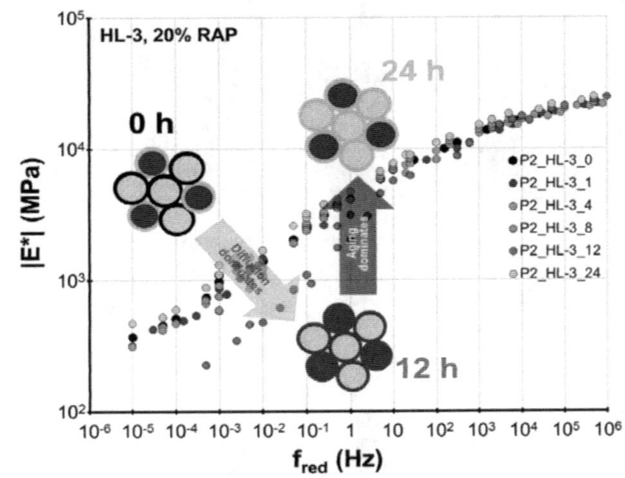

Figure 6. Master curve for absolute value of dynamic modulus (|E*|) for surface asphalt mix from Plant 2

Modulus exhibits a decrease up to 12 hours of silo storage, particularly at lower frequencies, corresponding to higher temperatures. After 12 hours, modulus starts to increase indicating dominance of aging processes during silo storage. At zero hours of storage, RAP binder is still stiff enough that its contribution to the mix is mainly as a filler. As storage time increases, stiff aged binder of RAP gets less viscous and interacts more easily with virgin binder, i.e. they diffuse into each other further. Therefore, more of RAP binder becomes available and binder content of the asphalt mix increases. Thus, part of the previously occupied space with fillers (stiff RAP) become part of the volume occupied by binder, i.e. VFA increases. These observations are in line with similar trends noted on impact of mixing temperature on improved compactability and evolution of volumetric properties of RAP-containing mixes [10,11].

Changes in mix performance parameters such as rutting depths, moisture susceptibility based on stripping inflection point (SIP) and fatigue cracking resistance with silo storage time also confirmed progress of diffusion. Figures 7 to 9 summarize rutting and fatigue cracking performance of the four mix samples as a function of silo storage duration, collected from the two plants. Plant 1 data on rutting depths for both the surface and the base courses suggest a decreasing trend within the errors of the Hamburg Wheel Tracking test. Rutting depths from Plant 2, on the other hand, demonstrate a clear decrease as a function of silo storage duration. It should be noted that the decrease in rutting depths in

Plant 2 data are due to diffusion progress up to 12 hours, as $|E^*|$ data clearly indicated dominance of diffusion process (see Figure 6). Figure 8 presents impact of diffusion progress on stripping inflection point (SIP) of asphalt mixes (this data set is limited to Plant 2 only). Data suggests that with further progress of diffusion between RAP and virgin binders, better adhesion of the binder blend to the aggregates is achieved. Fatigue cracking data, on the other hand, present an apparently different behaviour between the samples of the two plants. While both surface and base courses of Plant 2 reach an almost a plateau after 8 hours of storage, Plant 1 samples exhibit a continuous evolution in fatigue resistance properties during silo storage. Figure 9 suggests that while diffusion is still in progress between RAP and virgin binders in Plant 1 samples, it has achieved a completed stage in Plant 2 mix samples after about 8 hours of storage. As mentioned earlier, parameters such as mixing temperatures, silo temperature, operation of silo (stagnant vs continuous discharge) could have been contributing factors to the behaviour observed in Plant 1 fatigue performance.

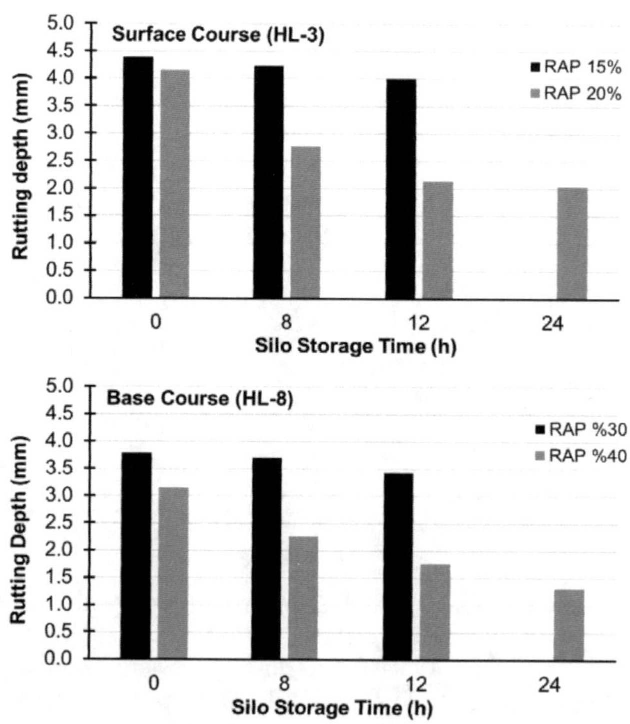

Figure 7. Rutting performance of asphalt mixes as a function of silo storage time

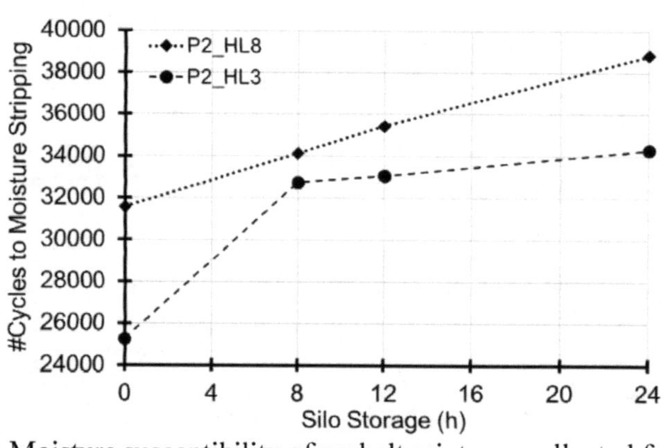

Figure 8: Moisture susceptibility of asphalt mixtures collected from Plant 2

Susceptibility was estimated based on stripping inflection point (SIP) from measuring rutting depths. RAP content of these mixtures was 20% and 40% for the surface (HL3) and base (HL8) courses, respectively.

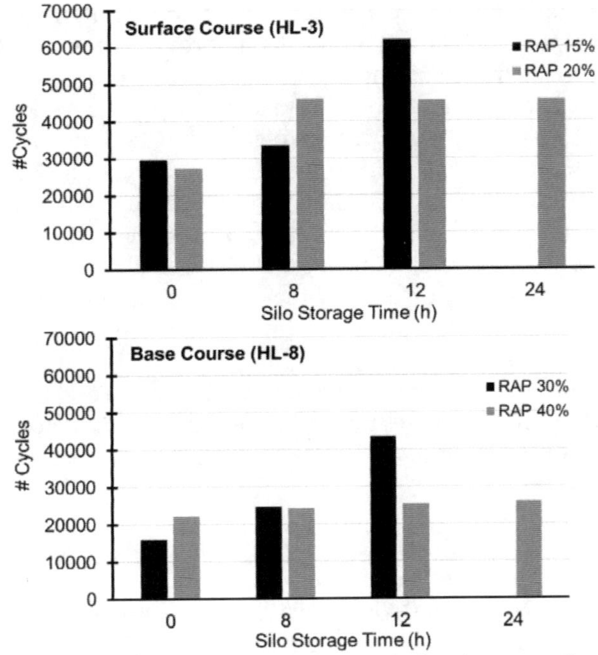

Figure 9: Fatigue performance of the RAP-containing asphalt mixes as a function of silo storage time

4. Concluding remarks

Managing the stiffness of the incorporated RAP, achieving proper cohesion within the binder blend and having proper adhesion to aggregates is important to optimize the quality and performance of a RAP-containing mix. After establishing surface contact between aged binder of RAP and new virgin binder through mechanical mixing, diffusion process is the dominant phenomenon driving the blending between the binders.

Studies carried out by Imperial Oil and ExxonMobil Technology Centres demonstrate that diffusion between RAP and virgin binder follows Arrhenius behaviour; therefore, temperature has a significant impact on progress of diffusion between the aged and virgin binder layers. Higher temperatures and longer contact time facilitate progress of diffusion toward completion. Proper control and tracking of thermal history of manufactured mix can significantly help maximizing the progress of diffusion, while minimizing the influence of evaporation, oxidation and absorption. A proper understanding of diffusion distance distribution, thus the required incubation time for optimal blending, in the final asphalt mix and providing enough driving force to facilitate the diffusion progress are key to optimize the quality of blending between virgin and aged RAP binders. At typical manufacturing temperatures, simulations suggested that only a 20% blending might be achieved right after manufacturing stage if the average diffusion distance is about 100μm.

Tracking evolution of complex viscosity in laboratory-produced mix samples suggested that average diffusion distance might be about 800 μm in an asphalt mix. Such average diffusion distance in asphalt mix samples suggested that approximately 12 hours is needed for diffusion to achieve its final state, i.e. full blending between RAP and virgin binders at an average manufacturing temperature of 150°C.

In both binder and laboratory-produced mix studies, the role of silo storage was emphasized as a platform to promote progress of diffusion. Simulations had indicated that silo storage can help achieving up to 100% of ideal blend viscosity if enough time and driving force, i.e. temperature, is provided. This hypothesis was verified through a plant trial on surface and base asphalt mix samples. Mix samples containing up to 40% RAP were stored in asphalt plant silos and material were examined at different storage times. Dynamic modulus results indicated that diffusion is the dominant process up to 12 hours of storage, after which aging processes take over. This duration is consistent with the performed predictive simulations on laboratory-manufactured asphalt mixes. Progress of diffusion also had noticeable manifestation on reduction of rutting depths, improved resistance to moisture damage and fatigue cracking of the examined mix samples.

The overview of the diffusion studies spanning from laboratory-scale binders blending up to the field verification of asphalt mix samples indicates that RAP-containing asphalt mixes are dynamic systems. Evolution of diffusion between RAP and virgin binders can have significant impact on the binder contribution from RAP and volumetric properties of the mix. This dynamic nature poses a challenge to mix designers, i.e. RAP-containing mixes may not possess the expected properties right after manufacturing stage. It is also possible that unexpected changes occur in mix properties as diffusion between RAP and virgin binders progresses over the service life of a mix. On the other hand, the performance results from the conducted plant trial indicate that progress of diffusion can provide opportunities to optimize performance behaviours of an asphalt mix. Selection of Binder grade and binder content to design a rutting-resistant or a fatigue-resistant mix are directionally opposite to each other. However, utilizing RAP coupled with a maximized progress of diffusion between aged and virgin binders can help optimizing for both rutting resistance and fatigue cracking resistance at the same time.

References

[1] Daniel J and Lachance A 2005 Mechanistic and Volumetric Properties of Asphalt Mixtures with RAP *Transportation Research Board: Journal of the Transportation Research Board* 1929

[2] Bonaquist R 2007 Can I Run More RAP *Hot Mix Asphalt Technology* 11-13

[3] Ozer H, Al-Qadi IL, Carpenter SH, Aurangzeb Q, Roberts GL and Trepanier J 2009 Evaluation of RAP Impact on Hot-Mix Asphalt Design and Performance *Journal of Association of Asphalt Paving Technologists*, 78, 317-351

[4] McDaniel RS, Shah A, Huber GA and Copeland A 2012 Effects of Reclaimed Asphalt Pavement Content and Virgin Binder Grade on Properties of Plant Produced Mixtures *Road Materials and Pavement Design* 13(S1), 161-182

[5] Zaumanis M and Mallick RB 2015 Review of Very High-Content Reclaimed Asphalt Use in Plant-Produced Pavements: State of the Art *International Journal of Pavement Engineering* 16, 39-55

[6] Zaumanis M, Mallick RB and Frank R 2014 100% recycled hot mix asphalt: A review and analysis *Resources, Conservation and Recycling* 92, 230-245

[7] Liu S, Shukla A and Nandra T 2017 Technological, Environmental and Economic Aspects of Asphalt Recycling for Road Construction *Renewable and Sustainable Energy Reviews*, 75, 879-893

[8] Aurangzeb Q, Al-Qadi IL, Ozer H and Yang R 2014 Hybrid Life Cycle Assessment for Asphalt Mixtures with high RAP Content *Resources, Conservation and Recycling*, 83, 77-86

[9] Hansen K and Copeland A 2017 Annual Asphalt Pavement Industry Survey on Recycled Materials and Warm-Mix Asphalt Usage: 2016 *7th Annual Survey Information Series* 138

[10] Ma T, Huang X, Zhang Y and Wang H 2016 Influence of Preheating Temperature of RAP on Properties of Hot-Mix Recycled asphalt Mixture *Journal of Testing and Evaluation*, 44, 762-769

[11] Madrigal DP, Iannone A, Martinez AH and Giustozzi F 2016 Effect of Mixing Time and Temperature on Cracking Resistance of Bituminous Mixtures Containing Reclaimed Asphalt Pavement Material *Journal of Materials in Civil Engineering* 29, 0417058

[12] Zhang K and Muhunthan B 2017 Effects of Production Stages on Blending and Mechanical Properties of Asphalt Mixtures with Reclaimed Asphalt Pavement *Construction and building Materials* 149, 679-689

[13] Mogawer WS, Austerman AJ and Bonaquist R 2012 Determining the Influence of Plant Type and Production Parameters on Performance of Plant-Produced Reclaimed asphalt Pavement Mixtures *Transportation Research Board: Journal of the Transportation Research Board*, 2268, 71-81

[14] Mogawer W, Bennert T, Daniel JS, Bonaquist R, Austerman A and Booshehrian A 2012 Performance Characteristics of Plant Produced High RAP Mixtures *Road Materials and Pavement Design*, 13, 183-208

[15] McDaniel RS, Shah A and Huber G 2012 Investigation of Low- and High-Temperature Properties of Plant-Produced RAP Mixtures *Federal Highway Report*, FHWA-HRT-11-058

[16] Rinaldini E, Schuetz P, Partl M, Tebaldi G and Poulikakos L 2014 Investigating the Blending of Reclaimed Asphalt with Virgin Materials using Rheology *Electron Microscopy and Computer Tomography, Composites: Part B*, 67, 579-587

[17] Bressi S, Cavalli MC, Partl MN, Tebaldi G and Dumont AG 2015 Particle Clustering Phenomena in Hot Asphalt Mixtures with High Content of Reclaimed Asphalt Pavements *Construction and Building Materials*, 100, 207-217

[18] Navaro J, Bruneau D, Drouadaine I, Colin J, Dony A and Cournet J 2012 Observation and evaluation of the degree of blending of reclaimed asphalt concretes using microscopy image analysis *Construction and Building Material*, 37, 135-143

[19] Noferini L, Simone A, Sangiorgi C and Mazzotta F 2017 Investigation on Performances of Asphalt Mixtures made with Reclaimed Asphalt Pavement: Effects of Interaction between Virgin and RAP Bitumen *International Journal of Pavement Research and Technology,* 10, 322-332

[20] Cavalli MC, Partl MN and Poulikakos LD 2017 Measuring the binder film residues on black rock in mixtures with High Amounts of Reclaimed Asphalt *Journal of Cleaner Production*, 149, 665-672

[21] Bowers BF, Moore J, Huang B and Shu X 2014 Blending Efficiency of Reclaimed Asphalt Pavement : An Approach Utilizing Rheological Properties and Molecular Weight Distribution, *Fuel*, 135 63-68

[22] Kriz P, Grant DL, Veloza BA, Gale MJ, Blahey AG, Brownie JH, Shirts RD and Maccarone S 2014 Blending and diffusion of reclaimed asphalt pavement and virgin asphalt binders *Road Materials and Pavement Design*, 15, 78-112

[23] Kriz P, Grant D, Gale MJ, Lavorato S and Pahalan A 2014 Reclaimed Asphalt Pavement- Virgin Binder Diffusion in Asphalt Mixes *Canadian Technical Asphalt Association*. Winnipeg, MB, Canada

[24] Yousefi Rad F, Roohi Sefidmazgi N and Bahia H 2014 Application of Diffusion Mechanism: Degree of Blending Between Fresh and Recycled Asphalt Pavement Binder in Dynamic Shear Rheometer *Transportation Research Board: Journal of the Transportation Research Board*, 2444, 71-77

[25] Oliver J 1975 Diffusion of Oils in Asphalt *Australian Road Research Board*, Report No. 9

[26] Carpenter S and Wolosick J 1980 Modifier Influence in the Characterization of Hot-Mix Recycled Material *Transactions Research Rec*, 777, 15-22

[27] Noureldin AS and Wood LE 1987 Rejuvenator Diffusion in Binder Film for Hot-Mix Recycled Asphalt Pavement *Transportation Research Record*, 1115, 51-61

[28] Zhao S, Huang B, Shu X and Woods ME 2016 Quantitative Evaluation of Blending and diffusion in High RAP and RAS Mixtures *Materials and Design*, 89, 1161-1170

[29] Huang B, Li G, Vukosavljevic D, Shu X and Egan BK 2005 Laboratory Investigation of Mixing Hot-Mix Asphalt with Reclaimed Asphalt Pavement *Transportation Research Board: Journal of the Transportation Research Board*, 1929, 37-45

[30] Zhao S, Baoshan H, Shu X and Woods ME 2015 Quantitative Characterization of Binder Blending: How Much Recycled Binder is Mobilized During Mixing, *Transportation Research Board: Journal of the Transportation Research Board*, 2506, 72-80

[31] Karlsson R and Isacsson U 2003 Application of FTIR-ATR to Characterization of Bitumen Rejuvenator Diffusion *Journal of Materials in Civil Engineering*, 15, 157-165

[32] Karlsson R and Isacsson U 2003 Laboratory studies of diffusion in bitumen using markers *Journal Materials Science*, 38, 2835-2844

[33] R. Karlsson, U. Isacsson & J. Ekblad, Rheological Characterization of Bitumen Diffusion, Journal of Materials Science, 42, 101-108 (2007).

[34] Ding Y, Huang B and Shu X 2017 Utilizing Fluorescence Microscopy for Quantifying Mobilization Rate of Aged Asphalt Binder *Journal of Materials in Civil Engineering*, 29, 04017243

[35] Ding Y, Huang B, Shu X and Zhu Z 2018 Fluorescence Microscopy for Determining Diffusion Coefficient between Virgin and Aged Asphalt Binders *Journal of the Association of Asphalt Pavement Technologists*

[36] Cong P, Hao H, Zhang Y, Luo W and Yao D 2016 Investigation of diffusion of rejuvenator in aged Asphalt *International Journal of Pavement Research and Technology*, 9, 280-288

[37] Elseifi MA, Al-Qadi IL, Yang SH and Carpenter SH 2008 Validity of Asphalt Binder Film Thickness Concept in Hot-Mix Asphalt *Transportation Research Record: Journal of the Transportation Research Board*, 2057, 37-45

[38] Garrick NW and Nunna R 1992 Effects of Hot Storage on the Properties of Asphalt Concrete Mixtures, *Conneticut Department of Transportation*

[39] Jacques C, Daniel JS, Bennert T, Reinke G, Norouzi A, Ericson C, Mogawer W and Kim YR 2016 Effect of Silo Storage Time on the Characteristics of Virgin and Reclaimed Asphalt Pavement Mixtures *Transportation Research Record: Journal of the Transportation Research Board*, 2573, 76-85

[40] Kandhal PS and Wenger ME 1974 Storage of Bituminous Concrete in Inert Gas *Highway research Record*, 468, 61-71

[41] Middleton S, GoodKnight J and Eaton JS 1967 The effects of Hot Storage on an Asphaltic Concrete Mix *Proceedings of the association of Asphalt Paving Technologists*, 36, 180-205

[42] Parr WK and Brock JD 1972 Statistical Study of Effect of Hot Storage on the Composition and Properties of Asphaltic Concrete *Proceedings Association of Asphalt Paving Technologists*, 41, 1-20

[43] Tuttle L 1966 Investigation of the Effects of Holding Hot Bituminous Mixtures in Storage hoppers for Short and extended Periods of Time *Proceedings of 11th Conference of Canadian Technical Asphalt Association*, 11, 237-248

IOP Conf. Series: Materials Science and Engineering 1075 (2021) 012007 doi:10.1088/1757-899X/1075/1/012007

Evaluating the mechanical performance of asphalt mixtures in cyclic torsional shear test

S Yokota[1,*], S Shibuya[1], T N Lohani[1], O Kamada[2] and Y Hisaro[3]

[1]Graduate School of Engineering, Department of Civil Engineering, Kobe University, Japan
[2]Technical Research Institute, Kajima Road Co., Ltd, Japan
[3]Engineering Department, Hanshin Expressway Engineering Co., Ltd, Japan
[*] Corresponding author: yokota@kajimaroad.co.jp

Abstract. In this paper, the mechanical performance of asphalt mixtures was examined in cyclic torsional shear test, in which the relationship between the averaged shear strain, γav, and the number of cyclic loading, N, was obtained by using a short, about 25mm height, specimen. In these tests, the effects of some influencing factors such as the type of asphalt mixture, the thickness of specimen and the temperature were examined. In each test, four indices; i.e., the slope of plastic flow line, the slope of stripping line, the number of load applications at the inflection point for stripping and the averaged shear strain at the inflection point for stripping, were defined for the relationship between γ_{av} and N. As a result of this study, we clarified that various performances such as rutting resistance, fatigue failure resistance and the stripping resistance of asphalt mixture can be appropriately evaluated by using the cyclic torsional shear test.

1. Introduction

In Japan, the importance of maintenance management of asphalt pavement is increasing. In order to efficiently repair asphalt pavement, it is necessary to properly evaluate the properties of existing asphalt pavement. Although there are many tests that directly evaluate the properties of existing asphalt pavement, it is extremely difficult to properly evaluate thin asphalt pavement like a pavement on bridge. For this reason, the authors have developed the cyclic torsional shear test and are studying methods of characterizing asphalt mixtures. This test can be appropriately evaluated not only for specimens about 40 to 60 mm height applied at the earthwork pavement but also for thin specimens about 30 mm height, collected in the pavement on bridge.

In this paper, we describe the influence on the test results when the type of asphalt mixture, temperature and specimen thickness were changed.

2. Cyclic torsional shear test

2.1. Test apparatus

The cyclic torsional shear test apparatus consists of four main components, a vertical loading unit, a torque loading unit, temperature control unit and a data acquisition and control unit. Figure 1 shows the test apparatus. The left side in the figure is the data acquisition and control unit. The center is the loading device which is the main device and the right side is the temperature control unit.

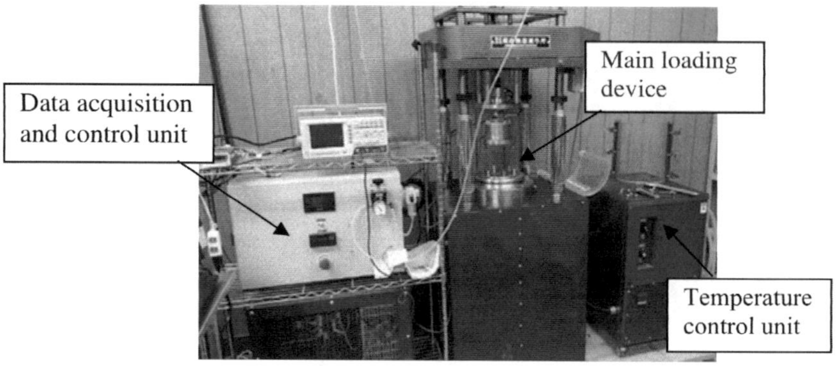

Figure 1. Cyclic torsional shear test overview

2.2. Testing procedure

For the cyclic torsional shear test, cylindrical specimens of asphalt mixture produced in the laboratory or cylinder cores taken from the site can be used. The diameter of the specimen is 100 mm. First, the each face of the specimen is bonded to the top loading cap and the bottom loading cap using an epoxy adhesive (Figure 2a) Next, set the specimen in the loading unit and place an acrylic cylindrical water tank around the specimen (Figure 2b). Then, the hot water adjusted to the test temperature is put into the aquarium from the temperature control unit. After about an hour, which is sufficient time for the specimen temperature to reach test temperature, the cyclic torsional shear test is started. This test is prescribed torque is repeatedly loading by rotating the bottom loading cap. At this time, the vertical displacement is kept at zero. The repeatedly shear loading is continued till either the specimen ruptured or the torsion angle reached 45 degrees.

a. Loading cap and specimen

b. Test situation

Figure 2. Specimen unit

2.3. Testing condition

Table 1 shows the cyclic torsional shear test conditions. In this study, three torques of 14Nm, 28Nm and 42 Nm were set in order to confirm the influence of the difference of the loading torque on the test result. The loading waveform was a haversine wave with a loading time of 0.1 sec and a pausing time of 0.7 sec, and displacement in the vertical direction of the specimen was set to 0. The test temperature was set at 60 ° C in consideration of the maximum pavement temperature in summer in Japan and the 50 ° C and 40 ° C tests were also confirmed in order to check the effect of temperature.

Table 1. Test conditions

Item	Contents	Remarks
Torque	14Nm, 28Nm, 42Nm	
Torque wave	Haversine wave	A loading time of 0.1 sec and a pausing time of 0.7 sec.
Vertical displacement of specimen	Kept 0	
Test temperature	60°C, 50°C, 40°C	

2.4. Evaluation index

In the cyclic torsional shear test, average shear strain, γ_{av}, for each number of load applications is obtained from the following equation

$$\gamma_{av} = \frac{2R\theta}{3H} \tag{1}$$

where γ_{av} is average shear strain, θ is torsion angle (rad), R is specimen radius (m) and H is specimen height (m).

In this test, the relationship between the number of cyclic loading and the average shear strain as shown in Figure 3 is obtained [1]. First, the average shear strain rapidly rises immediately after the start of loading, and then gradually increases as the specimen stripping, fatigue and plastic deformation progress. And as it progresses to some extent, the average shear strain rapidly increases again and the specimen breaks. This deformation behavior is similar to those observed in Hamburg wheel-tracking test on asphalt mixtures (ASSHTO 2008). In this study, tangent lines of the curves shown in Figure 3 are called consolidation line, Plastic flow line, and stripping line [2, 3]. Then, the point of intersection between plastic flow line and stripping line is called the inflection point of stripping. The cyclic torsional shear test evaluates the characteristics of the asphalt mixture with the following indicators; the number of cyclic loading at the inflection point for stripping, the average shear strain at the inflection point for stripping, the slope of the plastic flow line, and the slope of stripping line. The slope of the plastic flow line shows that the plastic flow is high when the slope is small. And the slope of stripping line shows tenacity until failure sufficiently progresses, and the tenacity is strong when the slope is small.

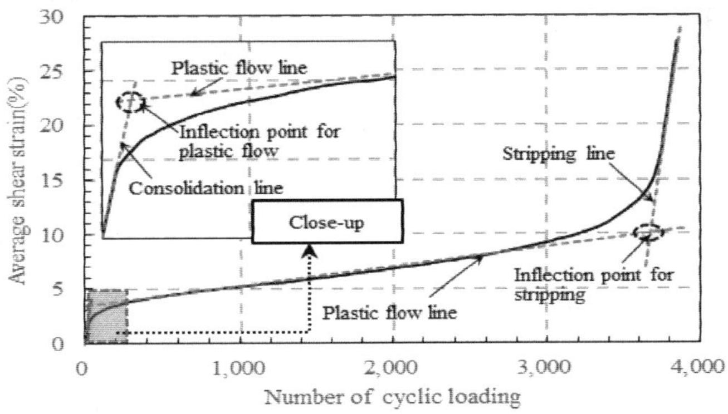

Figure 3. Relationship of average shear strain with number of cyclic loading

2.5. Asphalt mixtures

The properties of the asphalt mixture are shown in Table 2. In this study, two types of asphalt mixture were used.

Table 2. Properties of asphalt mixture

Types of asphalt mixture		Dense-graded		Asphalt mortar
	sieve (mm)			
Gradation (%)	19	100		-
	13.2	98.6		-
	4.75	62.1		-
	2.36	42.4		-
	0.6	24.8		-
	0.3	17.5		100
	0.15	8.5		25.0
	0.075	5.6		20.3
Asphalt content (%)		5.4	5.4	11.1
Binder type		Straight asphalt	Polymer-modified binder	Polymer-modified binder
Density (g/cm^3)		2.383	2.384	2.112
Air void (%)		4.8	4.6	6.4
Stability (kN)		6.4	10.5	3.3
Flow value (1/10cm)		25	34	36
Dynamic stability (cycles/mm)		340	7500	-

3. Test results

3.1. Loading torque

Figure 4 shows the relationship between the number of cyclic loading and the average shear strain according to the cyclic torsional shear test results using a dense-graded asphalt mixture. The thickness of the specimen is 50 mm. In the Polymer-modified binder loading torque of 14 Nm, although the number of loading operations was up to 300 thousand times, destruction of the specimen was not observed and the test was interrupted.

As a result of the test, when the loading torque (shear stress) increased, the number of cyclic loading at the inflection point of stripping was reduced. In addition, compared with the difference of binder, it can be said that a Polymer-modified binder has more the number of cyclic loading at the inflection point of stripping and better fatigue fracture resistance compared with a Straight asphalt. On the other hand, the average shear strain at the inflection point of stripping is nearly constant regardless of the type of binder when the asphalt mixture is the same.

Table 3 shows each evaluation index obtained from the cyclic torsional shear test result. When Polymer-modified binder was used and the loading torque is 14 Nm, the specimen was not destroyed. Focusing on the slope of the plastic flow line, the Polymer-modified binder is smaller than that of Straight asphalt. That is, it can be seen that the plastic flow resistance of the Polymer-modified binder is excellent. This trend is similar to the dynamic stability obtained in the general wheel tracking test [4] for evaluating the plastic flow resistance.

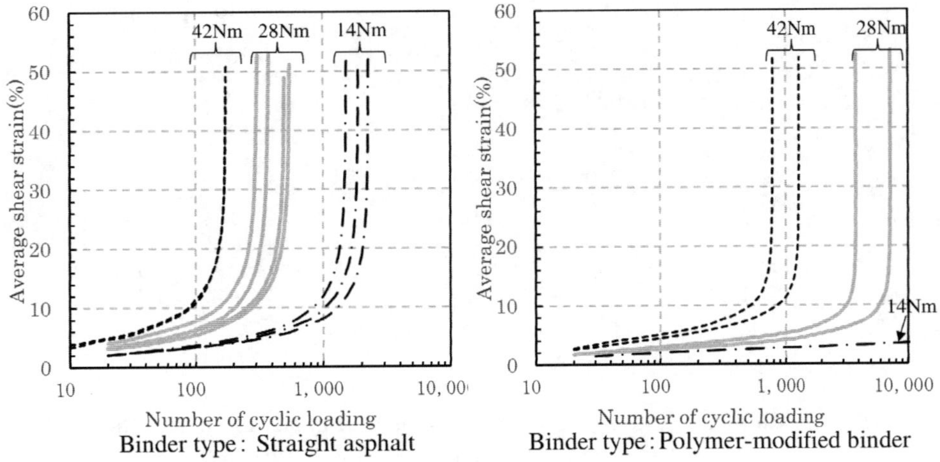

Figure 4. Relationships of average shear strain with number of cyclic loading for dense-graded asphalt mixture

Table 3. Evaluation indices for dense-graded asphalt mixtures

Binder type	Torque	Inflection point for stripping		slope of plastic flow line (%/times)×10^5	slope of stripping line (%/times) ×10^5
		Number of cyclic loading (times)	Average shear strain (%)		
Straight asphalt	14Nm	1,830	15	685	58,275
	28Nm	410	16	3,275	150,538
	42Nm	166	16	7,573	367,434
Polymer-modified binder	14Nm	The specimen was not destroyed			
	28Nm	5,467	10	142	72,039
	42Nm	1,022	13	927	188,384

3.2. Temperature

Figures 5 and 6 show the relationship between the test temperature and the number of cyclic loading at the inflection point of stripping or the slope of the plastic flow line by the repeated torsional shear test using a dense-graded asphalt mixture. The thickness of the specimen was 50 mm which was the same as the examination of the loading torque. In both straight asphalt and polymer-modified asphalt, when the temperature rises, the number of cyclic loading at the inflection point of stripping decreases and the slope of the plastic flow line increases, because plastic flow resistance decreases. It was found that the torsional shear test can clearly evaluate the temperature sensitivity characteristics of the asphalt mixture.

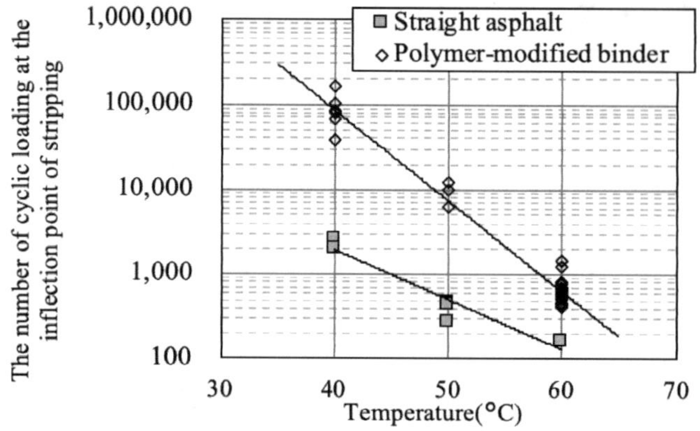

Figure 5. Relationships of temperature with the number of cyclic loading at the inflection point of stripping for dense-graded asphalt mixture

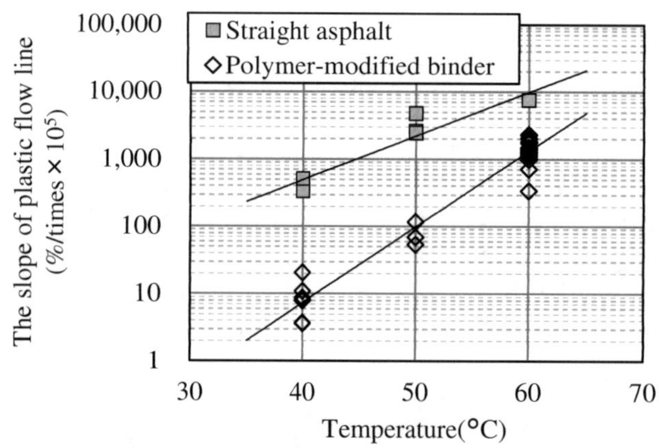

Figure 6. Relationships of temperature with the slope of the plastic flow line for dense-graded asphalt mixture

3.3. Thickness of specimen

3.3.1. Test results

Cyclic torsional shear tests were conducted on specimens with different thicknesses of dense-grade asphalt mixture and asphalt mortar used polymer modified asphalt. Figure 7 shows the relationship between the thickness and the number of cyclic loading at the inflection point of stripping, and Figure 8 shows the relationship between the thickness and the slope of plastic flow line. Although values differ for each asphalt mixture when the thickness of a specimen increases, the number of cyclic loading at the inflection point of stripping decreases and the slope of plastic flow line tends to increase.

By the way, although the results are omitted in this paper, there was no effect of thickness on the average shear strain and the slope of stripping line.

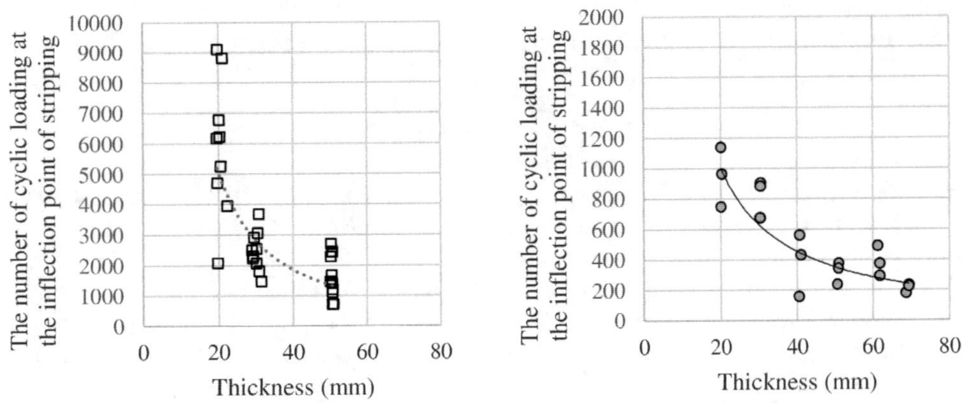

Polymer-modified Dense-graded Asphalt mortar

Figure 7. Relationship between thickness and the number of cyclic loading at the inflection point for stripping

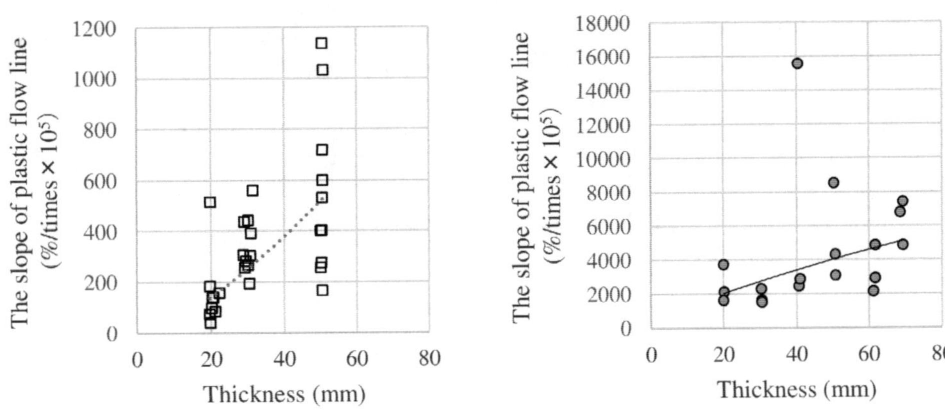

Polymer-modified Dense-graded Asphalt mortar

Figure 8. Relationship between thickness and the slope of plastic flow line

3.3.2. Study on thickness correction

When evaluating various asphalt mixtures in the cyclic torsional shear test, it is desirable to be able to evaluate with the same index when the thickness is different. In this study, we tried to obtain the thickness correction value from the cyclic torsional shear test results with different thickness. That is, the reference thickness was 50 mm, and the correction value was obtained from the ratio of the values of the thicknesses of the specimens. A specific derivation method is as follows.

1. The average value of the test results at the thickness of the specimen of 50 mm is obtained. Let this value be the test result Y_{50} of the reference thickness

2. The correction factor α is obtained by the following equation. Then, the average value of the correction factor of each index is obtained.

$$\alpha = \frac{Y_{50}}{Y} \tag{2}$$

where α is correction factor, Y_{50} is test results with a reference thickness of 50 mm, Y is test results of arbitrary thickness.

3. Find the relationship between the specimen thickness ratio (H / H_{50}) and the correction factor. Then, a correction coefficient for an arbitrary thickness (H) is obtained by the following equation.

$$\alpha = a_1 \times \left(\frac{H}{H_0}\right)^{a_2} \tag{3}$$

where H_0 is Reference thickness (50 mm), H is arbitrary thickness, a_1, a_2 are Regression factor.

The relationship between the specimen thickness ratio (H/H_0) of each evaluation index and the correction coefficient (α) is shown in Figures 9 and 10.

The correction factor for the number of cyclic loading at the inflection point of stripping increases with increasing thickness, and it can be expressed well by the proposed regression equation. Also, polymer-modified dense-graded asphalt mixture and asphalt mortar have almost the same correction factor. Even if the properties of asphalt mixture are different, it is possible to use the same correction factor.

The correction factor of the slope of plastic flow line decreases with increasing thickness, and it can be expressed well by the proposed regression equation. However, there are differences in correction factors depending on the asphalt mixture type. Whether the same correction factor can use for evaluating the flow resistance should be studied in the future.

By the way, although omitted in this paper, it was found that the average shear strain and the slope of stripping line had no effect of thickness. For this reason, correction of thickness is not required for these indexes.

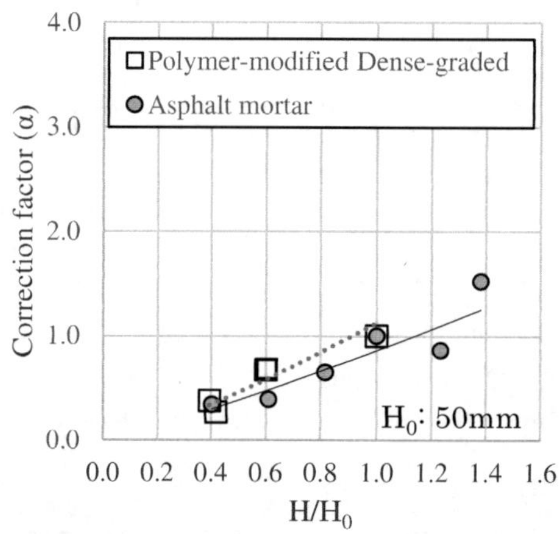

Figure 9. Relationship between (H/H₀) and correction factor

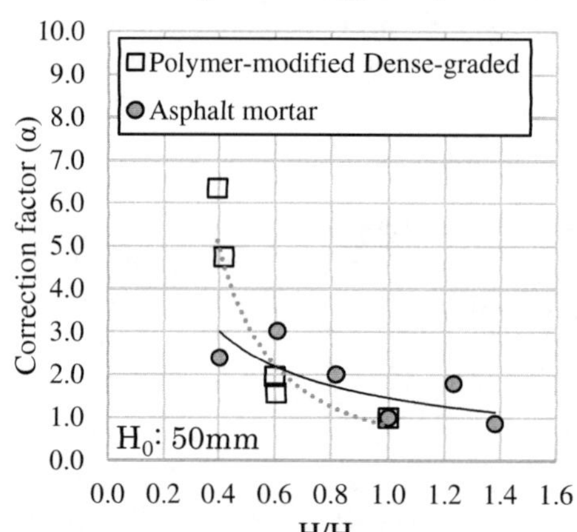

Figure 10. Relationship between (H/H_0) and correction factor

IOP Conf. Series: Materials Science and Engineering 1075 (2021) 012007 doi:10.1088/1757-899X/1075/1/012007

4. Summary

The findings obtained in this study are as follows.

1. Amongst the polymer-modified dense-graded asphalt mixture, and straight-asphalt dense-graded asphalt mixture, the polymer-modified dense-graded asphalt mixture exhibited the highest plastic flow resistance. This trend was similar for the magnitude of dynamic stability.
2. The higher the temperature, the lower the plastic flow resistance and the fatigue fracture resistance, implying that the cyclic torsional shear test enables us to properly evaluate some basic behavior of asphalt mixtures.
3. The test results of the cyclic torsional shear test were influenced by the thickness of specimen. Accordingly, the correction factor regarding the specimen thickness was proposed.

References

[1] Yoshida et al. 2013 Evaluation of properties related to durability of asphalt mixture based only on cyclic torsional shear test (in Japanese) *The 30th Japan Road Conference*, 3027, 2013.10.

[2] Fuke et al. 2010 An attempt to evaluate the durability of asphalt mixture based on cumulative deformation accompanying cyclic torsional shear (in Japanese) *Japan Society of Civil Engineers 2010 Annual Meeting*, v-044

[3] Yoshida et al. 2012 Interpretation and application of repeated torsional shear test on asphalt mixtures *Advances in Transportation Geotechnics II*, pp414-419

[4] Japan Road Association 2007 Method of wheel tracking test, *Pavement survey/test method manual [third volume]*, B003, 2007.6.

IOP Conf. Series: Materials Science and Engineering 1075 (2021) 012008 doi:10.1088/1757-899X/1075/1/012008

Prediction of unconfined compressive strength of cement stabilized pavement materials

R P P K Premarathne[1,*] and A Sawangsuriya[2]

[1]School of Engineering and Technology, Asian Institute of Technology, Pathum Thani 12120, Thailand
[2]Bureau of Road Research and Development, Department of Highways, Bangkok 10400 Thailand
[*] Corresponding author: piyumee.premarathne@gmail.com

Abstract. This paper evaluated and justified suitable prediction equations for unconfined compressive strength (UCS) of cement stabilized pavement materials. Equations by multiple linear regression analysis were also proposed. Data including pavement recycling base and subbase layers were collected from the Bureau of Material, Analysis, and Inspection, Department of Highways, Thailand. Multiple linear regression was conducted using the SPSS software. Three equations obtained from multiple linear regression and past researches were compared. Two criteria were considered in the selection of equations in order to check the accuracy and precision of each equation. First criterion was based on the predicted UCS vs. actual UCS graph. The slope and interception values of each graph were evaluated. Second criterion used two indices: ranking distance (RD) and ranking index (RI). For a good correlation, both these indices shall approach zero. By considering the statistical analysis outputs from the two criteria, a relationship between UCS and California Bearing Ratio – water cement ratio [CBR/ (w/c)] proposed by Sunitsakul in 2012 was justified as a suitable equation to predict the UCS of cement stabilized pavement materials. The study also confirmed that the relationship proposed by Sawangsuriya in 2016 for fine-grained soil with plasticity index was not appropriate for the prediction of UCS because the majority of the samples were non-plastic.

1. Introduction

Cement stabilization involves blending cement with the marginal material to increase its mechanical properties through chemical reactions [1]. An unconfined compressive strength (UCS) test is often adopted in practice to evaluate the strength of cement stabilized materials used in highway construction. Due to the time and construction constraints, many researchers proposed a number of prediction equations for the UCS. With the availability of prediction equations, it is a challenge for engineers and practitioners to select suitable equation for their highway construction practices. To predict the UCS of cement stabilized pavement materials prior to the testing, existing prediction equations must be evaluated and justified.

The objective of this study was to evaluate and justify suitable equations for predicting the UCS of cement stabilized pavement materials. Selection of the prediction equations for the statistical analysis was performed based on the available data in the proposed equations and the applicability of proposed equation for pavement materials. Such prediction equations will help engineers and practitioners to estimate the UCS of cement stabilized pavement materials for highway construction practices. This study was limited to cement stabilized pavement materials only.

2. Literature review

Process of soil stabilization helps improving the durability and strength of in situ soils [2]. Cement stabilized soil or soil cement layers are semi-rigid and exhibit increased bearing capacity and durability. Soil cement provides strong and uniform support for pavement layers as well as firm and stable working platform for pavement construction. Use of Portland cement as a stabilizing agent helps increasing the inter-particle friction within the soil mass while reducing the moisture susceptibility of existing in situ material [3]. Previous studies have shown that soil stabilization is an economical approach in strengthening the earth for construction requirements and in diminishing the amount of soil exchange [5]. Soil cement can improve engineering properties like the Unconfined Compressive Strength (UCS) of stabilized material. Hardened soil cement mixtures also resist adverse environmental effects while reducing plasticity index, increasing shrinkage limit, meeting strength thresholds, and improving resilient moduli [1].

Factors including cement content, parent soil mineralogy and gradation, construction practices, and curing method can affect the performance of cement stabilized soils [6]. Many factors including the physical and chemical properties of the soil, the amount of cement, the porosity, and the moisture content at the time of compaction affect the behavior of soil cement [7, 8]. Over the years, engineers have attempted different methods to stabilize soils that are subjected to fluctuations in strength and stiffness properties as a function of fluctuation in moisture content. The engineering properties vary widely due to heterogeneity in soil composition, difference in micro- and macro-structure of soils, heterogeneity of geologic deposits, as well as difference in physical and chemical interactions between the soil and candidate stabilizers [1].

When cement stabilized material is used as a compacted layer over a soil with a low bearing capacity, the system tends to fail under tensile stresses at the base of the improved layer [9]. Thus, it seems to be more reasonable to use tensile strength as a direct measure of cement stabilized material's strength. However, research has shown that the tensile strength of soil-cement typically ranges between 9% and 14% of the UCS [7, 10-12]. Because of those reasons, the UCS has been adopted as the most convenient test for the determination of the soil-cement strength and for the investigation of the effect of different variables on the soil-cement strength [13].

The UCS of soil is the load per unit area at which an unconfined cylindrical specimen of soil will fail in the simple compression test. The strength and stiffness of cement stabilized pavement materials are often evaluated through UCS tests. Preparation and curing of the samples used for UCS testing should be carried out in accordance with ASTM D 1632 in which it is recommended to moist cure for soil cement stabilized material samples. Testing of the cured samples should be performed following ASTM D 1633 in which it is needed to immerse the samples in water for two hours prior to testing [3]. It has been used systematically in most experimental programs to verify the effectiveness of the cement stabilization or to access the importance of influencing factors on the strength of cemented soils [9].

Table 1 shows the recommended tolerance limits for soaked compressive strength specified by ACI [14].

Table 1. Range of compressive strength in soil cements [14]

Soil Type	AASHTO Classification	Soaked Compressive Strength (psi)	
		7 Days	28 Days
Sand and gravelly	A-1, A-2, A-3	300-600	400-1000
Silty	A-4, A-5	250-500	300-900
Clayey	A-6, A-7	200-400	250-600

Table 2 shows the UCS criteria recommended by U.S. Army [4]. The lowest cement content in the mixture design that meets the requirements should be used as the design cement content.

Table 2. Unconfined compressive strength criteria [4]

Stabilized Layers	Minimum 7-Day Unconfined Compressive Strength (psi)	
	Flexible Pavement	Rigid Pavement
Base Course	750	500
Subbase, select material or subgrade	250	200

2.1. Prediction of unconfined compressive strength of cement stabilized pavement materials

Several equations were proposed by previous research to predict the UCS of cement stabilized pavement materials. The Equation 1 proposed by Sunitsakul in 2012 [15] involving water to cement (w/c) ratio, soaked California Bearing Ratio (CBR), and UCS at seven days of curing period of soil-cement is illustrated in Figure 1.

$$UCS = 0.427[\frac{CBR}{\frac{w}{c}}]^{0.578} \quad (1)$$

where

UCS Unconfined compressive strength at 7 days curing time (MPa)

CBR Soaked CBR at 95 % of maximum dry density based on modified Proctor compaction and is measured at 0.1 inch penetration (%)

w/c Water to cement ratio

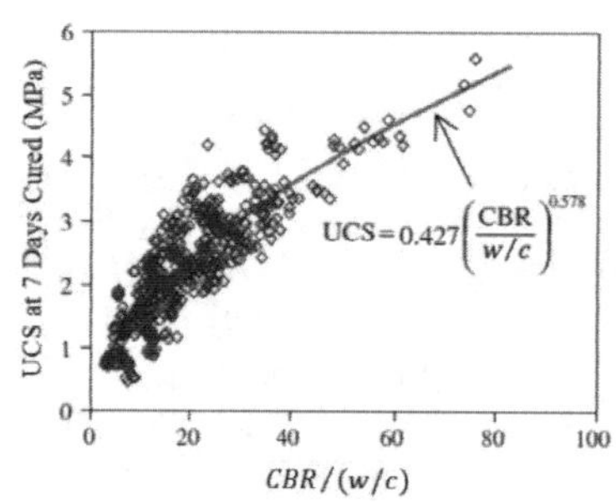

Figure 1. UCS model for cement stabilized bases [15]

Some aspects of cement content for stabilizing pavement materials were studied by Sawangsuriya in 2016 [16] as shown in Figure 2 along with the prediction Equations 2 and 3 for the UCS at seven days curing time for mixtures having varied cement contents. The coefficient of the proposed equations reflects the properties of the materials (e.g. gradation, plasticity index etc.)

$$UCS = CBR(\frac{f}{c})^{-0.6} \quad (2)$$

$$UCS = 12CBR(\frac{f}{c})^{-0.6} \quad (3)$$

where

UCS Unconfined compressive strength at 7 days curing time (ksc)

CBR Soaked CBR at 95 percent of maximum dry density based on modified Proctor compaction and is measured at 0.1 inch penetration (%)

f Fine content by weight (%)

c Cement content by weight (%)

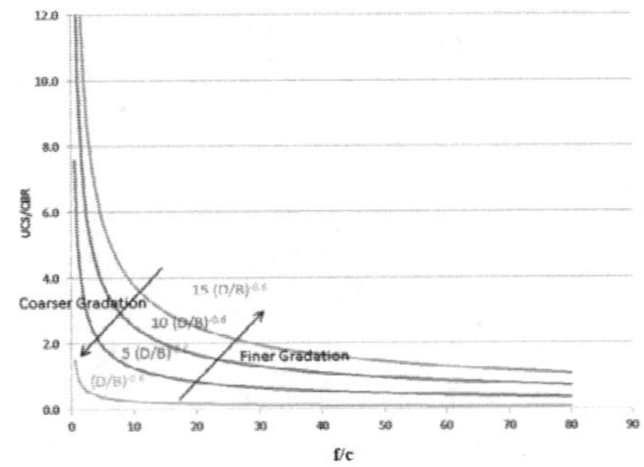

Figure 2. Relationship between UCS/CBR and f/c [16]

The performance of lateritic soil cement mixtures was investigated by Jaritngam et al. [17] through an experimental study. The experimental results in terms of UCS were increased significantly with the increasing cement content as shown in Figure 3. A multiple regression Equation 4 based on cement content (c), curing time (D), and dry density (γ_{dry}) were proposed to predict the UCS of lateritic soil-cement. Good correlations between the experimental data and the proposed prediction equation was observed as follows.

$$UCS = -108.468 + 7.607c + 0.674D + 47.656\gamma_{dry} \quad (4)$$

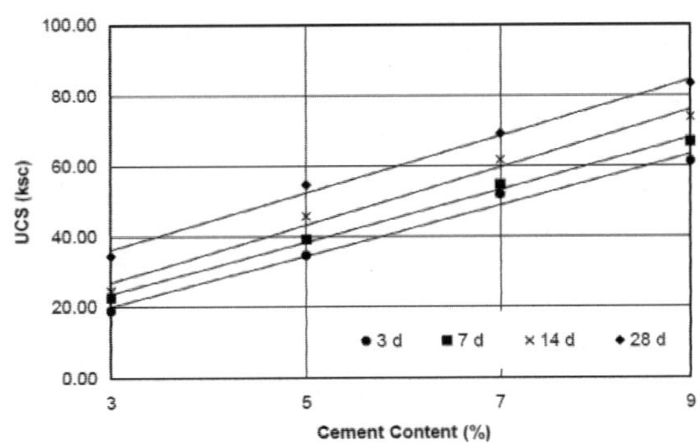

Figure 3. Effect of cement content and curing time on UCS [17]

3. Methodology

3.1. Statistical analysis on selected prediction equation

Data required for this study was obtained from the Bureau of Material, Analysis, and Inspection, Department of Highways (DOH), Thailand. Most of the mixtures had no plasticity while 10 out of total 55 mixes had plasticity. Ordinary Portland cement had been used for the cement stabilization process. Cement content of the mixes varied from 2% through 7% with respect to the dry weight of soil. Data obtained from the DOH were unconfined compressive strength (ksc), cement content (%), gravel content (%), fine content (%), California Bearing Ratio (%), maximum dry density (gm/ml), optimum moisture content (%), liquid limit (%) and plasticity index (%). UCS prediction equations for cement stabilized pavement materials were selected based on availability of data and their applicability to pavement materials. Selected equations were evaluated to determine suitable prediction equations. The best results were obtained from the prediction equation having the highest value of correlation coefficient (R). A logical assessment of the validity of the above correlation was evaluated using two different indices: ranking distance (RD) [18] and ranking index (RI) [19].

$$RD = \left\{\left[1 - \bar{y}\left(\frac{Predicted\ value}{Measured\ value}\right)\right]^2 + \sigma_y^2\left(\frac{Predicted\ value}{Measured\ value}\right)\right\}^{0.5} \quad (5)$$

$$RI = \bar{y}\left|\ln\left(\frac{Predicted\ value}{Mesured\ value}\right)\right| + \sigma_y\left|\ln\left(\frac{Predicted\ value}{Measured\ value}\right)\right| \quad (6)$$

where

RD	Ranking distance
RI	Ranking index
$\bar{y}$	Mean value
σ_y	Standard deviation

For a good correlation, both these indices shall approach zero. The $\bar{y}$, σ_y, RD, and RI of predicted to actual values of UCS should be satisfactory and acceptable, suggesting the validity of the proposed equation for estimation of UCS.

Multiple Linear Regression Analysis (MLRA) was used to develop regression equations for the prediction of UCS of cement stabilized pavement materials using the SPSS (Statistical Package for the Social Sciences) software. Regression models were developed and coefficients for each independent variable were obtained from MLRA.

4. Results and discussion

The unconfined compressive strength (UCS) was the dependent variable. Cement content (C), fine content (f), California Bearing Ratio (CBR), and maximum dry density (MDD) were the independent variables. Optimum moisture content (OMC) was excluded from the analysis as it failed to show a significant linear relationship (R^2 = 2.045E-8) and a correlation (Pearson linear correlation coefficient = 0) with UCS. Cement content showed the strongest linear relationship (R^2 = 0.72) and the highest correlation with UCS (Pearson linear correlation coefficient = 0.85).

Three prediction equations could be obtained from MLRA. As they have significant independent variables and higher R^2 values, they were statistically analyzed and then compared with Equations 1, (2), (3), and (4). Three equations obtained from multiple linear regressions can be presented as follows.

$$UCS = 8.518 + 3.834c - 0.323f + 0.069CBR \quad (7)$$

$$UCS = -58.332 + 3.791c - 0.379f + 33.169MDD \quad (8)$$

$$UCS = -52.005 + 3.787c - 0.319f + 28.882MDD + 0.048CBR \quad (9)$$

where

UCS	Unconfined compressive strength (ksc)
c	Cement content (%)
f	Fine content (%)
CBR	California bearing ratio (%)
MDD	Maximum dry density (gm/ml)

Table 3. Summary of multiple linear regression analysis

Equation	Independent Variables	R	R^2	Significance
Eq. 3	c f CBR	0.91	0.82	0.00
Eq. 4	c f MDD	0.91	0.83	0.00
Eq. 5	c F MDD CBR	0.92	0.84	0.00

Figures 4 and 5 show the variation of dependent variable UCS with independent variable CBR/ (w/c) for Equations 1, 2 and 3 respectively. For each graph, the dash line indicate variation of UCS/CBR values predicted by the equation and the solid line indicates variation of actual UCS/CBR values with f/c. A best fit curve was fitted for each data set to compare predicted values and actual values. A detailed summary of graphs is given in Table 4.

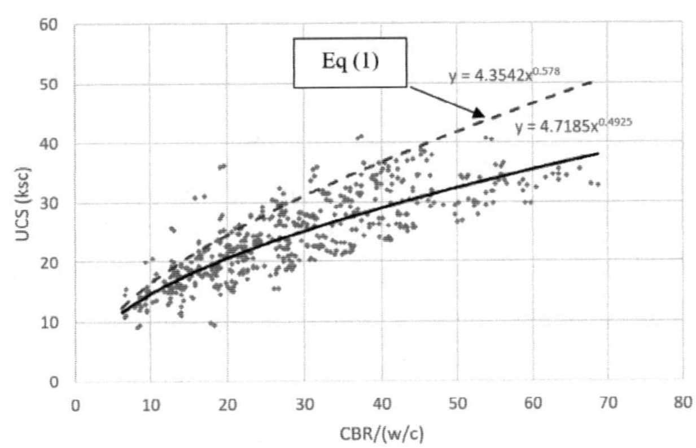

Figure 4. Variation of UCS with CBR/ (w/c)

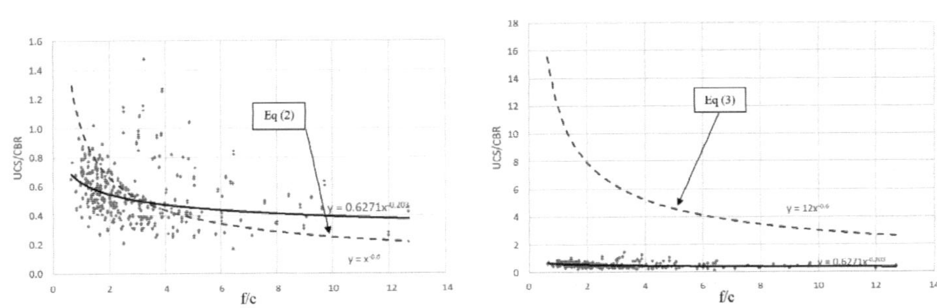

Figure 5. Variation of UCS/CBR with f/c for Equations 2 and 3

Table 4. Detailed summary of graphs

Equation	Independent Variable/ Variables	Dependent Variable	Proposed Fitting Parameters		Best Fitting Parameters	
			Coefficient	Exponent	Coefficient	Exponent
(1)	CBR/ (w/c)	UCS	4.35	0.58.	4.72	0.49
(2)	f/c	UCS/CBR	1	-0.6	0.63	-0.2
(3)	f/c	UCS/CBR	12	-0.6	0.63	-0.2

A common and simple approach to evaluate models is to regress predicted vs. actual values (or vice versa) and compare slope value and interception value against the 1:1 line [20]. Table 5 shows the summary of predicted UCS vs. actual UCS for Equations 1, 2 and 3 and equations proposed by this study (7), (8), and (9).

Table 5. Detailed summary of predicted UCS vs. actual UCS

Equation	Slope Value of Predicted UCS vs. Actual UCS Plot	Intercept Value of Predicted UCS vs. Actual UCS Plot
(1)	0.99	5.29
(2)	1.45	-5.53
(3)	17.38	-66.31
(7)	0.83	4.11
(8)	0.84	3.98
(9)	0.82	4.39

Equation 3 shows the highest deviation from slope value = 1 and interception value = 0 in the predicted UCS vs. actual UCS graph. The nearest slope value to slope value =1 is given by equation (1). It has the 4th highest interception value of 5.29. Equation 8 proposed by this study gives the lowest difference in interception value = 0 and interception value = 3.98. Calculated RD and RI values for each equation are given in Table 6.

Table 6. Average RD and RI values

Equation	Ranking Distance (RD)	Ranking Index (RI)
(1)	0.3	0.59
(2)	0.43	0.89
(3)	40.75	232.57
(7)	0.1	0.19
(8)	0.09	0.18
(9)	0.09	0.18

For the average RD and average RI values, the lowest average RI (RI = 0.18) and lowest value for RD (RD = 0.09) were given by Equations 8 and 9. It showed that there is no significant improvement in the prediction equations when the number of independent variables is increased from three to four. Equation 3 gives the highest average RD (RD ~ 40.8) and highest average RI (RI ~ 232.6) values. This is because Equation 3 is recommended for the use of fine-grained materials with plasticity index. Majority of the samples had no plasticity except for ten mixtures which was negligible (Figure 5). Therefore, it can be justified that Equation 3 was not suitable for the prediction of UCS of non-plastic or a material with no plasticity index (PI).

5. Conclusion

Three UCS prediction Equations 1, 2 and 3 suggested by previous research were selected based on the availability of data and applicability to pavement materials. They were analyzed to come up with a suitable UCS prediction equation. To check the accuracy and precision of each prediction equation, predicted UCS vs. actual UCS graphs were plotted. The slope value and interception value of each graph were considered to test the significance of slope value and intercept value. Two indices (e.g. ranking distance (RD) and ranking index (RI)) were obtained. For a good correlation, both these indices shall approach zero. The RD and RI of predicted to actual values of UCS should be satisfactory and acceptable, suggesting the validity of the proposed equation for estimation of UCS. Depending on the results of the analysis, it can be concluded that the relationship between UCS and CBR/ (w/c) proposed by Sunitsakul in 2012 [15], Equation 1 was a very good prediction equation for the prediction of UCS of cement stabilized pavement materials. Results also showed that the Optimum Moisture Content (OMC) does not have a linear relationship (R^2 = 2.045E-8) and no correlation (Pearson Linear Correlation Coefficient = 0). Cement content has the strongest linear relationship (R^2 = 0.72) and the highest correlation with UCS (Pearson Linear Correlation Coefficient = 0.85). Study confirmed that Equation 3 proposed by Sawangsuriya in 2016 [16] was not suitable for the prediction of UCS as the majority of the samples were non-plastic.

For further studies, additional material properties such as porosity, initial shear modulus, and curing days suggested by Consoli in 2011 [8], and Jaritngam in 2012 [17] can be taken into account. More test materials can also be investigated to validate prediction equations. Further soil types can be investigated for any possible correlations.

References

[1] Little DN and Nair S 2009 Recommended Practice for Stabilization of Subgrade Soils and Base Materials (Washington, DC: The National Academies Press)

[2] Winterkorn HF and Fang HY 1991 *Foundation Engineering Handbook* (New York: Springer) pp 88-143

[3] International ASTM 2017 Standard Test Methods for Compressive Strength of Molded Soil-Cement Cylinders (West Conshohocken, PA: ASTM International)

[4] McCarthy JE 1994 *Soil Stabilization for Pavements* (Washington, DC: Departments of the Army and the Air Force)

[5] Kukko H 2000 Stabilization of clay with inorganic by-products *J. of Mater. in Civ. Eng.* pp 307-309

[6] Mitchell JK, Ueng TS and Monismith CL 1972 Behavior of Stabilized Soils under Repeated Loading – Performance Evaluation of Cement Stabilized Soil Layers and Its Relationship to Pavement Design (Berkeley: University of California)

[7] Clough GW, Sitar N, Bachus RC and Rad NS 1981 Cemented sands under static loading *J. of the Geotechnical Eng. Div* vol 107 pp 799-817

[8] Consoli NC, Rosa AD, Corte MB, Lopes LDS Jr and Consoli BS 2011 Porosity-cement ratio controlling strength of artificially cemented clays *J. of Mater. in Civ. Eng.* pp 1249-1254

[9] Consoli NC, Vendruscolo MA and Prietto PDM 2003 Behavior of plate load tests on soil layers improved with cement and fiber *J. of Geotechnical and Geoenvironmental Eng.* pp 96-101

[10] Ingles OG and Metcalf JB 1972 *Soil Stabilization Principles and Practice* vol 11 (Washington, DC: The National Academies Press)

[11] Dass RN, Yen SC, Das BM, Puri VK and Wright MA 1994 Tensile stress-strain characteristics of lightly cemented sand *Geotechnical Test. J.* pp 305-314

[12] Consoli NC, Prietto PDM, Carraro JAH and Heineck KS 2001 Behavior of compacted soil-fly ash-carbide lime mixtures *J. of Geotechnical and Geoenvironmental Eng.* pp 774-782

[13] Consoli NC, Foppa D, Festugato L and Heineck KS 2007 Key parameters for strength control of artificially cemented soils *J. of Geotechnical and Geoenvironmental Eng.* pp 197-205

[14] Committee ACI 2009 *Report on Soil Cement* (Farmington Hills, MI: American Concrete

Institute)

[15] Sunitsakul J, Sawatparnich A and Sawangsuriya A 2012 Prediction of unconfined compressive strength of soil–cement at 7 days *Geotechnical and Geological Eng.* pp 263-268

[16] Sawangsuriya A 2016 Some aspects of cement content for stabilized pavement materials *3rd Thai Geotechnical Conf.* (Bangkok: The Engineering Institute of Thailand)

[17] Jaritngam S, Yandell WO and Taneerananon P 2013 Development of strength model of lateritic soil-cement *Eng. J.* pp 69-78

[18] Cherubini C and Orr TA 2000 Rational procedure for comparing measured and calculated values in geotechnics *The Int. Symp. on Coastal Geotechnical Eng. in Practice* pp 261-265

[19] Briaud JL and Tucker LM 1988 Measured and predicted axial response of 98 piles *J. of Geotechnical Eng.* pp 984-1001

[20] Piñeiro G, Perelman S, Guerschman JP and Paruelo JM 2008 How to evaluate models: observed vs. predicted or predicted vs. observed *Ecological Modelling* pp 316-322

The Effectiveness and Efficiency of Implementing Fast Set Concrete in Reparation and Reinstatement of Concrete Pavements

Teoh Sze Tean[1]

[1]ECO2 BUILDER PTE LTD, 12 Woodlands Square #07-75 Singapore 737715
WOODSQUARE TOWER 1

[*] Corresponding author: teoh@ecotwo.sg

Abstract. The aim of this research is to determine the average compressive strength of fast set concrete with different formulations and their viabilities to be adopted for reparation and reinstatement works in concrete pavements especially when time is critical consideration. Fosroc's Patchroc Rapid Setting Patch (RSP) cement was used as one of the components of the concrete. Conventionally, the recommended thickness for fast set concrete is only up to 75mm. With the technical supports from Fosroc (UK), we made a breakthrough by manipulating the design mix ratio and allow the fast set concrete remains its optimum performance with thickness up to 400mm. For the testing methodology, ten batches of concrete cube samples were produced, and each batch consists of ten numbers of samples. We are working on different mix design to determine the design with highest strength. All samples are tested in laboratory to ascertain their compressive strengths at 3 hours, 4 hours, 5 hours, 6 hours, 1 day, 3 days, 7 days, 28 days, 60 days and 90 days. Findings and analysis from several successful fast set concrete projects in Singapore Changi Airport were carried out and the data will be presented in this paper as well. Based on the most recent results, we are able to produce fast set concrete that achieve initial compressive strength more than 40MPa at 3 hours and the strength is increasing gradually with time. The projects using fast set concrete in aircraft pavement reparation and reinstatement were effective and efficient with minimum disturbance to the aircraft operation. From these results, we can conclude that fast set concrete could bring remarkable benefits and contributions to reparation and reinstatement works on operational area. This is because fast set concrete able to gain high early strength and the time for closure can be reduced drastically.

Keywords: fast set concrete, effective and efficient, pavement repair, pavement reinstatement

1. Introduction

Structures built by concrete including builds, bridge decks, road pavements, and aircraft pavements will deteriorate over time due to usage, loading and environmental factors. When concrete starts to deteriorate, defects such as cracks, spalling, bond splitting and others will occur and jeopardize the structural integrity of the concrete structures. If these defects are left unattended, severe, and fatal accidents may occur and result in lots of losses including lives. Therefore, concrete reparation is a

IOP Conf. Series: Materials Science and Engineering 1075 (2021) 012009 doi:10.1088/1757-899X/1075/1/012009

crucial field in the construction industry in order to extend and preserve the structural life span of the concrete structures.

However, concrete reparation is not a simple task as it seems. In operational area such as aircraft pavements in airport, concrete reparation is subjected to numerous of constraints especially for airport with high intensity of air traffic namely Singapore Changi Airport. The operation of Changi Airport covers over 100 airlines which are flying to 400 cities and 100 countries. There are about 7,400 flights depart or land in Changi Airport. Besides runway, most of the aircraft maneuver areas are constructed in concrete pavements such as taxiway entrances, parking bays and etc. Thus, during reparation works are carried out on the concrete pavement, time required for reparation, areas of closure, duration of closure and the disruption to the airport operation shall be kept at the minimum.

Therefore, fast set concrete is being used in concrete reparation works in airport operation areas where time is a critical concern [1,2]. Fast set concrete is a type of concrete which can gain early strength rapidly and achieve adequate amount of strength in shorter period of time as compared to ordinary concrete. On the other hand, the usage of fast set concrete reduces the overall time required for concrete curing. With such advantages of fast set concrete, reparation works on the defective concrete can be completed in a timely manner and minimize the disturbance to airport operations.

The aim of this research is to study the effectiveness and efficiency of implementing fast set concrete in reparation and reinstatement of concrete pavements especially for operational area such as airport. In order to study the effectiveness and efficiency of the fast set concrete, the compressive strength of the fast set concrete in the relation of time shall be determined.

2. Research Methodology

Experimental: 10 batches of concrete samples were produced by using fast set concrete and each batch of concrete consists of 10 concrete cube samples. One of the main components of the fast set concrete used in this research is Fosroc's Patchroc Rapid Setting Patch (RSP). All of the concrete sample were set to a third party laboratory, Admaterials Technologies Pte Ltd, to determine the compressive strengths of the concrete samples at time interval of 3 hours, 4 hours, 5 hours, 6 hours, 1 day, 3 days, 7 days, 28 days, 60 days and 90 days.

Before the experiment was carried out, a correlational study was conducted with technical supports from Fosroc (UK) to study the mix ratio and design mix of the fast set concrete. Despite the conventional fast set concrete is recommended to be casted not more than 75mm thick, a breakthrough was made by discovering a design which allow the fast set concrete to be casted up to 400mm thick while remain its structural characteristics.

3. Result and Findings

As the results, the fast set concrete is able to achieve high early strength as reflected in the results shown below (figure 1 and table 1). At time interval of 3 hours, the fast set concrete had already gained an average compressive strength of 43.53 MPa and the strength is increasing gradually over time. At 6 hours, it has already achieved compressive strength of 50.49 MPa. At 28 days, the compressive strength has developed to 64.85Mpa and continues to increase to 93.11Mpa and 96.26Mpa at 60 days and 90 days respectively.

Table 1. Average Compressive Strength achieved by Fast Set Concrete Over Time.

Compressive Strength (MPa) / Durations	Batch 1	Batch 2	Batch 3	Batch 4	Batch 5	Batch 6	Batch 7	Batch 8	Batch 9	Batch 10	**Avr. Compressive Strength (MPa)**
3 hours	38.9	36.1	43.8	54.5	42.8	46.9	45.0	40.4	48.8	38.1	**43.53**
4 hours	40.8	38.0	48.5	55.1	49.0	47.0	50.5	40.2	47.2	41.6	**45.79**
5 hours	44.2	50.7	54.1	53.6	51.3	48.4	48.3	41.5	53.5	45.3	**49.09**
6 hours	41.9	47.2	54.8	54.9	55.8	46.4	48.2	49.0	59.6	47.1	**50.49**
1 day	52.7	55.4	57.4	61.8	56.3	50.0	60.3	55.7	63.3	53.5	**56.64**
3 days	57.0	58.7	62.9	66.7	58.5	56.6	59.0	57.0	70.7	53.6	**60.07**
7 days	62.5	70.9	67.0	69.5	61.2	65.6	61.8	59.8	72.7	57.5	**64.85**
28 days	77.9	82.3	95.8	80.9	91.0	89.2	92.0	80.8	94.7	84.6	**86.92**
60 days	83.4	85.0	98.0	95.2	95.1	97.4	95.2	90.5	101.7	89.6	**93.11**
90 days	90.7	90.0	99.3	101.3	98.8	100.0	96.5	93.2	102.3	90.5	**96.26**

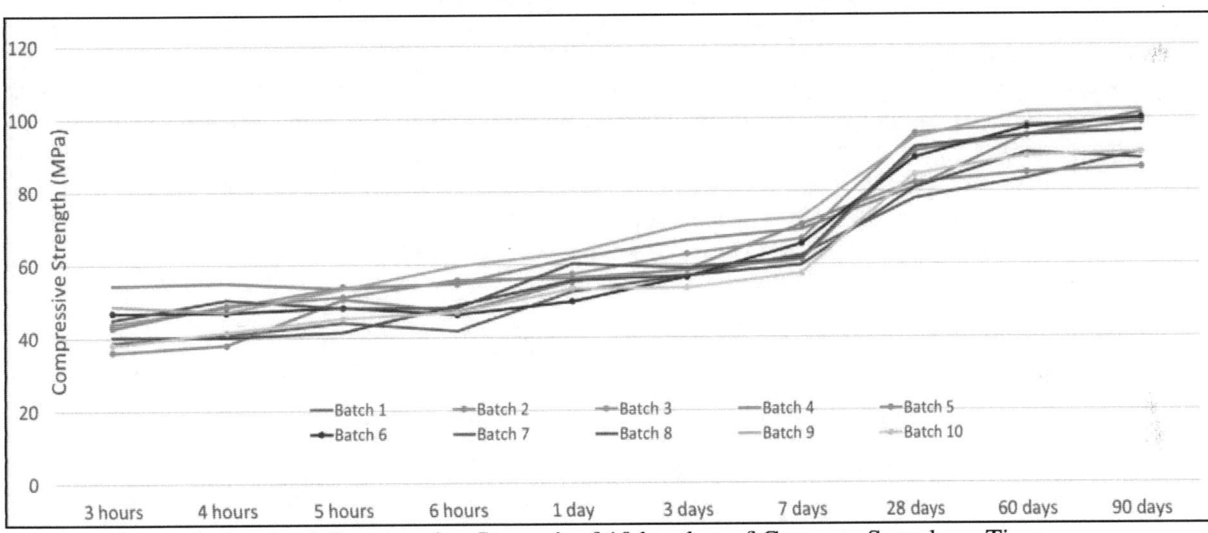

Figure 1. Graph of Compressive Strength of 10 batches of Concrete Sample vs Time.

Based on the results shown above, fast set concrete is able to achieve high early strength with average compressive strength of 43.53 MPa in 3 hours which is as good as ordinary mix of grade 40 concrete which can only reach compressive strength about 40 MPa at 28 days.

Maturity method can be implied to determine the strength of the concrete when the fast set concrete is casted on site and testing equipment is absent [3]. Maturity method is an approach to determine the concrete strength with the relation of temperature and time. The basis of this method uses the fundamental concept of heat releases while the concrete its developing its properties and gaining strength due to chemical reactions between cement, water and other components in the concrete.

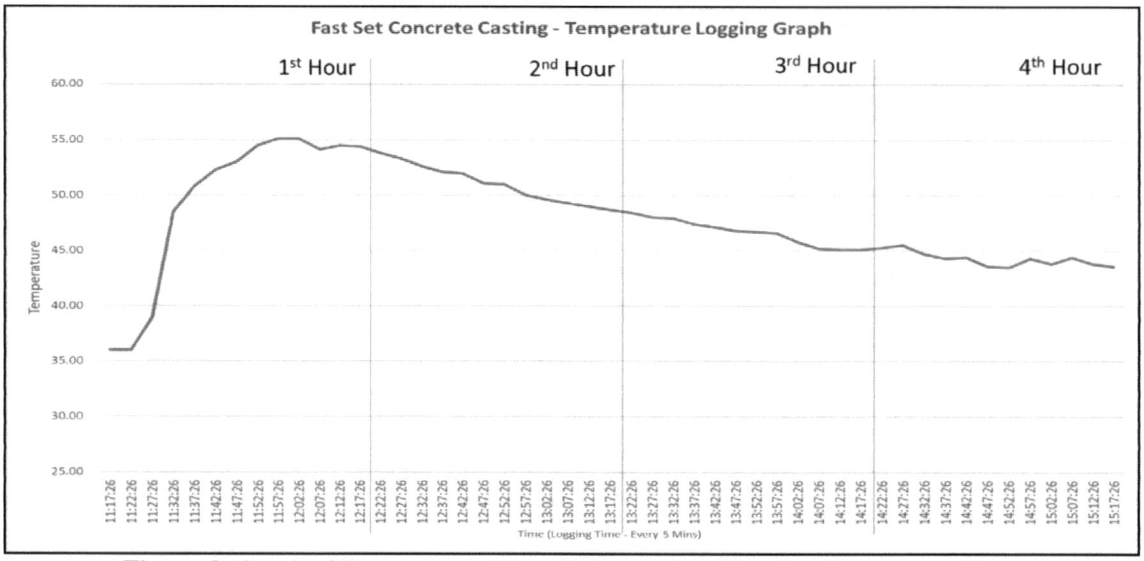

Figure 2. Graph of Temperature / Heat Releases from Fast Set Concrete vs Time.

From figure 2 above, it can be observed that the fast set concrete starts to release a significant amount of heat and reach the optimum temperature within the first hour of concrete casting. It is due to the concrete is gaining strength and reactions take place rapidly and allow the concrete to achieve high early strength at the first hour.

If fast set concrete is being implemented as the major concrete repair in airport operational areas, the duration of the working hours and closures can be reduced drastically. Most importantly, the repair area can be opened and resume its operation as soonest at 3 to 6 hours.

4. Conclusion

Repairing defective concrete pavement with fast set concrete is an effective and efficient method. The characteristics and properties of fast set concrete enable it to develop sufficient amount of early strength and allows the defective concrete pavements to be repaired at soonest as several hours. As the result, the disruption to the operational areas is kept as the lowest rate and risks are minimized also. Nevertheless, repairing defective concrete pavement with fast set concrete may be costly and it requires specialized contractors or builder to execute the works. However, by considering the benefits of fast set concrete and the way it can enhance the effectiveness and efficiency in concrete repairing works, fast set concrete is still one of the best choices to be utilized and implemented. Other than airports, fast set concrete can also be used to repair concrete pavements or structures in other operational areas namely expressway, bus bays, bridge decks etc.

5. References

[1] Mays, Ged., **Durability of Concrete Structures: Investigation, Repair, Protection**, first edition, E&FN Spon, New York, 2002, 288 pp.

[2] Mo Li, Victor C. Li, **High-Early-Strength Engineered Cementitious Composites for Fast, Durable Concrete Repair—Material Properties**, 2011

[3] National Ready Mixed Concrete Association, **Maturity Methods to Estimate Concrete Strength**, 2006

IOP Conf. Series: Materials Science and Engineering 1075 (2021) 012010 doi:10.1088/1757-899X/1075/1/012010

Sustainable pavement: A review on the usage of pavement as a mitigation strategy for UHI

SFI Al-humairi[1,2,*], AH Alias[1], NA Haron[1], S Hassim[1], and F Mohd Jakarni[1]

[1]Department of Civil Engineering, Faculty of Engineering, Universiti Putra Malaysia, 43400 UPM Serdang, Selangor, Malaysia
[2]Alliance Geotechnical Group, TX 77535, USA
[*] Corresponding author: sarmadfismael@gmail.com

Abstract. Urban Heat Island (UHI) phenomenon is currently affecting many millions of people worldwide. The higher temperatures are experienced in urban areas as compared to the surrounding countryside. The UHI effect has been a great threat to human habitation, and the way to mitigate this problem has been a global concern over decades. Artificial and manmade pavements have been recognised as one of the major origins that deteriorate the thermal environments for living. On the one hand, commonly used pavement materials could lead to a reduced latent heat flux and an increased sensible heat flux to the environment. Following this, there is a series of experiments and numerical analysis that has been conducted by researchers. The selection of pavement type requires an understanding of a local context of an area and characteristics of the pavements. This current study emphasises on reviewing the features, factors and impacts of different type of pavements, as well as determining the suitability of different types of pavements. It is expected that the outcome from this study can assist in further understanding of pavements and its effects on UHI, as well as can ultimately assist in reducing the effect of UHI on human population.

1. Introduction

The rapid urbanisation is documented as a main cause of replacement in most of the green surfaces, vegetation or natural soil by artificial land covers, such as asphalt mixtures and cement concrete pavements [1]. Dark surface, higher heat storage capacity and large thermal inertia of concrete material contribute to a large solar radiation which will be stored in day time and emitted into the atmosphere at night [2,3]. This heat release has a significant impact on the thermal climate. The high air temperature of city centres is termed as the urban heat island (UHI) effect as compared to the sub-urban rural areas [4].

The local climate change is considered a global issue and has been a major concern for researchers. Many studies had reported on the UHI influence on residents, in which it was found that UHI deteriorated thermal comfort, increased the concentration of pollutants and threatened human health [5]. Apart from that, it was found that UHI would increase the energy demand during high temperatures days, in which an increase of one degree of outside thermal air would result in 0.45 to 4.6 per cent of increment in electricity usage [6,7].

To counterbalance the influence of urban heat island, cool pavement technology has been developed to reduce the emission of heat flux, hence, mitigating the urban heat island phenomenon [8]. Cool pavement is proposed by United States Environmental Protection Authority (USEPA) to achieve lower surface temperature and reduce heat island effects [9]. The cooling pavement term refers to the paving materials that remain cool, reflect more solar energy and enhance water evaporation as compared to the conventional pavement [10,11].

The aim of this review paper is to present and define conventional pavement types, to explain consequences, such as the increasing urban temperature, as well as discuss the influence of UHI on the residents and service life of the structure. Apart from that, this paper will clarify the strategies to mitigate the UHI effect with an emphasis on new cool pavements technology, including discussion on the benefits, costs, strengths, and policy of cool pavement.

2. Conventional pavements

Pavements cover about 40 per cent of the urban area of cities[10,12]. The dark colour for pavement material contributes to a high percentage of heat absorption, storage, as well as re-radiation to the atmosphere [13]. Most studies indicated that pavement had a temperature of at least 60 degrees Celsius in summer [14]. Pavement can be generally classified into three types: flexible, rigid or a combination of flexible and rigid. A flexible pavement consists of three essential layers: 1) surface layer, 2) base and 3) subbase layers, with all these layers installed on the subgrade. Nevertheless, the rigid pavement does not have a base layer; thus, concrete is used instead of asphalt in the surface layer. In some cases, an upper layer of a rigid pavement is layered with a thin asphalt to generate rigid-flexible composite pavement [15]. The asphalt surface layer provides a comfortable driving and long service life, while a rigid base supports a pavement structure that bears the great traffic loads.

2.1. Asphalt concrete pavement

Asphalt pavement or known as flexible pavement consists of a mixture of asphalt, air voids and various sizes of aggregates depending on the design to handle any load from light to heavy vehicles. Studies had shown that over 80 per cent of paved surfaces are made of asphalt concrete as compared to other paving types [1,16,17]. Under continuous heavy load, paired with moisture and rising temperature, this type of pavement is subjected to a number of damages, including rutting, cracking and corrugation[18].

As the pavement aged, the environmental conditions will assist in oxidising the binder in the surface layer, exposing the aggregates and consequently, increasing the albedo in the asphalt concrete, which can range from 0.09 to 0.18 [6,12–14]. Apart from the environmental conditions, it is observed that the asphalt concrete exposure to abrasion will increase the albedo significantly. The albedo increases rapidly in the first year after placement before staying in a steady state value [20,21].

2.2. Cement concrete pavement

Portland cement concrete pavement (PCCP) basically comprises cement, water, as well as fine and coarse aggregates. Reinforcement steel bar is utilised in some form to mitigate crack prevention that constitutes the basic problem of concrete pavement design. As a result, PCCP has a higher bearing capacity as compared to the flexible pavement. In contrast with asphalt concrete, the albedo of cement concrete will decrease over time due to the exposure of dirt and tire wear, which will darken the colour of the pavement surface.

3. Heat mitigation strategy

To reduce the possible effects of global warming and urban heat island phenomenon, cooling strategies are proposed to contribute to improving the urban thermal environment and thermal comfort of a city dwellers. The climate condition, the area intended and expectations of the economic effects should be considered to choose an appropriate strategy. Cooling strategy can be divided into 3, which were green surface, cool roofs and cool pavement [22].

3.1. Green and blue surface

Research had found that adding 10 per cent of vegetation to the urban areas would have a significant impact in improving the UHI effect by reducing the air temperature of up to 0.5 to 0.8 degrees Celsius [23]. Water ponds, river and water sprinkle can be considered as significant strategies to mitigate heat urban in areas that could provide the required water easily [24]. Nonetheless, although green and blue infrastructures have been common in mitigating the UHI effect, using these strategies without proper consideration to gray surfaces may not be significant to cooling the climate. This is because pavements are still by and large the basic infrastructure in developing cities, as well as constitute a large percentage of the cities` area. Therefore, apart from green and blue surfaces, it is equally important to use pavement technologies as heat mitigation strategy.

3.2. Cool roofs

Reflective or cool materials can be used to provide cool roofs in urban buildings. Studies' results reported that roof surface covered with albedo can result in a reduced electricity demand in the building, as well as a reduce sensible heat emission to the atmosphere [6,25]. Moreover, the changes of the cooling surface temperature can increase the materials' life time of the roof due to the decreasing chemical interactions as compared to the conventional roof [11,26].

3.3. Cool pavement

Cool pavements are basically an implementation of light coloured materials on a structure surface[27–29], which will then reduce the air-surface and the pavement temperature. The cooling techniques used by cool pavements include the installation of albedo [12], heat evaporation[10], as well as heat storage and harvesting [30].

Multiple studies had displayed that the technique of cool pavements undoubtedly reduced the surface temperature of pavements [31,32]. For example, research had demonstrated that cool pavement stayed cooler in the sun than traditional pavements. It is also noted that cool pavement had several benefits as compared to conventional pavement, such as energy savings and emission reductions[33], improved comfort and health[5], a long-life and safe road [18], improved water quality and noise reduction[34,35], as well as increased driver safety and decreased billions of tons of carbon dioxide emissions [6].

4. Cool pavement strategies to reduce urban heat island

Cool pavement technologies tend to store less heat resulting to lower surface temperatures. Reflective pavements, evaporation pavements and heat-harnessing pavement can be classified as cooling pavements. The surface of cooling pavements remains cooler than conventional pavements due to either reflecting more solar radiation, as well as decreasing the pavement temperature and sensible heat emits evaporation of the water existing in the top and bottom pavement layers or suppression of the surface heat absorbed to extraction of the heat from the pavement as a form of renewable energy for other usages and cooling the pavement[10,11]. Viewed from above, the UHI issue cannot be limited to only colour materials or rough surface of the pavement but developed widely to also include open-pore structures [36].

Besides several benefits of cool pavements in achieving environmental profit, another benefit is related to an improved construction performance and long-live service of pavements. The high temperature of asphalt pavement in summer can cause rutting distress and aging, while increasing surface temperature in cement concrete pavements produced risk by thermal cracking [37].

The difference in economic factors of cities influences directly on the cost of cool paving technologies that will be affected significantly to choose the best type. For these reasons, it is proposed that a comprehensive study for all pavement types should be conducted. This study depends on documented work of literature by a committee to make a decision to be then applied in the project. Nowadays, conventional pavements which have been developed with a light-coloured surface commonly use a comparison to other cool pavements due to requirements of a little cost and high bearing capacity of heavy loading [21].

4.1. Reflective pavements

The utilisation of a reflective pavement can be considered as one of the ways to mitigate the UHI effect, since it is the most cost-effective and can significantly decrease the surface temperature of the pavement [30]. Slurry seal is considered a typical treatment, which is applied over the existing pavement to get reflective pavements, while existing conventional asphalt pavement is replaced with yellow thin layer asphalt through the mixture of asphalt emulsion binder with infrared reflective pigments and fine aggregates to present much higher reflectivity [1]. In United States, a chip seal is applied as a reflective coating with a higher albedo in addition to its function to a treatment of a bituminous surface. The chip seal is typically used on the roads with lower traffic volumes, such as rural roads, walking sides and parking lots. [38]. In another case, white concrete pavements are employed instead of grey concrete pavements by adding pigments of titanium dioxide (TiO2) with a concrete mixture [6].

There is a number of benefits through the usage of reflective pavements. First, the reflective pavements can reduce the surface temperature and consequently, reduced sensible heat is emitted to the atmosphere [39]. This will enhance thermal comfort and mitigated the urban heat island [40]. Second, preventing glare and visual discomfort to provide driving safety will consequently minimise eye-related disorders and road accidents [19]. In contrast, since the application of solar reflective coatings will change the original surface texture of the pavement, further studies are needed to determine the suitability of the application of this technique. It is reported that in the warm-weather climates the reflective pavements with a high albedo value reduces the pavement surface temperature but will give an output of discomfort thermal for the humans in the ground level [41].

4.2. Evaporation and permeable pavements

Permeable pavement technology has been developed to significantly reduce air temperature [42]. It allows water to filtrate through the pores to supply the sub-layers and storage in the top surface. This in turn will provide a cool surface that will affect the air temperature on the level ground. Generally, there is a number of benefits of permeable pavements, which includes an increased skid resistance and a reduced tire-pavement noise [43], a higher sound absorption [41], an increase in snow and ice melting rates [42], and an improvement of the quality of water supplied to the groundwater [37]. This type of pavement is usually used in places with light traffic loads, low speeds, commercial areas, and parking surfaces [34]. Permeable pavements can be classified into porous pavers, pervious pavers and water retaining pavements.

4.2.1 Porous pavers

Porous paver is a type of concrete or plastic geocells that form a cellular grid system with a similar performance as compared to rigid pavement [10]. The surface-air cooling of porous pavers depends on the evaporation capability through a channel for the water stored in the pavement layers. The cellular grid system has internal holes filled with materials, such as grass, concrete or fired-clay bricks that have a high capacity to hold moisture [10]. When grass is used to fill the holes, the roots aid to transport moisture from deeper soil to cool the surface-air of the pavement. On the contrary, when concrete or fired-clay bricks are used, this allows the water to pass around the pavement into the base or sub-base layers as in permeable interlocking concrete pavements. The rough surface and the low thermal inertia cause a high thermal if the water is not available. This type of pavers is usually used for pedestrian walkways, sidewalks, pathways, and driveways.

4.2.2. Pervious pavers

Various development strategies pervious pavements have a lesser impact on the environment. Sustainable materials and eco-friendly strategies play an important role to become a suitable candidate application on roads. Pervious pavement layer is made of asphalt or concrete with a high ratio of air voids that allows the water to pass directly through the pavement into sublayers of crushed aggregate, then it is infiltrated into the ground. Pervious pavements are designed to accept precipitation; hence, during application, side drains are not required, so the construction cost is reduced. The aggregate physical characteristics have a significant influence on the properties of pervious concrete and controlled porous proportion on the water permeability through pavement layers. The pavement layer typically is thicker than the traditional concrete to increase the bearing capacity of heavy weight. The usage of pervious concrete in low volume roads, such as parking lots[35], sidewalks and roads without heavy load is due to weaker durability and mechanical properties of the structure[44].

Evaporation process can keep pervious pavements cooler than conventional pavements about one to two days after sprinkling under hot conditions. Cooling is negligible with a loss of water and more absorption of solar irradiation because pervious pavements have a rough surface and a lower thermal inertia than conventional pavements[32]. Continuously, maintenance of pavements is important to sustain a structure and keep life cycle performance. This is done by using a high-pressure washing with water and/or vacuum to remove dust and other foreign particles from the upper layer of the pavement [43].

Finally, after the permeable pavement types are explained, it can be concluded that due to the absence of water in air voids, it reduced the amount of water used for water evaporation. Consequently, the water evaporation rate is suddenly reduced, while the temperature rate increases rapidly [45]. During hot days the permeable pavement temperature is found to be higher than another type of impermeable pavement under dry condition. An open-graded asphalt mixture absorbs more solar radiation and increases the top layer temperature during the day. Irrigation water reduces surface pavement temperature and then, it reduces the significant effect of mitigating the near air surface temperature. The sprinkling assist is observed to cool the pavement surface and directly decrease 10 degrees Celsius after wetting the surface[42].

4.2.3. Water-retaining pavements

Any pavement of water-holding and water-retentive is indicated for the longest period by a water-retaining pavement. It is developed to solve the problem of fast infiltrate water through the pervious concrete. Water-retentive pavements have a porosity similar to permeable pavements, but the water permeability of the water-retentive pavements is lower than the other one[10]. The component essentially for this type of pavement is asphalt or cement. The different fillers are added to the mixture to assist in retaining the amount of water. After raining, the water-retaining pavements are expected to stay cooler than permeable pavements. The high capillary force assists the water to be sucked from the sub-base of the pavement because of a lack of water in the upper surface. This type of pavement illustrates a better efficiency in cooling the air-surface when the water is held in the surface layer that had a range of 25 millimeters [10,46].

4.3. Heat storage and harvesting

Heat-harvesting pavements cannot only reduce surface temperature, but also extract absorbed heat as renewable energy for other usages [11]. For this type of pavement, the usage of asphalt pavements as the surface material can perform better than the concrete pavements. This is because the black colour of asphalt mixture can absorb solar irradiation better as compared to concrete's solar absorption coefficient. Technique design of harvesting pavement is formed by combining an asphalt solar collector, installing pipes with a circulating fluid inside and installing a generator beneath the pavement. Water is usually used as the fluid in the pipework to extract the pavement heat [10,47]. Apart from water, natural air can be used in the pipework to harness heat from the pavement [10,48]. In case of water fluid, the pipe systems can be eliminated and a multiplayer asphalt pavement system can keep the water inside the structure. This multiplayer asphalt pavement system circulates in the porous middle layer between impermeable layers.

Due to the temperature difference between the fluid circulating into pipes and the pavement, the pavement absorbs formed solar radiation. This solar radiation is transferred to the fluid by counterbalance and then, decreases pavement temperature to provide a comfortable near surface air and a sustainable structure. The harvested energy can be stored in a roadside water tank for other usage or for snow-melting systems in winter time [11]. To apply this system, more studies regarding durability and scrutiny on the power output are needed. In addition, continuous maintenance activities that add extra cost should be done as most studies had also been conducted in laboratories.

4.4. Innovative methods recently

i. Evaporation-enhancing permeable pavement is developed as a new permeable pavement to contribute mitigation of the UHI effect. In addition, storm water in cities, in which the groundwater is high is managed. The evaporation-enhancing permeable pavements have a good capillary capacity, which causes the evaporation to stay for a long period as compared with the conventional permeable pavement that only lasted for the first two days after raining[49].
ii. A new technology is applied to a reduced temperature of pavement surface and sub-layers [37]. Implanting of steel rods with high thermal conductivity in different levels is carried out to accelerate the flow of the absorbing heat by the top layers, and the heat is transferred to the bottom layers to decrease heat accumulation. Then, the influence on the air-surface is decreased in the ground level. Additionally, the structure will be strong to increase rutting resistance and long-term service.

5. Factors affecting cool pavement selection

5.1. Albedo

One of the important parameters in assessing the temperature in pavement is albedo. Albedo can be defined as a ratio of solar radiation reflected to the radiation incident on a surface [50]. A higher value of surface albedo reflects more proportion of the solar radiation, therefore, reducing pavement surface temperature. There are two methods to measure albedo. First, albedo is measured by using an albedometer or a double pyranometer, in which the accuracy of the results is high depending on the size of the sample and weather conditions[20,51]. The other test is carried out in a laboratory using a spectral reflectance obtained from a spectrophotometer to obtain an approximate estimate of the albedo value[20,19].

Albedo affects local microclimate, energy consumption and pedestrians. Most of the previous studies focused on the influence of an increased albedo of the inside and the outside environment. Part of solar radiations reflects from a pavement surface back to the sky without influencing the thermal atmosphere, while the remaining part influences the surrounding buildings. As a result, it is recommended not to use pavements with albedo as a surface in cities with high buildings [52]. A number of research is done in analysing the impact of pavements on the ground level related to the thermal comfort for pedestrian [23].

5.2. Cost

Economic factors and local availability of materials have a direct impact on the cost of projects. Thus, managing the project cost requires a customisation of optimal pavement designs for site-specific conditions. For instance, asphalt pavements are favoured because of their low initial costs and short project lengths. Moreover, asphalt pavements are preferred in some cases which expected to do additional activities related to infrastructure, such as water, power, sewage, and pipeline after the initial construction [11]. Conversely, although rigid pavement is less prone to damages due to environmental conditions and heavy-vehicle loading, the rehabilitation work for this type of pavement is more expensive than asphalt concrete [53].

Prior studies had also noted the importance of cost in selecting cool pavement technologies [54]. Rehabilitation and maintenance of pavements are considered a requirement to sustain the quality and functionality of the pavements. Assessing the phases of cool pavement before the overall implementation includes production, construction, maintenance, use, and post-construction maintenances of the pavement to avoid extra costs. Therefore, many studies have been conducted to investigate the costs of cool pavement technologies [16,11]. Nonetheless, there is no any international standard since the cost depends on the pavement type and its location. When reflective and permeable payments are compared, lower cost and easy potential installation in reflective pavements makes reflective pavement preferred for marketing than other applications. In addition, extra cost of permeable pavement is used to dislocate the clogging dirt in the cavity because the open porous structure is prone to clogging. [30].

5.3. Strength

Pavements are critical to land transportation use (roads, walking, commercial plazas, and parking). The large proportion of these paving should be able to carry a heavy loading. Rigid or flexible pavements are built and cured until they were strong enough to carry traffic. A rigid base can greatly improve the bearing capacity of a pavement structure [15]. Structural design and material composition can influence the pavement behaviour. In asphalt pavement case under heavy loads, the void spaces of porous asphalt deteriorate with time due to deformation, resulting in a reduction in air voids of the surface pavement and consequently, reducing the efficiency of the pavements in allowing rainwater to pass through the pavement to the underlying layers [44].

Due to strength, heat-harvesting pavements cannot be used under high traffic loadings without specific engineered techniques to support the pavement structure [11,55]. Besides that, in the case of permeable pavements, the strength of pervious concrete decreases with the increase in porosity and permeability [44]. Thus, it can be seen that the majority of existing cool pavement is used for low vehicle volumes, such as parking lots and commercial areas with low speeds and light traffic loads. Further studies are needed to ensure that the structural design and material properties of cooling pavements could be used effectively on roads with high loads and speed.

6. Conclusion

The local climate changes contribute to the urban heat island phenomenon. Urban climate change has a significant impact on electricity consumption, indoor and outdoor ambient temperature, citizens comfort, threatening human health, economic impacts, as well as passive impact on the paving life service. To counterbalance the effect, this paper has reviewed cooling strategies, including green, blue and grey surfaces to mitigate the UHI effect. It is discovered that apart from green and blue surfaces, focus must be given to grey surfaces and cool pavements since grey surfaces has a significant effect on the ambient temperature and covered most of the roads of the cities.

Cool pavements can also be achieved by modifications of the materials or designs for the existing pavements. To select either reflective pavement surface, evaporative pavements, or heat-harvesting pavements, the effect of the urban heat needs to be observed. Benefits of using cool pavements to battle the urban heat island effect, strength, cost, pros and cons during implementation the project should also be considered.

The cool pavement techniques were applied in some countries, such as Toronto[24], Greece[6], USA[35], Malaysia[2], China[4], and Japan[56] in this study. In short, it could be concluded that reflective pavements were appropriate for climate areas with hot dry summer and for the open area with long-hour insolation. However, evaporative pavements may be more suitable for tropical regions with sufficient water available over the year. In conclusion, it is recommended to conduct more practical studies of cool pavements, in addition to a combination of the methods under variable conditions to address the gaps and reducing the UHI effect.

References

[1] Gilbert HE *et al.* 2017 Energy and environmental consequences of a cool pavement

campaign *Energy Build* **157** 53–77

[2] Benrazavi RS, Kamariah BD, Ujang N, and Benrazavi NS 2016 Effect of pavement materials on surface temperature in tropical climate *Sustain. Cities Soc.* **22** 94–103.

[3] Levermore G, Parkinson J, Lee K, Laycock P and Lindley S 2018 The increasing trend of the urban heat island intensity *Urban Clim.* **24** 360– 368

[4] Sulaiman NM *et al.* 2018 A critical review of Urban Heat Island phenomenon in the context of Greater Kuala Lumpur, Malaysia *Sustain. Cities Soc.* **39** 99–113

[5] Giridharan R and Emmanuel R 2018 The impact of urban compactness, comfort strategies and energy consumption on tropical urban heat island intensity: A review *Sustain. Cities Soc.* **40** 677–687

[6] Kyriakodis GE and Santamouris M 2018 Using reflective pavements to mitigate urban heat island in warm climates - Results from a large scale urban mitigation project *Urban Clim.* **24** 326–339

[7] Santamouris M 2015 Regulating the damaged thermostat of the cities - Status, impacts and mitigation challenges *Energy Build* **91** 43–56

[8] Akbari H and Kolokotsa D 2016 Three decades of urban heat islands and mitigation technologies research *Energy Build* **133** 834–842

[9] C US EPA, OAR, OAP 2016 *Using Cool Pavements to Reduce Heat Islands* 3–5

[10] Mathi M, Selvaraj S and Priya V 2018 The Urban Heat Island Effect Its Causes and Mitigation with Reference to the Thermal Properties of Roof Coverings *Int. Conf. Urban Sustain* **197** 522–538

[11] Qin Y 2015 A review on the development of cool pavements to mitigate urban heat island effect *Renew Sustain Energy Rev.* **52** 445–459

[12] Akbari H, Menon S and Rosenfeld A 2009 Global cooling: Increasing world-wide urban albedos to offset CO 2 *Clim. Change* **94** 275–286

[13] Hendel M, Parison S, Grados A and Royon L 2018 Which pavement structures are best suited to limiting the UHI effect? A laboratory-scale study of Parisian pavement structures *Build. Environ.* **144** 216–229

[14] Higashiyama H, Sano M, Nakanishi F, Takahashi O and Tsukuma S 2016 Field measurements of road surface temperature of several asphalt pavements with temperature rise reducing function *Case Stud. Constr. Mater.* **4** 73–80

[15] Li S, Huang Y and Liu ZH 2016 Experimental evaluation of asphalt material for interlayer in rigid-flexible composite pavement," *Constr. Build. Mater.* **102** 699–705

[16] Pomerantz M, Akbari H, Chen A, Taha H and Rosenfeld AH 1997 Paving materials for heat island mitigation (Berkeley: Lawrence Berkeley National Lab)

[17] Li H *et al.* 2019 Study on the gradient heating and healing behaviors of asphalt concrete induced by induction heating *Constr. Build. Mater.* **208** 638–645

[18] Yinfei D, Jiaqi C, Zheng H and Weizheng L 2018 A review on solutions for improving rutting resistance of asphalt pavement and test methods," *Constr. Build. Mater.* **168** 893–905

[19] Santamouris M, Synnefa A and Karlessi T 2011 Using advanced cool materials in the urban built environment to mitigate heat islands and improve thermal comfort conditions *Sol. Energy* **85** 3085–3102

[20] Sen S and Roesler J 2016 Aging albedo model for asphalt pavement surfaces *J. Clean. Prod.* **117** 169–175

[21] Tran N, Powell B, Marks H, West R and Kvasnak A 2009 Strategies for Design and Construction of High-Reflectance Asphalt Pavements *Transp. Res. Rec. J. Transp. Res. Board* **2098** 124–130

[22] Lai D, Liu W, Gan T, Liu K and Chen Q 2019 A review of mitigating strategies to improve the thermal environment and thermal comfort in urban outdoor spaces *Science of the Total Environment* **661** 337–353

[23] Wang Y, Berardi U and Akbari H 2016 Comparing the effects of urban heat island mitigation

strategies for Toronto, Canada *Energy Build* **114** 2–19

[24] Taleghani M and Berardi U 2018 The effect of pavement characteristics on pedestrians' thermal comfort in Toronto *Urban Clim.* **24** 449–459

[25] Doulos L, Santamouris M and Livada I 2004 Passive cooling of outdoor urban spaces. The role of materials *Sol. Energy* **77** 231–249

[26] Taha H 1997 Modeling the impacts of large-scale albedo changes on ozone air quality in the South Coast Air Basin *Atmos. Environ.* **31** 1667–1676

[27] Taleghani M 2018 The impact of increasing urban surface albedo on outdoor summer thermal comfort within a university campus *Urban Clim.* **24** 175–184

[28] Taha H 2008 Urban Surface Modification as a Potential Ozone Air-quality Improvement Strategy in California: A Mesoscale Modelling Study *Boundary-Layer Meteorol.* **127** 219–239

[29] Kleerekoper L, Taleghani M, van den Dobbelsteen A and Hordijk T Urban measures for hot weather conditions in a temperate climate condition: A review study *Renew. Sustain. Energy Rev.* **75** 515–533

[30] Qi JD, He BJ, Wang M, Zhu J and Fu WC 2019 Do grey infrastructures always elevate urban temperature? No, utilizing grey infrastructures to mitigate urban heat island effects *Sustain. Cities Soc.* **46** 101392

[31] Battisti A, Laureti F, Zinzi M and Volpicelli G 2018 Climate mitigation and adaptation strategies for roofs and pavements: A case study at Sapienza University Campus *Sustain.* **10** 3788

[32] Qin Y and Hiller JE 2016 Water availability near the surface dominates the evaporation of pervious concrete *Constr. Build. Mater.* **111** 77–84

[33] Mastrapostoli E, Santamouris M, Kolokotsa D, Vassilis P, Venieri D and Gompakis K 2016 On the ageing of cool roofs: Measure of the optical degradation, chemical and biological analysis and assessment of the energy impact *Energy Build.* **114** 191–199

[34] Weiss PT, Kayhanian M, Gulliver JS and Khazanovich L 2019 Permeable pavement in northern North American urban areas: research review and knowledge gaps *Int. J. Pavement Eng.* **20** 143–162

[35] Kumar K, Kozak J, Hundal L, Cox A, Zhang H and Granato T 2016 In- situ in fi ltration performance of different permeable pavements in a employee used parking lot e A four-year study *J. Environ. Manage.* **167** 8–14

[36] Zhong R, Leng Z and Sun Poon C 2018 Research and application of pervious concrete as a sustainable pavement material: A state-of-the-art and state-of-the-practice review *Constr. Build. Mater.* **183** 544–553

[37] Yinfei D, Zheng H, Jiaqi C and Weizheng L 2018 A novel strategy of inducing solar absorption and accelerating heat release for cooling asphalt pavement *Sol. Energy* **159** 125–133

[38] WSDOT - What is a Chip Seal? [Online] Available: https://www.wsdot.wa.gov/Regions/Eastern/ChipSeal/ [Accessed: 21- Mar-2019]

[39] Jiang J, Jin Y, Bao T and Ou X 2019 Sensible heat discharging from pavements with varying thermophysical properties *Sustain. Cities Soc.* **45** 431–438

[40] Kolokotsa DD, Giannariakis G, Gobakis K, Giannarakis G, Synnefa A and Santamouris M 2018 Cool roofs and cool pavements application in Acharnes, Greece *Sustain. Cities Soc.* **37** 466–474

[41] Taleghani M 2018 Outdoor thermal comfort by different heat mitigation strategies- A review *Renew. Sustain. Energy Rev.* **81** 2011–2018

[42] Wang J, Meng Q, Tan K, Zhang L and Zhang Y 2018 Experimental investigation on the influence of evaporative cooling of permeable pavements on outdoor thermal environment *Build. Environ.* 140 184– 193

[43] Xie N, Akin M and Shi X 2019 Permeable concrete pavements: A review of environmental

benefits and durability *J. Clean. Prod.* **210** 1605–1621

[44] Chandrappa AK and Biligiri KP 2016 Pervious concrete as a sustainable pavement material-Research findings and future prospects: A state-of- the-art review *Constr. Build. Mater.* **111** 262–274

[45] Hassn A, Chiarelli A, Dawson A and Garcia A 2016 Thermal properties of asphalt pavements under dry and wet conditions *Mater. Des.* **91** 432–439

[46] Nemirovsky EM, Welker AL and Lee R 2013 Quantifying Evaporation from Pervious Concrete Systems: Methodology and Hydrologic Perspective *J. Irrig. Drain. Eng.* **139** 271–277

[47] Bobes-Jesus V, Pascual-Muñoz P, Castro-Fresno D and Rodriguez- Hernandez J 2013 Asphalt solar collectors: A literature review," *Appl. Energy* **102** 962–970

[48] García A and Partl MN 2014 How to transform an asphalt concrete pavement into a solar turbine *Appl. Energy* **119** 431–437

[49] Liu Y, Li T and Peng H 2018 A new structure of permeable pavement for mitigating urban heat island *Sci. Total Environ.* **634** 1119–1125

[50] Wu H, Sun B, Li Z and Yu J 2018 Characterizing thermal behaviors of various pavement materials and their thermal impacts on ambient environment *J. Clean. Prod.* **172** 1358–1367

[51] Li H, Harvey J and Kendall A 2013 Field measurement of albedo for different land cover materials and effects on thermal performance *Build. Environ.* **59** 536–546

[52] Chen J, Wang H and Zhu H 2017 Analytical approach for evaluating temperature field of thermal modified asphalt pavement and urban heat island effect *Appl. Therm. Eng.* **113** 739–748

[53] Plati C 2019 Sustainability factors in pavement materials, design, and preservation strategies: A literature review *Constr. Build. Mater.* **211** 539–555

[54] Grimmond CSB and Oke TR 2002 Turbulent Heat Fluxes in Urban Areas: Observations and a Local-Scale Urban Meteorological Parameterization Scheme (LUMPS) *J. Appl. Meteorol.* 41 792–810

[55] Taylor P, Northmore AB and Tighe SL 2014 International Journal of Pavement Engineering Performance modelling of a solar road panel prototype using finite element analysis 37–41

[56] Furumai H, Kim J, Imbe M, and Okui H 2008 Recent application of rainwater storage and harvesting in Japan *IWA Conf.* 1–7

IOP Conf. Series: Materials Science and Engineering 1075 (2021) 012011 doi:10.1088/1757-899X/1075/1/012011

Investigation on effects of aggregate roughness on bond strength of aggregate-bitumen systems

V Mishra[1]* and D Singh[2]

[1]Department of Civil Engineering, Indian Institute of Technology Bombay, Powai, Mumbai – 400076, India
[2]Department of Civil Engineering, Indian Institute of Technology Bombay, Powai, Mumbai – 400076, India, Phone: +91-22-2576-7304
*Corresponding author: vinamramishra89@gmail.com

Abstract. It is critical to understand bonding-behavior of aggregate with bitumen. The surface chemistry and roughness of aggregates play an important role to create sufficient bond with bitumen. Usually, asphalt-bond-strength (ABS) test is used to directly quantify bonding-behavior of aggregate-bitumen system, further, surface-free-energy (SFE) approach has also been used by some researchers. AASHTO T-361 recommends to use aggregates polished with 280-silicon-carbide (SiC) powder to determine bond-strength with bitumen. However, increase or decrease in polishing-level can alter aggregate-roughness and may affect bond-strength. Further, there can be numerous roughness-scale for aggregates such as, arithmetic (R_a), root-mean-square (R_q) and box-roughness (S_a), which needs detailed study. Therefore, one of the strong motivations of this study was to evaluate effects of aggregate-roughness on bond-strength using ABS approach. Similarly, aggregate SFE may change with change in its roughness, thereby affecting aggregate-bitumen bonding. The present study evaluates effects of aggregate-roughness on aggregate-bitumen bond-strength
using ABS and SFE approaches. Basaltic substrates and viscosity-grade binder (VG30) were selected for this study. To generate different roughness-levels, aggregate substrates were polished with 220SiC and 1000SiC powders. At each polishing-level, R_a and S_a parameters were evaluated using zeta-profilometer. The ABS and SFE measurements were conducted on combination of aggregate and bitumen. The results showed that bond-strength increased with increase in aggregate-roughness, indicating better bonding for rough aggregates. Further, S_a parameter which remained unaddressed in earlier research studies was recommended to quantify aggregate roughness rather than R_a parameters, because it measures roughness over area, which is one of the major contributions of present study.

1. Introduction

Presence of moisture in bituminous mixes results into bonding failure of aggregate-bitumen system. Failure of aggregate-bitumen bonding can lead to cracking, potholes, rutting and raveling distresses in bituminous pavements. Thus, it is crucial to achieve better adhesive bond between aggregate and bitumen. Surface texture of aggregate plays a vital role in developing bond strength between aggregate and bitumen [1-3]. In general, rough textured aggregates make better bond with bitumen [1]. However, aggregate texture changes during crushing, attrition, and abrasion mechanism during production process and during service life. For example, as the pavement ages, the protruding aggregate become polished due to wearing action of vehicular movements, which lowers friction and skid resistance between pavement and vehicle tires. The reduced friction under wet surface conditions aggravates risk of hydroplaning which is undesirable considering serviceability criteria. Further, polished surface is more likely to undergo bonding failure [1]. Therefore, timely measurement of surface roughness is critical. Further, change in aggregate texture may affect its surface chemistry which influences its bonding ability with bitumen. Therefore, the effect of polishing on the surface chemistry of aggregates also needs to be understood.

The bond strength between aggregate and bitumen can be indirectly evaluated using Surface Free Energy (SFE) approach based on contact angle measurements [4-7]. Koc and Bulut [8] evaluated the SFE components of aggregates polished with silicon carbide (SiC) powder (200, 400, 600, and 1000) and 5-micrometer alumina oxide grit. However, surface roughness values were assumed to be less than 1 μm after polishing with 600SiC grade abrasive powder. Similarly, Kittu et al. [7] evaluated the roughness of unpolished, 400SiC and 600SiC grade polished aggregates using 2-D profilometer. Both studies have reported strong dependence of contact angle on aggregate texture. However, the effect of polishing levels on bond strength was not studied. Recent studies have used Asphalt Bond Strength (ABS) test approach which considers aggregate roughness to determine bond strength of aggregate-bitumen system in accordance with AASHTO T-361 [1, 9-10]. However, standard method recommends aggregates polished with 280SiC grit powder without consideration for magnitude and effect of varying roughness. Hence, it was observed that limited studies were conducted on effect of varying aggregate roughness on bond strength with bitumen. Therefore, the present was undertaken to estimate the effect of aggregate roughness on aggregate-bitumen bond strength.

2. Objective and Scope of study

As discussed, quantification of aggregate roughness was primarily done using 2D profilometer and its effect on bond strength was rarely studied. Roughness of aggregates was usually measured using 2D profiler with arithmetic scale (R_a) [7-8]. However, direction of roughness measurement i.e. along the X-direction (horizontal) or Y-direction (vertical) is not clearly stated which may affect roughness measurements. Further, use of different scale of roughness measurements, for example, root-mean-square (R_q) and box-roughness (S_a) may result in different roughness values. The R_a roughness scale measures roughness over line along horizontal axis (R_a-H) or vertical axis (R_a-V), whereas, S_a measure roughness over the scanned area. Therefore, it will be interesting to study whether the measurement of roughness depends on the choice of scale used. Further, it is important to investigate the impact of higher and lower polishing levels on bonding potential of aggregate-bitumen system.

Therefore, the present study is motivated to determine the roughness of aggregate under different polishing levels using 3D zeta profilometer and identify the effects of aggregate roughness on aggregate-bitumen bond strength. The present study considers measurement of aggregate roughness using line roughness (R_a) and area roughness or box roughness (S_a) scales. Two different levels of aggregate roughness were generated by polishing with 220SiC and 1000SiC grade abrasive powders. Thereafter, the roughness was evaluated using zeta profilometer with different scales of measurement (i.e. R_a and S_a). Further, bond strength of aggregate and bitumen was evaluated using SFE and ABS test approach at different polishing levels.

Thus, the overall objective of the present study was to quantify varying levels of aggregate surface roughness and identify its effect on bond strength with bitumen. To achieve the desired goals, following sub-objectives were formulated:

1. Measurement of aggregate roughness at different polishing levels using 3D profilometer.
2. To estimate the effects of different levels of aggregate roughness on surface chemistry of aggregates and aggregate-bitumen bond strength.

3. Material and Methodology

Basalt aggregate and viscosity grade bitumen (VG-30) were used in the present study. These are the commonly used materials for road construction in India. To determine the SFE of bitumen, water (W), diiodomethane (D) and ethylene glycol (E) were selected as probe liquids due to its lower condition number (4.47). However, water (W), formamide (F) and ethylene glycol (E) were selected as probe liquids for measuring contact angles of aggregate [5]. The methodology adopted in the present study involves the measurement of aggregate roughness at different polishing stage, evaluation of aggregate surface chemistry and bond strength using SFE approach and evaluation of aggregate-bitumen bond strength using ABS test approach at different roughness levels. Brief discussion of the adopted methodology is presented in the following sub-sections.

3.1 Aggregate Roughness Measurement

The steps in measuring aggregate roughness involves cutting of aggregates using diamond saw cutter, polishing using required grades of SiC powder and cleaning of polished aggregates using ultrasonic water bath. The roughness of aggregate was measured using 3D Zeta profilometer. The main components of zeta profilometer are camera, standard objectives (5x, 10x, 20x and 50x), precision Z-drive mechanism, zeta optics module, control box, substrate holder, and computer system (figure 1a). When a beam of LED bombards the aggregate surface, the X, Y and Z dimensions of a point are measured (figure 1b). Thereafter, using zeta optics at any level of magnification 2D composite image and 3D true color image is obtained. The roughness of aggregate at different scales i.e. R_a along the horizontal axis (R_a-H) and vertical axis (R_a-V), and S_a is calculated using data analysis software [7, 11]. The R_a is the arithmetic scale which measures line roughness as sum of deviations of measured points from the average peak [11] and is presented in equation (1). On the other hand, S_a is the extension of line roughness scale R_a. It measures absolute value of surface roughness measured as the difference in height of each point and arithmetic mean of surface (equation (2)). The line roughness scale i.e. R_a-H and R_a-V, and area roughness scale i.e. S_a is presented in figure 1c-1e. The measurement of roughness is presented in figure 1f.

$$R_a = \frac{1}{n_r}\sum_{i=1}^{n_r}|z(i) - \bar{z}| \tag{1}$$

$$S_a = \frac{1}{A}\iint|Z\,(x,y)|dxdy \tag{2}$$

where,
(i)= local peak height of specimen at i^{th} measurement.
$\bar{z}$= average peak height.
n_r= number of measurements on the sampling length r.
A= scanned area

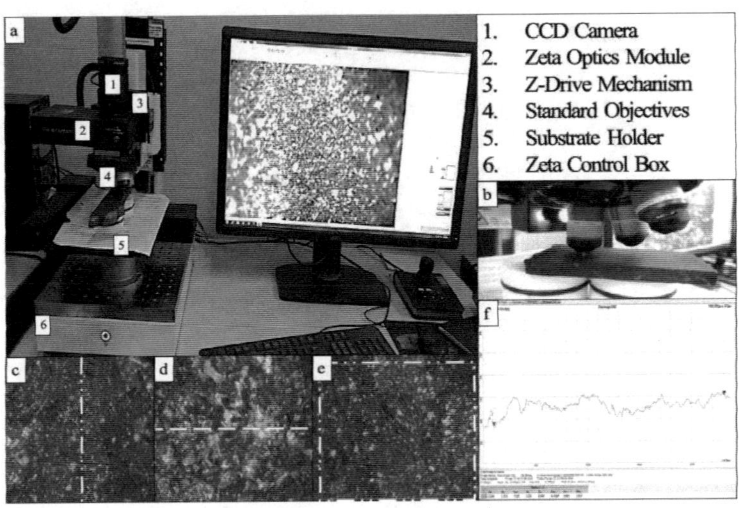

Figure 1. (a) Zeta Profilometer, (b) Roughness measurement using zeta profiler, (c) R_a-V profile over line, (d) R_a-H profile over line Roughness measurement, (e) S_a profile over surface, (f) Roughness measurement profile

3.2 Surface Free Energy (SFE) approach

The SFE of material (Υ) is its intrinsic property, consisting of Lifshitz-van der Waals (Υ^{LW}), Lewis acid (Υ^{+}) and Lewis base (Υ^{-}) components [12]. The $\Upsilon^{+}$and Υ^{-}components are together termed as polar component (Υ^{AB}) (equation (3) and (4)). Using contact angle approach, the components of SFE (i.e. Υ_S^{LW}, Υ_S^{+} and Υ_S^{-}) of any substrate 'S' can be determined using probe liquids 'L' according to acid-base theory using equation (5) [12].

$$\Upsilon = \Upsilon^{LW} + 2\sqrt{\Upsilon^{+}\Upsilon^{-}} \quad (3)$$

$$\Upsilon^{AB} = 2\sqrt{\Upsilon^{+}\Upsilon^{-}} \quad (4)$$

$$\Upsilon(1 + Cos\theta) = 2\sqrt{\Upsilon_S^{LW}\,\Upsilon_L^{LW}} + 2\sqrt{\Upsilon_S^{+}\Upsilon_L^{-}} + 2\sqrt{\Upsilon_S^{-}\Upsilon_L^{+}} \quad (5)$$

The SFE components of aggregate and bitumen are useful to determine their bond strength by evaluating energy parameters. Work of dry adhesion (W_{AB}), work of wet adhesion (W_{ABW}), cohesion energy (W_{BB}) and wettability ($W_{AB} - W_{BB}$) are regarded as energy parameters. Energy of adhesion in absence of moisture is termed as work of dry adhesion (W_{AB}), whereas, adhesion energy in presence of moisture is termed as work of wet adhesion (W_{ABW}) (equation (6) and (7)). Cohesion energy is the work required to create unit area crack within bitumen (equation (8)). Combining these energy parameters, the bond strength between aggregate and bitumen can be estimated by determining energy ratio (ER) parameter using equation (9) [6].

$$W_{AB} = 2\sqrt{\Upsilon_A^{LW}\Upsilon_B^{LW}} + 2\sqrt{\Upsilon_A^{+}\Upsilon_B^{-}} + 2\sqrt{\Upsilon_A^{-}\Upsilon_B^{+}} \quad (6)$$

$$W_{ABW} = \Upsilon_{AW} + \Upsilon_{BW} - \Upsilon_{AB} \quad (7)$$

$$W_{BB} = 2\,\Upsilon_B \quad (8)$$

$$ER = \left|\frac{W_{AB} - W_{BB}}{W_{ABW}}\right| \quad (9)$$

3.3 Asphalt Bond Strength (ABS) test approach

ABS test measures the tensile force required to break bonding between aggregate and bitumen in terms of pull-off tensile strength (POTS) as per AASHTO T-361. ABS test set-up comprises of metallic pull-off stub, pressure hose, reaction plate, piston, and pneumatic adhesion tester. The determination of aggregate-bitumen bond strength involves aggregate cutting and polishing with 280SiC grade powder [13]. To prepare test specimens, washed aggregates, binder and pull-off stubs are kept in oven at 150°C for not less than 30 min. Thereafter, the aggregates are kept in another oven maintained at application temperature (60°C). The binder sample is collected in pull-off stub which is then pressed firmly against the aggregate surface preheated at required application temperature. The steps involved in preparing sample is present in figure 2a-2d. The prepared samples were then kept for dry (room temperature, 24 h) and wet conditioning (40°C, 24 h). The POTS of samples after dry conditioning ($POTS_{dry}$) and wet conditioning ($POTS_{wet}$) can be evaluated using equation (10). Further, to rank different combinations of aggregate and bitumen, bond strength ratio (BSR) parameter can be evaluated using equation (11).

$$POTS = \frac{(BP * A_g) - C)}{A_{ps}} \quad (10)$$

$$BSR = \frac{POTS\ wet}{POTS_{dry}} \quad (11)$$

where, A_g is contact area of gasket with reaction plate in mm^2; A_{ps} is area of pull-off stub in mm^2; C is piston constant; BP is burst pressure in kPa.

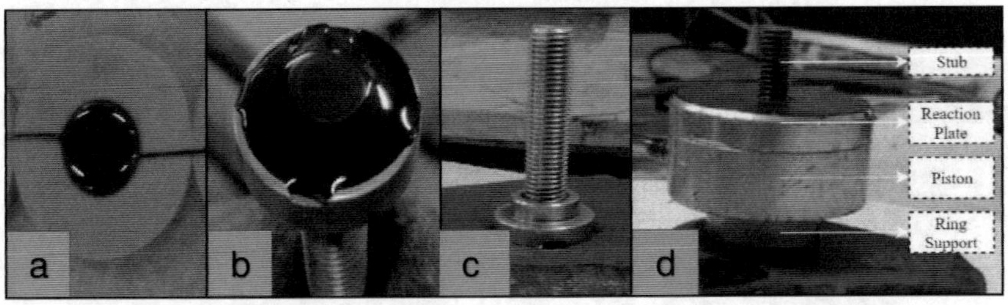

Figure 2. (a) Binder poured in silicon mold, (b) Binder collected on preheated stub, (c) Test Sample (d) Sample test arrangement.

4. Results and Discussion

4.1 Effect of polishing levels on aggregate roughness

The roughness of 220SiC and 1000SiC polished aggregates measured at different roughness scales (i.e. R_a-H, R_a-V and S_a) is presented in figure 3. As expected, roughness generated from 220SiC polishing level was found to be higher than roughness of 1000SiC polished aggregates. For example, S_a value for 220SiC polished aggregate was found to be 2.933 μm, which after progressive polishing to 1000SiC reduces it to 0.610 μm. The reduction in roughness is due to the higher grain size of 220SiC abrasive powder (63 micron) compared with 1000SiC (7 micron). In addition, p-values obtained using students t-test for roughness scales R_a-H, R_a-V and S_a were found to be 0.003, 0.000 and 0.000, respectively, therefore indicating significant difference in the roughness obtained using different polishing levels. Moreover, the difference in polishing levels is clearly evident from the 3D images obtained using zeta profilometer (figure 4). Further, it is to be noted that roughness measured using S_a scale evaluates higher values of roughness compared with R_a-H and R_a-V roughness scale (figure 3).

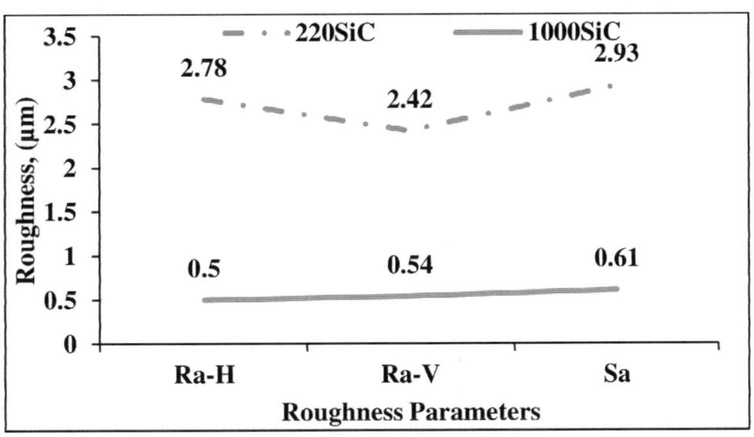

Figure 3. Roughness values of polished aggregates using different scales

As discussed, S_a scale is a surface roughness parameter which measures roughness of aggregate over area whereas R_a scale measures roughness of aggregate over line. Hence, higher value of S_a may be expected. However, no significant difference in the roughness measured using R_a-H, R_a-V and S_a was observed. But for a given scanning area, roughness measured over area seems to be more appropriate, hence, use of S_a parameter is recommended to indicate aggregate roughness. In addition, to substantiate the present study, further experimental study is required to collect roughness data of different aggregates with increased number of observations.

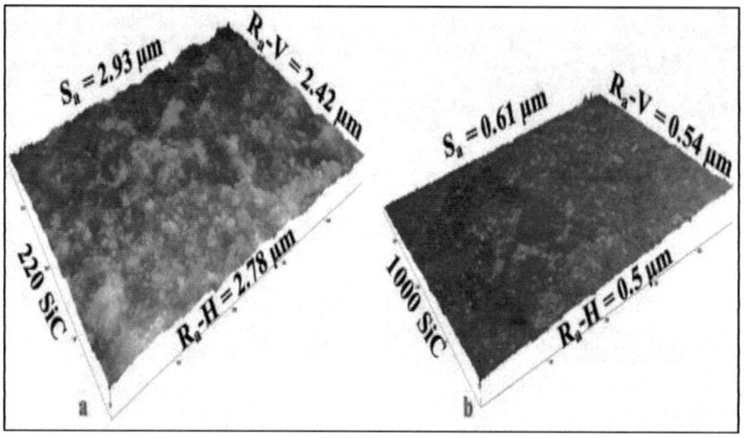

Figure 4. 3-D profile of polished aggregates (a) 220SiC (b) 1000SiC

4.2 Variation in contact angles due to change in aggregate roughness

Figure 5 shows the effect of texture on the contact angle of aggregate. The contact angle of aggregates was measured using water (W), formamide (F) and ethylene glycol (E). For each of the probe liquids, higher values of contact angle were observed for aggregate texture generated after polishing with 1000SiC grit powders when compared with texture generated after polishing with 220SiC grit powders. In addition, statistical analysis revealed that surface texture of aggregate has a significant effect (p-value<0.05) on contact angle of aggregate when measured using all the probe liquids (i.e. W, F, and E). Further, significant change in the contact angle may reciprocate in the form of change in SFE of aggregate. Therefore, it will be interesting to known, how significant is the change in the SFE of aggregates due to change in contact angles.

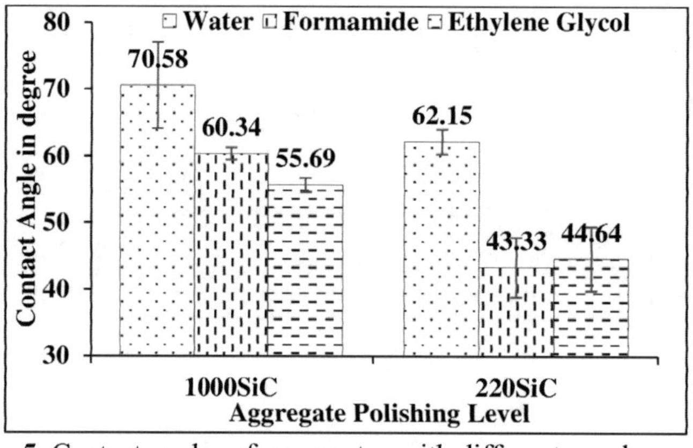

Figure 5. Contact angles of aggregates with different roughness levels

4.3 SFE of aggregate

The SFE components of basalt aggregate is presented in figure 6. It was observed that change in aggregate texture does not affect the acid and base SFE components of basalt aggregate. For example, the base SFE component of basalt aggregate was found to be similar i.e. 17.90 mJ/m^2 and 17.63 mJ/m^2 for 220SiC and 1000SiC polished surface, respectively (figure 6). However, significant increase in the dispersive components was observed with change in aggregate texture resulting from 220SiC polishing level to 1000SiC polishing level. Further, the total SFE of 220SiC polished aggregate (76.45 mJ/m^2) was found to be higher than SFE of 1000SiC polished aggregates (38.48 mJ/m^2). Statistical analysis using independent sample t-test showed significant difference (p-value=0.000) in the total SFE of

220SiC and 1000SiC polished aggregate due change in aggregate texture. Thus, it was observed that polishing of aggregates significantly reduces its SFE.

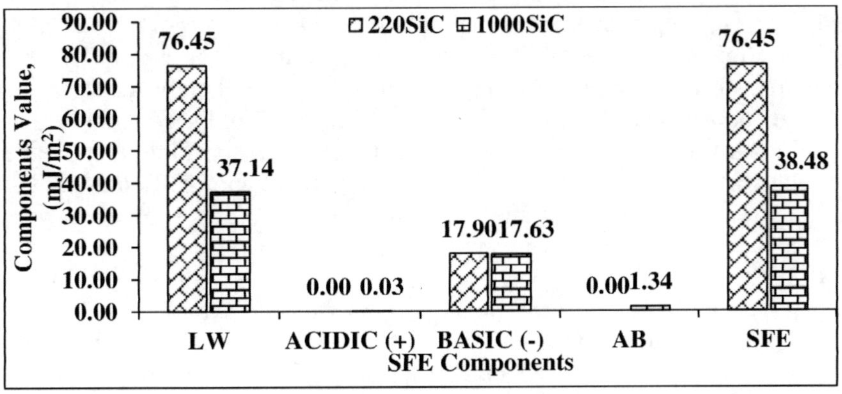

Figure 6. SFE components of aggregates

4.4 SFE of bitumen

The SFE components of bitumen was determined using sessile drop method with set of three different probe liquids namely, W, D and E. The γ^{LW}, γ^{+}, γ^{-} and, γ_b components of binder SFE was found to be 29.15 mJ/m^2, 0.00 mJ/m^2, 2.82 mJ/m^2, and 29.15 mJ/m^2, respectively. The determination of SFE components of bitumen is essential in determining energy parameters related to compatibility with aggregate.

4.5 Energy parameters and Energy Ratio

Using the SFE components of aggregate and bitumen, the energy parameters namely, W_{AB}, W_{ABW}, and wettability were evaluated (figure 7). Higher value of W_{AB} was observed for aggregates polished with 220SiC grade powder compared with aggregates polished with 1000SiC powder, which is favourable for better bond strength. In fact, dry adhesion in case of 220SiC polished surface was found to be 42% higher compared with 1000SiC polished surface. Subsequently, for a particular type of binder used in the present study, the wettability with variedly polished aggregates was evaluated. It was observed that wettability of rough textured aggregate (220SiC) is remarkably higher compared with smooth textured aggregate (1000SiC) (figure 7). Higher wettability indicates better bonding; thus rough textured aggregates were found to be favourable for better bonding compared with smooth textured aggregates.

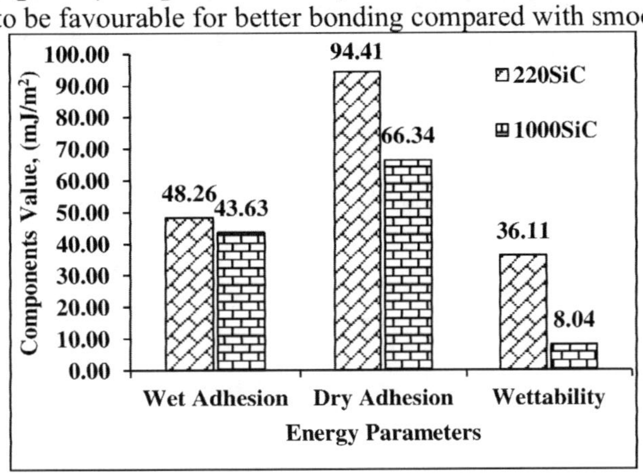

Figure 7. Energy parameters related to bond strength

Further, comparable values of WABW were observed for aggregates with polishing levels 220SiC and 1000SiC (figure 7). Additionally, to rank the combination of aggregate and bitumen system, ER parameter was evaluated. The ER of rough textured (220SiC) aggregate was found 0.75, while ER of smooth textured (1000SiC) aggregates was found to be 0.18, indicating higher bond strength of 220SiC polished aggregates. Higher ER value for 220SiC polished aggregates was expected due to better coating of binder on rough surface as indicated by higher value of wettability. In addition, ranking criteria characterizes 220SiC polished aggregate as "Fair", while 1000SiC polished aggregate as "Very poor" [6]. Further, similar level of aggregate polishing can be expected on bituminous pavements due to vehicular movement which is undesirable for aggregate-bitumen bonding. Hence, further study can be conceptualized to correlate the roughness and bond strength of laboratory polished aggregates with the aggregates polished due to vehicular movements.

4.6 Effect of aggregate roughness on POTS and Bond strength ratio (BSR)

Figure 8 shows the $POTS_{dry}$ and $POTS_{wet}$ of 220SiC and 1000SiC polished aggregates. The POTS of 220SiC polished aggregate was found to be higher than 1000SiC polished aggregate which was expected considering the higher S_a values of 220SiC polished aggregates (S_a=2.93) compared with 1000SiC polished aggregates (S_a=0.61). Increased roughness of aggregate promotes physical interlocking between aggregate surface and bitumen due to presence of summit and valley in aggregate surface (figure 4), resulting in increased bond strength. Further, statistical analysis revealed no significant decrease in the $POTS_{dry}$ (p-value=0.102) and $POTS_{wet}$ (p-value=0.237) values with increase in polishing from 220SiC to 1000SiC. However, it was noted that $POTS_{dry}$ reduces by 7.0% when the polishing level increase from 220SiC to 1000SiC. Similarly, $POTS_{wet}$ reduces by 9% when aggregate is polished with 1000SiC grade powder. Hence, wet conditioning regime is detrimental for bond strength of aggregate-bitumen system particularly for highly polished aggregates. Further, it is interesting that aggregate-bitumen combination with higher wettability has highest POTS values. Similar results, i.e. combinations with higher POTS corresponds to higher wettability was reported by Moya et al. [14].

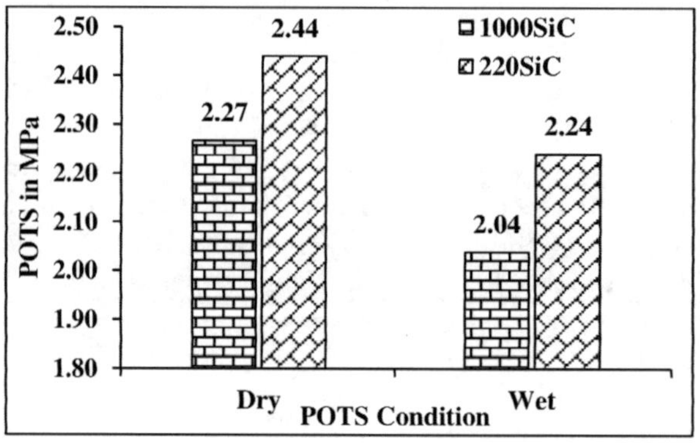

Figure 8. POTS of aggregate-bitumen system at varying levels of surface roughness

Further, the effect of polishing levels on bond strength was evaluated using BSR parameter (equation (11)). The BSR of 220SiC polished aggregate and 1000SiC polished aggregate was found to 0.92 and 0.90, indicating higher bonding potential of rough textured aggregates. As discussed, the increased bond strength of rough aggregates is due to several summit and valley in the rough textured aggregates generated after polishing with 220SiC grade abrasive powder compared with polishing after 1000SiC powder (figure 4).

5. Concluding Remarks

The present study was primarily focussed on measuring roughness of aggregate at different polishing levels (i.e. 220SiC and 1000SiC) and estimating its effects on bond strength with bitumen. The bond strength was evaluated using SFE and ABS test approaches. Based on the limited studies, following conclusions were drawn:

1. The polishing of aggregate significantly affects its surface roughness properties. However, no significant difference in roughness was observed while selecting different scales (i.e. R_a or S_a) for measurements.
2. Usually, R_a scale of measurement is suitable for 2D profilometer. However, with advance tools capable of measuring 3-dimensions at a point, measurement of roughness over area using S_a scale seems to be more logical.
3. Contact angle measurement was found to be significantly affected by change in aggregate roughness which indicates changes in the unbalanced forces on the surface of aggregate. The unbalanced forces govern the aggregate surface chemistry which was also found to vary significantly with change in aggregate texture.
4. No significant change in POTS values were observed with change in aggregate roughness. However, the POTS decreases with the increase in aggregate polishing level, which is undesirable for bond strength perspective. Hence, it is advisable to check for excessive polishing of aggregate surface in bituminous pavement.
5. Similarly, bond strength evaluated using ER and BSR parameters was found to higher for rough texture aggregates. This may be due to the presence of valley and peaks in rough textured aggregates as observed in 3D images which improved physical interlocking between aggregate and bitumen.

References

[1] Moraes R, Velasquez R, and Bahia H U 2011. Measuring the effect of moisture on asphalt aggregate bond with the bitumen bond strength test. Transportation Research Record, **2209(1)**, pp 70-81.

[2] Gubler R, Partl M N, Canestrari F, and Grilli A 2005. Influence of water and temperature on mechanical properties of selected asphalt pavements. *Materials and structures*, **38(5)**, pp 523 -532.

[3] Van Lent D Q, Molenaar A A A, and Van de Ven M F C 2009. Influence treatment in laboratory of stone surface on the surface roughness. *Journal of Testing and Evaluation*, **37(5)**, pp 417-423.

[4] Singh D, and Mishra V 2018. Different methods of selecting probe liquids to measure the surface free energy of asphalt. binders. *Construction and Building Materials*, **175**, pp 448 -457.

[5] Mishra V, and Singh D 2018. Effects of short term aging on aggregate bitumen compatibility. *Advances in Materials and. Pavement Performance Prediction (AM3P 2018), April 16- 18, 2018, Doha, Qatar* (p. 425). CRC Press.

[6] Little D N, Bhasin A, and Hefer A 2006. Using surface energy measurements to select materials for asphalt pavement. NCHRP project 9–37. *Transportation Research Board (TRB), Washington, DC, USA*.

[7] Kittu A T, Bulut R, and Puckette J 2014. Effects of surface roughness on contact angle measurements on a limestone aggregate. *Recent Developments in Evaluation of Pavements and Paving Materials*. pp 1-8.

[8] Koc M, and Bulut R 2014. Assessment of a sessile drop device and a new testing approach measuring contact angles on aggregates and asphalt binders. *Journal of materials in civil. engineering*, **26(3)**, pp 391-398.

[9] Habal A, and Singh D 2018. Influence of recycled asphalt pavement on interfacial energy and bond strength of asphalt binder for different types of aggregates. Transportation Research. Record, **2672(28)**, pp 154-166.
[10] Moraes R, Velasquez R, and Bahia H 2017. Using bond strength and surface energy to estimate moisture resistance of asphalt- aggregate systems. *Construction and Building. Materials*, **130**, pp 156-170.
[11] Gadelmawla E S, Koura M M, Maksoud T M A, Elewa I M, and Soliman H H 2002. Roughness parameters. *Journal of materials processing Technology*, **123(1)**, pp 133-145.
[12] Van Oss C J, Chaudhury M K, and Good R J 1988. Interfacial lifshitz- van der waals and polar interactions in macroscopic systems. *Chemical reviews*, **88(6)**, pp 927-941.
[13] TP 91 AASHTO 2011 Standard method of test for determining. asphalt binder bond strength by means of the binder bond strength (BBS) test.
[14] Aguiar-Moya J P, Salazar-Delgado J, García A, Baldi-Sevilla A, Bonilla-Mora V, and Loría-Salazar L G 2017. Effect of ageing on micromechanical properties of bitumen by means of atomic force microscopy. *Road Materials and Pavement Design*, **18(sup2)**, pp 203-215.

Acknowledgments
This research project was sponsored by Science and Engineering Research Board (SERB), Department of Science and Technology (DST), Government of India, under Project Reference No.: SERB No: SB/S3/CEE/013/2014. The ABS instrument used in the present research work was purchased through funding support from DST, India under project No. **DST/TSG/STS/2-15/76.** The authors would also like to acknowledge the Department of Mechanical Engineering, IIT Bombay for facilitating with Zeta Profilometer for aggregate roughness measurement.

Field Experiment for Accuracy Verification of the Devices Measuring International Roughness Index in TRUE Project: Five Years Report

K Tomiyama[1*], H Nakamura[2], H Mashito[3], M Jomoto[4] and S Kameyama[5]

[1]Division of Civil and Environmental Engineering, Kitami Institute of Technology, 165 Koen-cho, Kitami, 090-8507, Japan
[2]NIPPO Corporation, 6-70 Mitsuhashi, Nishi-ku, Saitama, 331-0052, Japan
[3]Toa Road Corporation, 315-126 Kaname, Tsukuba, 300-2622, Japan
[4]Taisei Rotec Corporation, 1456 Oaza-Kamiya, Konosu, 365-0027, Japan
[5]Hokkaido University of Science, 7-Jo 15-4-1 Maeda, Teine, Sapporo, 006-8585, Japan
[*] Corresponding author: tomiyama@mail.kitami-it.ac.jp

Abstract. A lot of devices measuring the International Roughness Index (IRI) has recently been developed and applied in response to the pavement inspection purposes. Consequently, the accuracy of measured IRIs obtained with different devices on the same road surface has become a practical concern for providing a consistent inspection result. Against this background, the specified non-profit organization Pavement Diagnosis Researchers Group (PDRG) in Japan has undertaken a project regarding the experiment to harmonize and compare Test methods for surface Roughness Under actual road Environment (TRUE project) since 2014. The TRUE project has so far conducted the roughness measuring experiments three times over five years at Hokkaido, Japan. This paper introduces the features of the TRUE project such as participated devices, test site selection, and data recording and reporting procedures. A benchmark testing for obtaining reference profiles is also provided to make the project reliable. In this paper, the results of experiments conducted in the last five years are statistically described as well. Finally, this paper introduces the potential of the TRUE project such as a new device grouping and the accuracy evaluation report for device certification for contributing to further improvement of a device participated in the project.

1. Introduction

Against the background that pavement infrastructures have aged, the Bureau of Public Road (BPR) of Ministry of Land, Infrastructure, Transport and Tourism (MLIT), Japan has issued a strategy for inspection of the whole road stock and has provided a tentative implementation guide for evaluating roughness by the International Roughness Index (IRI) since 2013. Then, BPR has officially published the guidelines for pavement inspection using IRI since 2016 [1]. According to the guidelines, road administrators are obliged to maintain their pavements effectively and efficiently with establishing a maintenance cycle consisted of inspection, diagnosis, treatment, and recording during the service period of a pavement. These strategies have accelerated the development of a large variety of roughness measuring devices corresponding to the needs for inspection purposes [2].

On this account, the reliability of the devices for obtaining the same result with different devices on the same profile has become an important matter to monitor the surface condition periodically in a life interval of pavements. Against this background, the specified non-profit organization Pavement Diagnosis Researchers Group (PDRG) in Japan has launched the experiment to harmonize and compare Test methods for surface Roughness Under actual road Environment (TRUE project) since 2014. A purpose of the project is to improve technologies of surface measurement under actual road environment by (a) supporting the experiment implementation, (b) analyzing the data obtained in experiments, and (c) reporting and publishing the outcomes of activities. PDRG has so far conducted roughness measuring experiments three times in the last five years as part of the TRUE project to verify the accuracy of devices measuring IRI. This paper provides the basic idea of the TRUE project and describe the results of three experiments. Not all of the devices used in Japan, but this study includes a number of them. The information presented in this paper is intended to provide a benchmark of the reliability for associating different devices.

2. Feature of TRUE project

2.1. Participated devices

One of the well-known experiments to harmonize roughness measurement devices was performed as part of the EVEN project by the Permanent International Association of Road Congresses - World Road Association (PIARC), which was conducted in the United States, Japan, and Europe [3-5]. Seven devices participated in the second experiment of the project conducted in Japan in 1998 [4]. After 16 years, in the TRUE project, thirty-four devices including high- and low- speed devices participated in the first experiment in 2014. Following the first experiment, twenty-eight devices took part in the second and third experiment held in 2016 and 2018, respectively. Table 1 summarizes the number of the participated devices. An example of the devices is shown in figure 1. This enthusiastic participation in every experiment infers a great concern and demand for the roughness inspection and its methods. In addition, after the profile measurement, the project investigated the structural properties of the pavements by a Falling Weight Deflectometer (FWD).

Table 1. Number of Participated Devices

Device Type \ Date	FY 2014 (Sep. 17-19, 2014)	FY 2016 (Sep. 14-16, 2016)	FY 2016 (Oct. 12-14, 2018)	Total
High-speed Devices	20	15	12	47
Low-speed Devices	14	13	16	43
Total	34	28	28	90

(a) (b) (c) (d)

Figure 1. An Example of Participated Devices: (a) Dipstick, (b) Inertial Profiler, (c) Mobile Mapping System (MMS), (d) Walking Profiler.

2.2. *Test Section*

The experiments of the TRUE project have been performed on prefectural roads with the cooperation of Hokkaido prefecture of Japan. Two test sites categorized into arterial (high volume) and residential (low volume) roads including two sections for each site have been established in consideration of the variety of roughness level. Each test section is 200 m long with 20 m and 5 m additional extents from the beginning and end points, respectively. Table 2 summarizes specifications of the test sites with the IRIs calculated from the reference profiles called the true profiles by the standard Golden-Car simulation. Only in the second experiment, one extra section was added in the Site 1 for further analyses, however, this paper does not deal with this section.

Table 2. Specifications of the Test Sites

Site		1		2	
Road Category		Arterial (High Volume) Road		Residential (Low Volume) Road	
Driving Speed (km/h)		40, 50, and 60 km/h		20, 30, and 40 km/h	
Length (m)		200		200	
Overview					
Section		No. 1	No. 2	No. 1	No. 2
IRI (mm/m)	FY 2014	2.6	1.8	6.3	4.5
	FY 2016	2.6	1.8	6.5	4.5
	FY 2018	2.8	1.9	6.7	4.7

2.3. *Data Recording and Reporting Procedures*

In the experiment, participated devices made three repeated runs for each test section. The high-speed devices were also required to obtain data at the following different operating speeds: 40 km/h, 50 km/h, and 60 km/h on Site 1 and 20 km/h, 30 km/h, and 40 km/h on Site 2. The participants were asked to report IRI and profile measurements with data format of ".csv" and ".xlsx", respectively, in accordance with specified formats of the project. The IRI data were reported with a fixed interval of 200 m and a minimum step available for each device. The TRUE project received profile and IRI measurements obtained with high-speed profilers (vans, trucks, cars, etc.) and hand-operated low-speed profilers, and IRI measurements obtained with response type systems after the experiment.

3. Reference "True Profile" Measurement

3.1. *Measurement procedure*

A reference profile called "true profile" for each test section was accurately measured by the same manner for every experiment. A true profile consists of the combination of measurements obtained with three different reference devices such as a Rod and Level, a static Dipstick, and a hand-operated profiler as shown in Figure 2 with basic three steps described in the following subsections.

3.1.1. Step 1. The static Dipstick measures the difference in height which is sensitive to the IRI calculation at an interval of 0.25 m.

3.1.2. Step 2. The Rod and Level measurements compensate longer wavelength ranges than the static Dipstick at an interval of 10 m.

3.1.3. Step 3. The hand-operated profiler interpolates shorter wavelength ranges than the static Dipstick at an interval of no longer than 0.05 m.

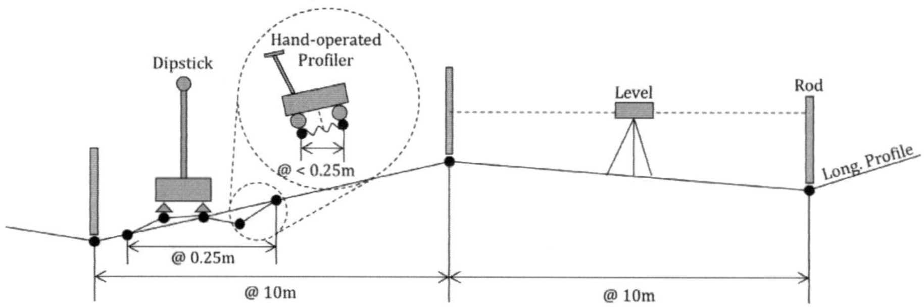

Figure 2. A Schematic Illustration of Measuring a True Profile

3.2. Benchmark testing

A benchmark device has been tested to validate the true profiles obtained with the above-mentioned procedure prior to the experiment. As a benchmark testing, we applied three levels of Rod and Level devices and total station (TS) as shown in table 3 to provide a ground truth. Figure 3 shows how do they work. In the testing, a longitudinal profile was measured for 105 m long at an interval of 0.05 m by using each device. Although the most reliable device in table 3 is Digital Level of level 1, it is inconvenient considering workability on residential roads. Thus, a benchmark device was selected from the remaining devices by assuming the Digital Level of level 1 as the reference.

Table 3. Candidate Devices for Benchmark Testing

Device	Level	Name/Resolution
Digital Level Bar-code Levelling Staff	1 1	DiNi 0.3 (Trimble)/ 0.01 mm
Digital Level Bar-code Levelling Staff	2 2	DL-502 (TOPCON)/ 0.1 mm
Auto Level Levelling Staff	3 -	AT-M3 (TOPCON)/ 1 mm
Total Station	1	TS15 1'' (Leica)/ 0.1 mm

Figure 3. Candidate Devices for Benchmark Testing: (a) Digital Level of Level 1 and Level 2 and (b) Auto Level of Level 3 and Total Station (TS) of Level 1.

Figure 4 indicates the result of benchmark testing compared in terms of (a) longitudinal profile, (b) amplitude gain, (c) profile agreement [6], and (d) IRI. As shown in figure 4 (a) for the profile comparison, Auto Level of level 3 had insufficient resolution for the elevation measurement as the profile shape was not smooth. This result negatively reflects to the amplitude gain, profile agreement for the roughness profile, and IRI as shown in figure 4 (b), (c), and (d), respectively. The testing result also indicates that TS is inaccurate on the profile agreement and IRI as shown in figure 4 (c) and (d), respectively. This negative result is due mainly to the heat haze on the ground. In contrast, Digital Level of level 2 realize well correlated result with Digital Level of level 1 as shown in figure 4 with less operation effort. Thus, in the TRUE project, we decided to use Digital Level of level 2 which has the resolution of 0.1 mm for elevation measurement as a benchmark device.

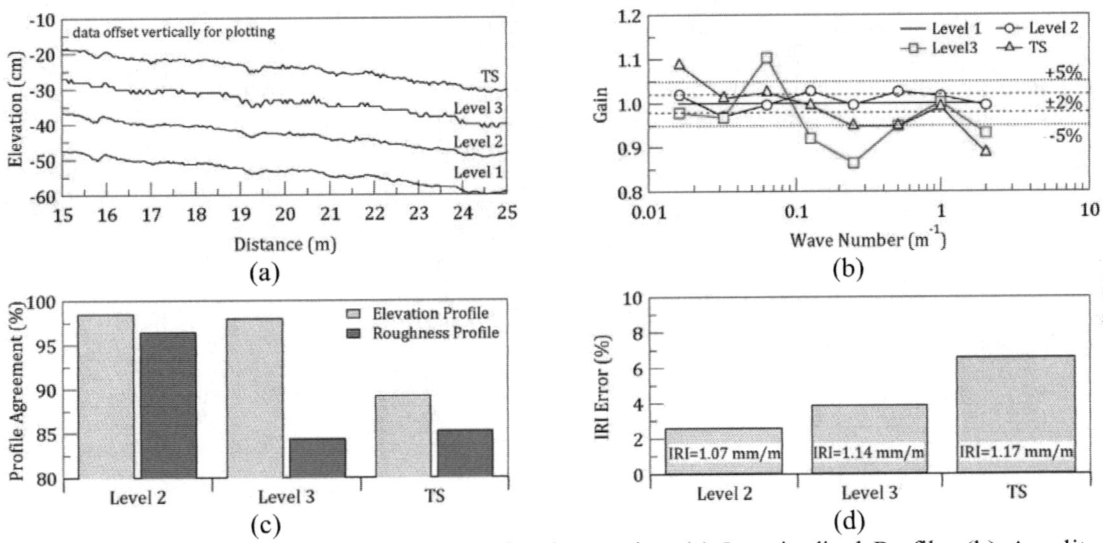

Figure 4. Result of the Benchmark Testing for Comparing (a) Longitudinal Profile, (b) Amplitude Gain, (c) Profile Agreement, and (d) IRI.

4. Statistical Description of Experiment Result

A standard practice for analyzing data in the TRUE project statistically describes the influence of operating speed, repeatability as precision, and reproducibility as well as portability of the measurements as bias in terms of the IRI. Table 4 summarizes the viewpoints of the analysis such as the type of error (random or systematic), comparison factor (within or between devices), comparison index (mean or standard deviation), and data averaging method.

4.1. Influence of Operating Speed

Since the influence of operating speed is assumed to be a systematic error within a device, it can be evaluated by the standard deviation of IRI data obtained by the different operating speeds. The high-speed devises made three repeated runs at the specific speeds on each site for assessing the influence of operating speed on the measurement result. The operating speeds were 40 km/h, 50 km/h, and 60 km/h on Site 1 and 20 km/h, 30 km/h, and 40 km/h on Site 2. Note that the operating speeds on Site 2 in 2018 were only 20 km/h and 30 km/h due to the environmental problem around the test site. Figure 5 shows the standard deviation of IRI obtained at different speeds on each site in each year. As shown in the figure, the mean of the standard deviation of IRI for the influence of operating speed increases with increasing IRI values, which is similar trend in 2014 and 2016. However, no relationship was observed in 2018 especially for Site 2. This result is caused by which a wide variety of devices participated in the experiment in 2018. The influence of operating speed was within 10% precision of the measured IRI values. For example, the influence tends to be less than 0.2 mm/m when the measured IRI is 2.0 mm/m.

Table 4. Viewpoints of Data Analysis

Term of Description	Type of Error	Comparison Factor	Comparison index
Influence of Speed (only for high-speed devices)	**Systematic** An ability to repeat the measures on different operation speeds	**Within** Deviation from the average obtained with repeated runs	**Standard Deviation**
Repeatability (Precision)	**Random** An ability to repeat the measures with a same profiler	**Within** Deviation from the average obtained with repeated runs	**Standard Deviation**
Reproducibility and Portability (Bias)	**Systematic** An ability to repeat the measures with a different profiler	**Between** Deviation from the average obtained with an expected value	**Mean Value**

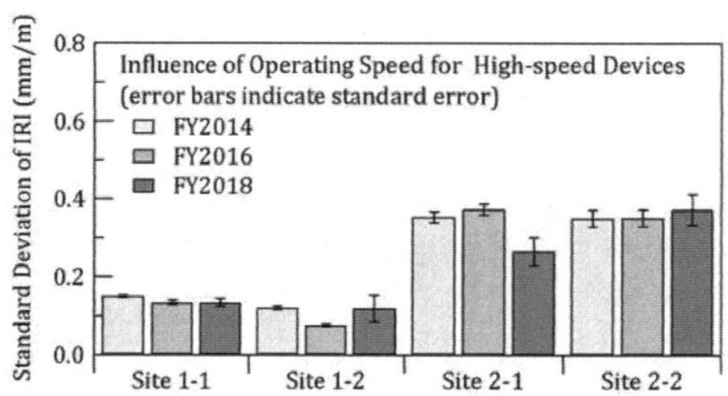

Figure. 5. Influence of Operating Speed for High-speed devices in FY 2014, FY 2016, and FY 2018.

4.2. Repeatability

Repeatability is the ability to obtain repeat measures with the same device at the same time [7], which is the random error corresponding to the precision within the device. Repeatability can be evaluated by the standard deviation of IRI measurements obtained by repeated runs. Figure 6 indicates the repeatability of IRI measurements obtained with the high- and low-speed devices. As shown in the figures, the standard deviation for repeatability increases with increasing IRI values, which is similar trend in 2014 and 2016. However, no relationship was observed in 2018 especially for Site 2 due to the participation of a wide variety of devices. For the high-speed devices shown in figure 6 (a), the repeatability was within 10% precision of the measured IRI values. For example, the the repeatability of high-speed devices tends to be less than 0.2 mm/m when the measured IRI is 2.0 mm/m. This trend is similar to the influence of operating speed. Thus, the repeatability corresponds to the influence of operating speed with respect to the within comparison factor.

As shown in figure 6 (b), the low-speed devices were more repeatable than the high-speed devices. This result can be easily anticipated because it is easier to trace the same line between repeated runs with low-speed devices than high-speed ones. The repeatability of low-speed devices except Site 1-2 in 2018 was within 1% precision of the measured IRI values. For example, the repeatability of low-speed devices tends to be less than 0.02 mm/m when the measured IRI is 2.0 mm/m.

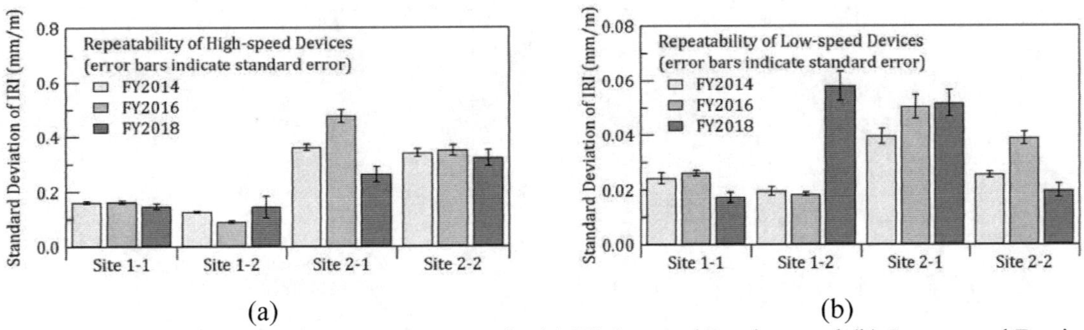

Figure 6. Repeatability of IRI Measurements in (a) High-speed Devices and (b) Low-speed Devices.

4.3. Reproducibility and Portability

Reproducibility is the ability to repeat the measures with a different device of the same basic design [7] which is the systematic error corresponding to the bias of measurements between devices. According to the literature [7], for profilers, reproducibility is not of as much interest as portability that is the ability to repeat the measures with completely different profiler design. The standard for portability is generally the true profile. On this account, this study evaluates the portability of IRI measurements as shown in figure 7. The figures indicate the percentile levels of the mean of measured IRIs for the repeated runs obtained with the participated devices corresponding to the reference IRIs obtained from the true profiles. The spread of a box-plot along horizontal axis denotes the reproducibility of measures. The degree of reproducibility increases with decreasing the spread. The deviation of a mean from a line of identity represents the portability of measures. The degree of portability increases with decreasing the deviation. This representation contributes to demonstrating the reliability of a candidate device in pavement inspection. For the high-speed devices, the 50th - 75th percentile devices were within the error of 10%, while some devices exceeded the error of 30%. On the other hand, the 25th – 50th percentiles of the low-speed devices were possible to achieve accurate measurement within the error of less than 10%.

5. Potential of the TRUE Project

5.1. Device Grouping

Roughness measurement devices have traditionally been categorized into the following four classifications on the basis of how directly their measures pertain to the IRI [8].

- (a) Class 4 – A roughness measure is not reproducible or stable with time, and can only be compared to IRI by subjective estimation.
- (b) Class 3 – A measure obtained from a response type system is calibrated to the IRI scale by correlation with reference measures from a Class 1 or 2 system.
- (c) Class 2 – A profile-based method is used that is reproducible and stable with time, and that is calibrated independently of other roughness measuring instruments.
- (d) Class 1 – A profile-based method similar to Class 2 is used. A profile-based measurement qualifies as a Class 1 measure if it is so accurate that further improvements in accuracy would not be apparent.

However, in recent years, a wide variety of roughness measuring devices have been developed and served for surface inspection activities. Thus, some newly developed devices are hard to be categorized into the traditional classifications. Against this background, we have proposed the new grouping criteria for roughness measurement devices by how the device measures IRI as shown in table 5 to promote the development of diverse devices.

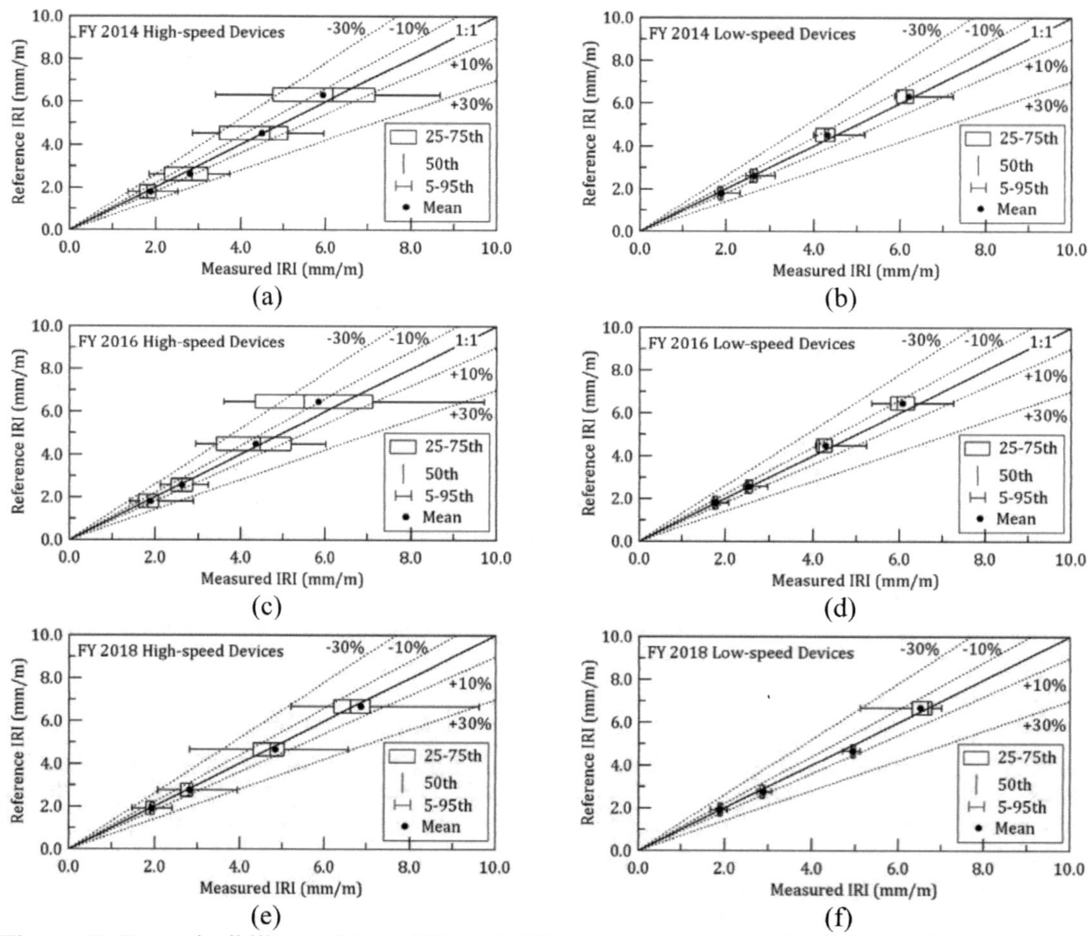

Figure 7. Reproducibility and Portability of IRI Measurements in (a) High-speed Devices of FY 2014, (b) Low-speed Devices of FY 2014, (c) High-speed Devices of FY 2016, (d) Low-speed Devices of FY 2016, (e) High-speed Devices of FY 2018, and (f) Low-speed Devices of FY 2018.

Table 5. New Grouping Criteria for Roughness Measurement Devices

<table>
<tr><th colspan="2">Group</th><th colspan="5">Requirements</th><th>Traditional Class</th></tr>
<tr><td rowspan="2">I</td><td>A</td><td rowspan="2">Subjective</td><td colspan="2">Visual Inspection/ Ride Experience</td><td colspan="2">On Vehicle</td><td rowspan="2">Class 4</td></tr>
<tr><td>B</td><td colspan="2">Visual Inspection</td><td colspan="2">By Walk</td></tr>
<tr><td rowspan="2">II</td><td>A</td><td rowspan="8">Profile-base Method</td><td colspan="2" rowspan="2">Static</td><td colspan="2">Measuring Elevation Directly*1</td><td rowspan="2">Class 1</td></tr>
<tr><td>B</td><td colspan="2">Measuring Elevation by Inclinometer*2</td></tr>
<tr><td rowspan="3">III</td><td>A</td><td rowspan="9">Dynamic</td><td rowspan="3">Low-speed</td><td colspan="2">Non-contact Sensor(s)*3</td><td rowspan="6">Class 2</td></tr>
<tr><td>B1</td><td rowspan="2">Contact Sensor(s)*4</td><td>Dedicated Device(s)</td></tr>
<tr><td>B2</td><td>Multi-purpose Device(s)*5</td></tr>
<tr><td rowspan="3">IV</td><td>A</td><td rowspan="6">High-speed</td><td colspan="2">Non-contact Sensor(s)*3</td></tr>
<tr><td>B1</td><td rowspan="2">Contact Sensor(s)*4</td><td>Dedicated Device(s)</td></tr>
<tr><td>B2</td><td>Multi-purpose Device(s)*5</td></tr>
<tr><td rowspan="3">V</td><td>A</td><td rowspan="3">Response Type</td><td colspan="2">Non-contact Sensor(s)*3</td><td rowspan="3">Class 3</td></tr>
<tr><td>B1</td><td rowspan="2">Contact Sensor(s)*4</td><td>Dedicated Device(s)</td></tr>
<tr><td>B2</td><td>Multi-purpose Device(s)*5</td></tr>
<tr><td>VI</td><td>-</td><td colspan="5">Otherwise</td><td>-</td></tr>
</table>

*1 e.g. Rod and Level, *2 e.g. Dipstick, *3 Lase, Ultrasonic, etc., *4 e.g. Wheel(s), *5 e.g. Smartphone Device(s)

IOP Conf. Series: Materials Science and Engineering 1075 (2021) 012012 doi:10.1088/1757-899X/1075/1/012012

5.2. Accuracy Evaluation Report for Device Certification

As early mentioned, one of the purposes of the TRUE project is to improve technologies of surface measurement devices. In order to achieve this purpose, the project certified the accuracy of the devices participated in the latest experiment conducted in 2018 by publishing the evaluation report shown in figure 8. The report includes the evaluation result of accuracy for a device in terms of the profile agreement and IRI with respect to the repeatability, portability, and influence of operating speed when the high-speed device. Detailed values of IRI are also reported as an appendix. This report contributes to further improvement of a device participated in the project by receiving the feedback of accuracy evaluation based on the well-established true profiles.

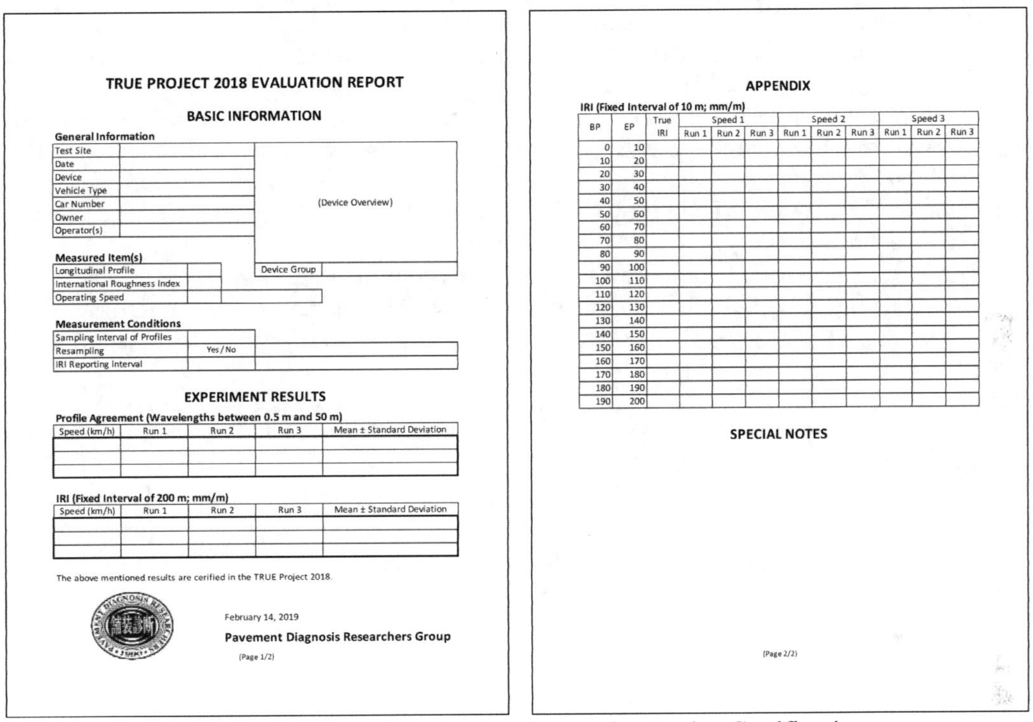

TRUE PROJECT 2018 EVALUATION REPORT

BASIC INFORMATION

General Information

Test Site		(Device Overview)
Date		
Device		
Vehicle Type		
Car Number		
Owner		
Operator(s)		

Measured Item(s)

Longitudinal Profile		Device Group	
International Roughness Index			
Operating Speed			

Measurement Conditions

Sampling Interval of Profiles		
Resampling	Yes / No	
IRI Reporting Interval		

EXPERIMENT RESULTS

Profile Agreement (Wavelengths between 0.5 m and 50 m)

Speed (km/h)	Run 1	Run 2	Run 3	Mean ± Standard Deviation

IRI (Fixed Interval of 200 m; mm/m)

Speed (km/h)	Run 1	Run 2	Run 3	Mean ± Standard Deviation

The above mentioned results are cerified in the TRUE Project 2018.

February 14, 2019

Pavement Diagnosis Researchers Group

(Page 1/2)

APPENDIX

IRI (Fixed Interval of 10 m; mm/m)

BP	EP	True IRI	Speed 1			Speed 2			Speed 3		
			Run 1	Run 2	Run 3	Run 1	Run 2	Run 3	Run 1	Run 2	Run 3
0	10										
10	20										
20	30										
30	40										
40	50										
50	60										
60	70										
70	80										
80	90										
90	100										
100	110										
110	120										
120	130										
130	140										
140	150										
150	160										
160	170										
170	180										
180	190										
190	200										

SPECIAL NOTES

(Page 2/2)

Figure. 8. Accuracy Evaluation Report for Device Certification.

6. Discussion and Conclusions

According to the pavement inspection guidelines using IRI published by the BPR in 2016 in Japan, a lot of devices measuring the IRI has recently been developed and applied in response to the inspection purposes. Consequently, the accuracy of measured IRIs obtained with different devices on the same road surface has become a practical concern for providing a consistent inspection result. In the light of this background, a project regarding the experiment to harmonize and compare Test methods for surface Roughness Under actual road Environment (TRUE project) has been launched since 2014.

The TRUE project has so far conducted the roughness measuring experiments three times at Hokkaido, Japan in September 2014, September 2016, and October 2018. This paper has provided the features of the TRUE project such as participated devices, test site selection, and data recording and reporting procedures. The features have been kept through the experiments conducted since the last five years. To make the project reliable, a benchmark device was tested to validate the procedure for obtaining true profiles prior to the experiment. As a result, in the TRUE project, Digital Level of level 2 which has the resolution of 0.1 mm for elevation measurement was selected as a benchmark device.

This paper has also statistically described the results of three experiments conducted in the last five years. The analysis of the data obtained in the experiments has focused on repeatability and reproducibility as well as portability of the IRI measurements obtained with the participated devices. The influence of operating speed on the IRI measurements has also been considered for the high-speed devices.

Finally, this paper has introduced the potential of the TRUE project that is a new device grouping and the accuracy evaluation report for device certification for contributing to further improvement of a device participated in the project.

References

[1] The Bureau of Public Road of Ministry of Land, Infrastructure, Transport and Tourism 2016 *Pavement Inspection Guidelines* (in Japanese)

[2] Tomiyama K, Nakamura H, Mashito H, Moriishi K, Jomoto M and Watanabe K 2018 Comparison of Different Roughness Measuring Devices in True Project *Proc. 8th Symposium on Pavement Surface Characteristics (SURF2018)*

[3] Schmidt B 2001 EVEN Project: Experiment to Compare and Harmonize Methods for Assessment of Longitudinal and Transverse Evenness of Pavement *Transp. Res. Rec.* No 1764 pp 221-231

[4] Kawamura A, Takahashi M and Inoue T 2001 Basic Analysis of Measurement Data from Japan in EVEN Project *Transp. Res. Rec.* No 1764 pp 232-242

[5] Descornet G, Berlemont B and Martin J M 2001 Study of Precision of Transverse Evenness Measurements in FILTER Experiment: Forum of European National Highway Research Laboratories Investigation on Longitudinal and Transverse Evenness of Roads Experiment *Transp. Res. Rec.*No 1764 pp 210-220

[6] Karamihas S M and Gillespie T D 2002 Development of Cross Correlation for Objective Comparison of Profiles, Publication *UMTRI-2002-36* The University of Michigan Transportation Research Institute

[7] Sayers M W and Karamihas S M 1998 *The Little Book of Profiling, - Basic Information about Measuring and Interpreting Road Profiles* The University of Michigan

[8] Sayers M W 1990 Profiles of Roughness *Transp. Res. Rec.* No 1260 pp 106-111

IOP Conf. Series: Materials Science and Engineering 1075 (2021) 012013 doi:10.1088/1757-899X/1075/1/012013

Dynamic Viscosity Standard of High Viscosity Modified Asphalt Considering Temperature and Load Coupling Effect

Jun Li[1], Yingtao Li[2], Mingliang Li[1], Haizhu Lu[2] and Dongwei Cao[3]

[1]Road Engineering Research Center, Research Institute of Highway Ministry of Transport, Xitucheng Road No.8, Beijing, 100088, P. R. China

[2]Jiangsu Expressway Maintenance Engineering Technology Co., Ltd., Tianyuan West Road No.168, Nanjing, 211106,P. R. China

[3]Zhonglugaoke(Beijing) Highway Technology Co., Ltd., Xitucheng Road No.8, Beijing, 100088, P. R. China

[*] Corresponding author: willow_lee@163.com

Abstract. The anti-ravelling and anti-rutting performances of porous asphalt pavement are significantly affected by dynamic viscosity of high viscosity modified asphalt. Climatic and traffic conditions vary greatly due to the vast territory of China, and dynamic viscosity standard has not been recommended up to now aiming at different temperature and load conditions. The influence of high viscosity additive dosage on key property indexes was firstly analysed for asphalt and asphalt mixture, and appropriate dosage was recommended. The dynamic stability changing law of porous asphalt mixture was revealed under different temperature, load and their coupling conditions. Multiple regression relation of them was established. Finally dynamic viscosity standard was proposed considering temperature/load coupling effect. The research results show that when high viscosity additive dosage exceeds 10% for 70# base asphalt, the increasing trend of dynamic viscosity becomes slowing down. Semi-logarithmic relations of dynamic stability with ravelling loss, Marshall stability were respectively found, and double-logarithmic relation of that with dynamic viscosity was found. The appropriate dosage of high viscosity additive is recommended as 12%. Based on multiple regression formula of dynamic stability with dynamic viscosity, temperature and load, the lower limit of 60℃ dynamic viscosity is recommended as 50000Pa•s for normal region, 100000Pa•s for heavy-load region, 200000Pa•s for high-temperature region, and 400000Pa•s for heavy-load & high-temperature region.

1. Introduction

The previous research shows that the asphalt binder with good performance can significantly enhance the performance of porous asphalt mixture in all aspects [1]. A large number of engineering practices also show that the performance of asphalt cement plays a decisive role in the durability of porous asphalt pavement. In the early stage, OGFC commonly used ordinary asphalt in the USA, but the road performances are poor. So some states once banned the use of such surfaces. The results of NCAT investigation in 1998 showed that OGFC using modified asphalt showed better road performance [2-4].

In Europe, the porous asphalt pavement is mostly made of modified asphalt. Fibre (mineral or lignin fibre) and anti-stripping agent (hydrated lime or amine) are added, and the road performance is generally good. Therefore, in recent years, the United States learned from Europe and developed a new generation of OGFC, promoting the use of modified asphalt and adding mineral fibre or lignin fibre to avoid asphalt flow and segregation of coarse and fine aggregates in the process of transportation and paving of the mixture.

In the research and application of porous asphalt pavement, Japan has fully learned from the experience of Europe and the United States. Modified asphalt was used at the beginning, but due to the inadaptability to the high temperature environment and traffic conditions in summer in Japan, the high temperature rutting problem still appeared in the early stage of porous asphalt pavement built with common modified asphalt. Later, modified asphalt with better adaptability and high viscosity was developed, which made the porous asphalt pavement widely used in Japan. Although the technical specifications of Japan stipulate that modified type I and modified type II bitumen can also be used, in fact, high viscosity bitumen is commonly used, which shows good application effect in use. The technical guidelines for porous asphalt pavement issued by Japan road association has put forward clear technical standards for high viscosity modified asphalt and its mixture [5].

Because of the different weather conditions, traffic conditions and test methods in foreign countries, China cannot directly use foreign technical standards. The climate conditions in Europe are relatively mild. The summer temperature in Japan is similar to that in southern China in terms of high temperature, but the traffic load standard in Japan is far lower than that in China. When constructing porous asphalt pavement in China, high performance asphalt should be selected according to the local climate and traffic conditions. By referring to the experience of foreign countries, the asphalt technical indexes and standards of economic and technical rationality are formulated to ensure the performance and durability of the asphalt pavement in heavy traffic and high temperature and rainy areas[6-14].

In 2004, Yulin expressway in Chongqing city constructed a 3 km test section of porous asphalt pavement. The performance comparison between SBS modified asphalt and high viscosity modified asphalt is specially arranged, in which the design void ratio of SBS modified porous asphalt pavement is 17% and that of high viscosity modified porous asphalt pavement is 20%. According to the follow-up observation in the later period, it was found that the porous asphalt pavement with SBS modified asphalt showed a certain degree of early damage after using less than 2 years, mainly manifested as pits and ruts, while the test section with high viscosity modified asphalt basically had no disease.

Figure 1. Disease of porous asphalt pavement with SBS modified asphalt(left: pits; right: ruts).

The engineering application of Yulin expressway fully proves the research results of laboratory tests. The performance of the asphalt mixture with common asphalt as binder is very poor, and the pavement is prone to serious early damage. When the void ratio of SBS modified asphalt mixture is large, there is the possibility of serious moisture damage. It is necessary to strictly carry out a comprehensive performance test or adopt technical measures such as adding hydrated lime. Modified asphalt with high viscosity can significantly improve the anti-moisture damage ability, anti-rutting ability and anti-ravelling damage of the porous asphalt mixture, and improve its durability [15]. It is a kind of asphalt binder with excellent performance and can be used for the porous asphalt pavement under high temperature and heavy traffic conditions.

2. Materials and test design

2.1. Materials design

Mitsubishi 70# A-grade matrix asphalt was adopted, and the test results of main technical indicators are shown in table 1. The test results of main technical indexes of high viscosity additives (HVA) are shown in table 2. The aggregate gradation composition of porous asphalt mixture is shown in table 3. The maximum nominal particle size of porous asphalt (PA) is 13.2mm, referred to as PA-13 for short, and its optimal oil stone ratio is 4.5%, and the designed voidage is 20%. Among the other constituent materials, coarse aggregate and fine aggregate are rolled from basalt, and the filler is limestone mineral powder, all of which meet the relevant provisions of JTG F40-2004 Technical specifications for construction of highway asphalt pavement.

Table 1. Main technical indexes of matrix asphalt.

Technical index	Unit	Test value	Technical requirement	Test method
Penetration degree	0.1mm	69	60~80	T 0604-2011
Softening point	℃	47	≥46	T 0606-2011
Ductility	cm	22	≥15	T 0605-2011
Solubility	%	99.6	99.5	T 0607-2011
Relative density	/	1.030	Actual measurement	T 0603-2011
	After TFOT test			T 0609-2011
Quality change	%	-0.02	±1.0	T 0609-2011
Penetration degree ratio	%	65.2	61	T 0604-2011
Ductility	cm	7	6	T 0605-2011

Table 2. Main technical indexes of high viscosity additive.

Technical index	Unit	Test value	Technical requirement	Test method
Shape	/	Granular, even, full	Granular, even, full	/
Single particle mass	g	0.015	≯0.03	/
Density	g/cm^3	0.983	0.90~1.00	GB/T 1033-2008
Melt index	g/10min	4.5	≮2.0	GB/T 3682-2000
Ash content	%	0.9	≯2	T 0614-2011

Table 3. Aggregate gradation composition of porous asphalt mixture.

Technical index	Quality passing rate through the following sieve pore(mm)									
	16	13.2	9.5	4.75	2.36	1.18	0.6	0.3	0.15	0.075
Aggregate gradation(%)	100	95.0	55.5	19.0	17.0	15.0	11.0	8.5	6.5	5.0

2.2. Test design

Four different HVA admixtures were selected, which were 0,5,10, 15 percentage respectively, and 0 was the control test. High viscosity modified asphalt and asphalt mixture with different HVA content were prepared and key performance indexes were tested. Asphalt test indexes including 60 ℃ dynamic viscosity and softening point. The test indexes of asphalt mixture include ravelling loss, Marshall stability and dynamic stability. By analysing the relationship between HVA content and each key performance index, the appropriate HVA content is recommended for the follow-up study.

The recommended HVA content was selected to form the PA-13 specimen, and the rut test was carried out under different test temperatures and loads. Test temperature was chosen as 55 ℃, 60 ℃, 65 ℃, 70 ℃ and 75 ℃. The loads were 0.5 MPa, 0.6 MPa, 0.7 MPa, 0.8 MPa and 0.9 MPa. By analysing the relationship between test temperature, load and dynamic stability, the functional formulas of them with dynamic stability are obtained by regression, respectively.

Using uniform design method, considering temperature, load, and dynamic viscosity, taking dynamic stability as the target variable, the influence rule of dynamic viscosity on dynamic stability under the coupling action of temperature and load was analysed. The functional formulae of temperature, load, dynamic viscosity with dynamic stability were obtained by further multiple regression.

Table 4. Factors and levels to be considered in uniform test.

Effect factors	Unit	Effect levels								
dynamic viscosity(Measured by HVA content)	%	0	2.5	5.0	7.5	10.0	12.5	15.0	17.5	20.0
temperature	℃	50	53	56	59	62	65	68	71	74
load	MPa	0.50	0.55	0.60	0.65	0.70	0.75	0.80	0.85	0.90

Table 5. Test design scheme.

Test number	Viscosity		Temperature(℃)	Load(MPa)
	HVA content (%)	Corresponding dynamic viscosity(Pa·s)		
1#	0	249.4	65	0.6
2#	2.5	376.3	53	0.85
3#	5.0	600.4	74	0.75
4#	7.5	1782.0	56	0.5
5#	10.0	21915.3	62	0.7
6#	12.5	63112.0	68*	0.9
7#	15.0	174978.7	50	0.65
8#	17.5	620190.2	71	0.55
9#	20.0	2146347.3*	59	0.8

3. Test results and discussion

3.1. Optimum dosage of HVA

3.1.1. Relationship between HVA dosage and asphalt properties
Dynamic viscosity and softening point of asphalt prepared with different HVA content were measured. The test results were shown in figure 2.

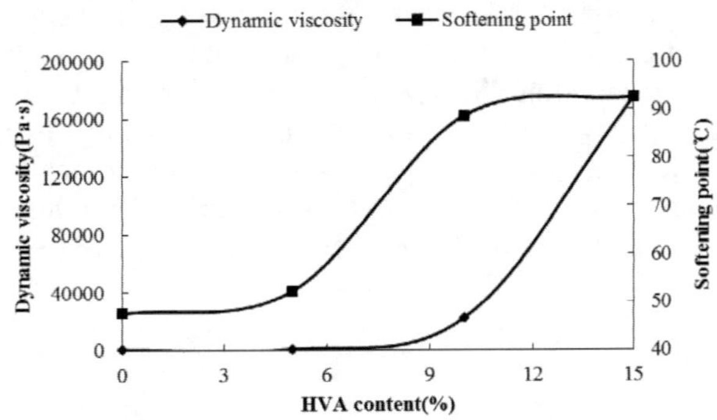

Figure 2. Relationship between HVA dosage and asphalt properties.

Figure 2 shows that the dynamic viscosity has a significant nonlinear relationship with the content of modifier. When the modifier content exceeds 10%, the dynamic viscosity increases rapidly. Also, The relation between softening point and modifier content is nonlinear. When the content of modifier is low, the increase is not so much.It increases sharply between 5% and 10%, and then tends to increase slowly after 10% dosage.

3.1.2. Relationship between HVA dosage and asphalt mixture performances
Ravelling loss, Marshall stability and dynamic stability of asphalt mixture prepared with different HVA content were measured. The test results were shown in figure 3.

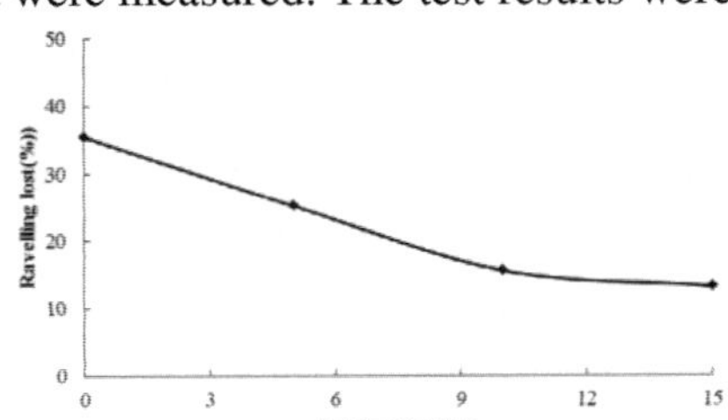

(a)Ravelling

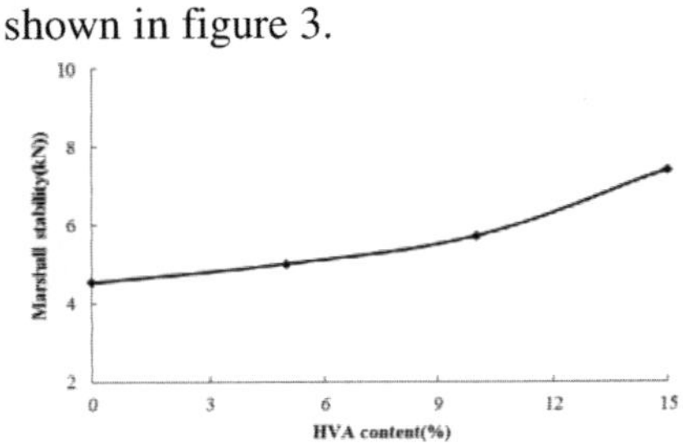

(b)Marshall stability

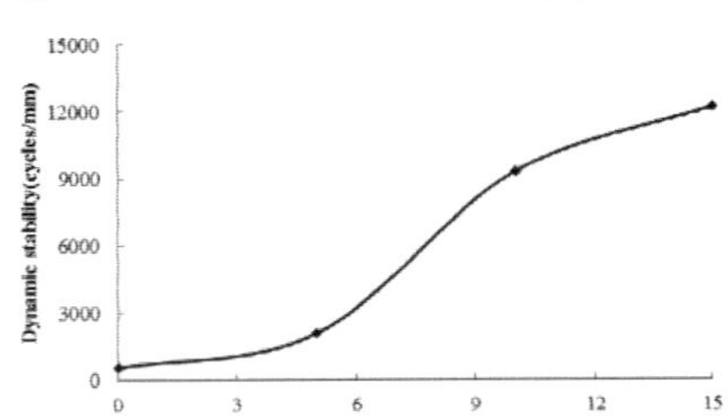

(c)Dynamic stability

Figure 3. Relationship between HVA dosage and asphalt mixture performances.

Figure 3(a) shows that the ravelling loss decreases with the increase of dynamic viscosity. When the modifier content increases from 0 to 15%, the ravelling loss decreases from 35.5% to 13.1%, indicating that the high viscosity asphalt has a significant contribution to the anti-ravelling performance of the porous asphalt mixture. One of the most common problems of porous asphalt pavement is the problem of particle ravelling, and improving the viscosity of asphalt can effectively solve this problem.

Figure 3(b) shows that the relationship between Marshall stability and modifier content is similar to that of dynamic viscosity, while figure 3(c) shows that the relationship between dynamic stability and modifier content is similar to that of softening point.

Considering that the content of modifier directly corresponds to dynamic viscosity, rutting is the main disease of asphalt pavement in high temperature and heavy load area. It is necessary to establish the relationship between dynamic viscosity and dynamic stability. But the nonlinear relationship between dynamic stability and dynamic viscosity is obvious. By taking logarithm of dynamic stability and dynamic viscosity, it is found that they have linear relationship. The regression diagram is shown in figure 4.

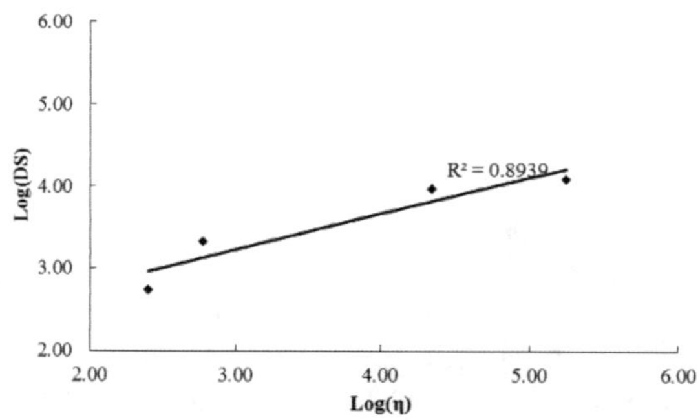

Figure 4. Relationship between dynamic viscosity and dynamic stability. DS means dynamic stability, η meas dynamic viscosity.

Figure 4 shows that after taking logarithms of dynamic stability and dynamic viscosity, the linear relationship between them is good, and the correlation coefficient R^2 reaches 0.8939. As an important evaluation index of asphalt pavement high temperature performance, the content of modifier is finally recommended to be 12% as the optimal content of 70# matrix asphalt based on its relationship with dynamic stability.

3.2. Independent effect of load and temperature on DS

3.2.1. Effect of load on DS

Set the temperature of rutting test for 60 ℃, and the loads respectively are 0.5 MPa, 0.6 MPa, 0.7 MPa, 0.8 MPa and 0.9 MPa. The dynamic stability index of porous asphalt mixture was tested to analyze the rutting resistance under different loads. The dynamic stability test results under different load conditions are shown in figure 5.

The pressure of the test wheel is controlled by the calibration pressure and the grounding area of the test wheel. For the load of 0.9 MPa, the test wheels with higher hardness rubber are required. For the load of 0.5 MPa, rubber test wheels with low hardness are required.

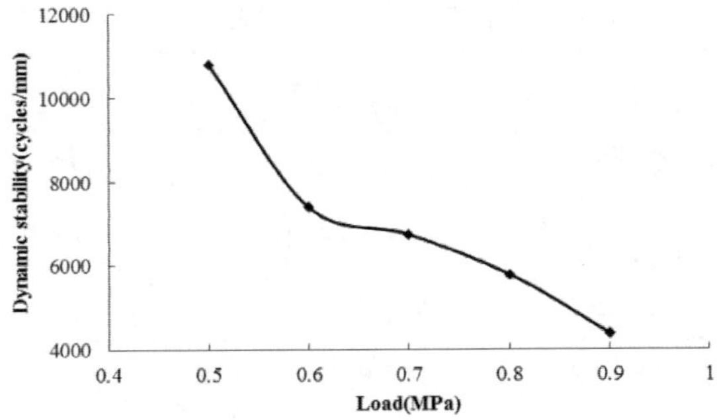

Figure 5. Relationship between dynamic stability and load

Figure 5 shows that with the increase of load, the dynamic stability index of rut test gradually decreases, but there is a mutation around 0.6 MPa. This shows that heavy traffic will make the high temperature stability of porous asphalt mixture worse, but high viscosity asphalt will weaken this effect, so that the dynamic stability of asphalt mixture can still reach 4,371 cycles /mm under the load of 0.9MPa.

3.2.2. Effect of temperature on DS

Set the load of rutting test for 0.7MPa, and the temperature respectively are 55 ℃, 60 ℃, 65 ℃, 70 ℃ and 75 ℃. The dynamic stability index of porous asphalt mixture was tested to analyze the rutting resistance under different temperature. The dynamic stability test results under different temperature conditions are shown in figure 6.

Although the pressure is set as constant in the test, the temperature will affect the hardness of the rubber test wheel, thus affecting the grounding area and pressure of the test wheel. Therefore, it is necessary to first keep the temperature constant, and then achieve the predetermined load by repeatedly adjusting the pressure and measuring the grounding area of the test wheel.

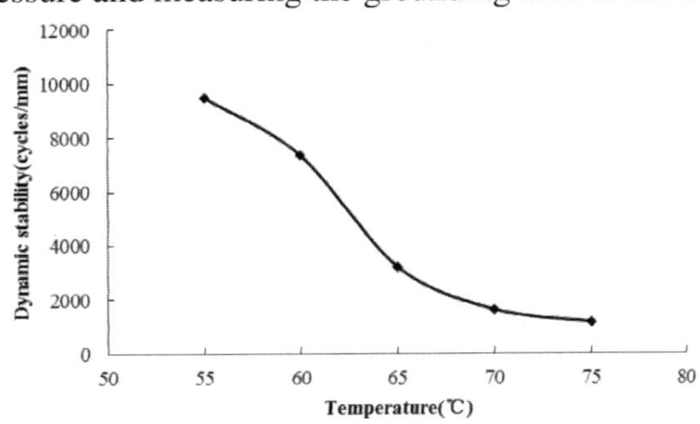

Figure 6. Relationship between dynamic stability and temperature.

Figure 6 shows that with the increase of temperature, the dynamic stability of rut test decreases. Compared with the rut test results under different load conditions, the dynamic stability of porous asphalt mixture is more sensitive to the change of temperature. The influence of temperature on the stability of high temperature is obviously more than that of load, so the damage of rutting caused by high temperature should be paid more attention.

In addition, after taking logarithms of the dynamic stability under different loads and temperatures, linear regression was performed respectively. Together with linear regression of dynamic viscosity, the summary is shown in table 6. This regression relationship lays a foundation for the subsequent analysis of the coupling effect of temperature and load on dynamic stability.

Table 6. Linear regression equation of dynamic viscosity, load, temperature.

Serial number	Effect factors	Equation	R^2
1	dynamic viscosity	$Log(DS) = 0.4419 Log\eta + 1.8992$	0.8939
2	Load	$Log(DS) = -0.8926 p + 4.4514$	0.9567
3	temperature	$Log(DS) = -0.0498 T + 6.7592$	0.9741

p indicates load and T indicates temperature.

3.3. Coupling effect of load and temperature on DS

The effects of single factors such as asphalt viscosity, load and temperature on the rutting resistance of porous asphalt mixture were analyzed, but a combination of high temperature and heavy load may be formed in engineering practice. On the other hand, engineers are often concerned about how to determine the dynamic viscosity requirements of asphalt under a specific climate and traffic load conditions, so that PA can meet the requirements of high temperature use, but not too wasteful and save costs. It is necessary to analyze the influence of temperature, load and dynamic viscosity of asphalt on the rutting resistance of pavement when the three factors change at the same time. Temperature and load are the external factors, asphalt dynamic viscosity is the internal factors.

Carry out this part of test according to the test scheme in table 5. The test results are shown in table 7. The equation 1 can be obtained by regression analysis of the test results in table 7. The correlation coefficient of this equation is 0.892.

Table 7. Rutting test results of uniform design scheme.

Test number	Dynamic viscosity (Pa·s)	Temperature (℃)	Load (MPa)	DS (cycle /mm)
1#	249.4	65	0.6	22.3
2#	376.3	53	0.85	2154.2
3#	600.4	74	0.75	33.2
4#	1782.0	56	0.5	7717.5
5#	21915.3	62	0.7	3244.8
6#	63112.0	68*	0.9	4470.1
7#	174978.7	50	0.65	12308.7
8#	620190.2	71	0.55	3881.7
9#	2146347.3*	59	0.8	10248.9

$$Log(DS) = 0.535 Log\eta - 0.63 p - 0.0653 T + 5.592 \quad (1)$$

According to the technical guide for drainage pavement (case) of Japan and relevant literature, the technical requirement for dynamic stability of porous asphalt mixture in rut test should be no less than 3000 cycles /mm. However, in practical engineering, rutting resistance often leaves a certain safety reserve, and its dynamic stability is more than 4000 ~ 6000 cycles /mm. The load of rut test in China is consistent with the standard wheel load of pavement design in China, which is 0.7MPa.

Based on the overload investigation results of expressway, the tire grounding pressure can reach 0.83 MPa when overload is 50%, 0.91 MPa when overload is 100%, and 0.99 MPa when overload is

200%. Rutting test temperature may be different in different countries, such as the UK using 45 ℃, in many parts of the Japanese using 60 ℃ except Hokkaido area using 45 ℃, but the principle is that it can reflect the region surface temperature in the summer.

Using equation 1, the dynamic viscosity which meets the design requirements of different dynamic stability under different load and temperature conditions can be calculated. See table 8 for details. Further, the limit dynamic viscosity of different typical regions are as shown in table 9.

Table 8. Dynamic viscosity calculated value meeting different design requirements.

Design load(MPa)	Design DS (cycles /mm)	Dynamic viscosity (Pa·s) under the following design temperature				
		55℃	60℃	62.5℃	65℃	67.5℃
0.7	4000	6575	26801	54112	109252	220582
	5000	9977	40671	82116	165794	334742
	6000	14029	57186	115460	233116	470666
0.8	4000	8622	35148	70965	143280	289286
	5000	13085	53339	107693	217434	439003
	6000	18398	74998	151422	305724	617262
0.9	4000	11308	46096	93069	187907	379388
	5000	17160	69953	141236	285157	575737
	6000	24128	98357	198585	400947	809518
1.0	4000	14830	60453	122056	246434	497555
	5000	22505	91740	185226	373974	755060
	6000	31643	128992	260437	525828	1061656

Table 9. The limit dynamic viscosity for typical regions.

Region type	load(MPa)	temperature(℃)	DS (cycles /mm)	Dynamic viscosity(Pa·s)
Normal region	0.7	60	5000	50000
Heavy-load region	0.9	60	5000	100000
High-temperature region	0.7	65	5000	200000
High-temperature & heavy-load region	0.9	65	6000	400000

4. References

[1] Huber G 2000 *Performance survey on open-graded friction course mixes* (Washington: National Academy Press)
[2] Qureshi N, Khurshid M and Watson D 2015 *Can. J. Civ. Eng.* **42** 1104-13.
[3] Chen X, Zhu H, Dong Q and Huang B 2016 *Int. J. Pavement Eng.* **29** 1-14
[4] Kline L 2010 *Comparison of open graded friction course mix design methods currently used in the united states*(South Carolina: Clemson University)
[5] Hiromistsu N 2004 *History of Porous Asphalt Pavement in Japan* (Japan: Taiyu Kensetsu Co., Ltd.).
[6] Li L, Geng H and Sun Y 2010 *J. Build. Mater.* **23** 12-7
[7] Chen Y, Tan Y and Chen K 2012 *J. Harbin Inst. Technol.* **44** 81-5
[8] Li L, Geng H and Sun Y 2015 *Mater. Struct.* **48** 2147-56
[9] Chen S, Zhu S, Qin X and Li Z 2013 *Adv. Mater. Res.* **753-755** 495-9
[10] Huang Z, Hu X, Wang J and Zhang J 2015 *J. Zhejiang Univ., Eng. Sci.* **49** 1448-54+86
[11] Zhang F and Hu C 2015 *Polym. Compos.* **38** 936-46
[12] Zhu S, Chen S, Qin X and Li Z 2012 *J. Wuhan Univ. Technol.* **34** 52-6
[13] Chen S, Dou H, Xing M and He R 2014 *J. Xi'an Univ. Archit. Technol. Nat. Sci.* **46** 380-4
[14] Zhang M and Hao P 2011 *IEEE 2011* 5357-60
[15] Jeong K, Hwang S, Lee S and Kim K 2011 *Ksce J. Civ. Eng.* **15** 1259-62

IOP Conf. Series: Materials Science and Engineering 1075 (2021) 012014 doi:10.1088/1757-899X/1075/1/012014

Performance study and application of porous ultra-thin wearing course for asphalt pavement maintenance

Mingliang Li[1], Haizhu Lu[2], Jun Li[1], Quan Mao[2] and Long Chen[3]

[1]Road Engineering Research Center, Research Institute of Highway Ministry of Transport, Xitucheng Road No.8, Beijing, 100088, P. R. China

[2]Jiangsu Expressway Maintenance Engineering Technology Co., Ltd., Tianyuan West Road No.168, Nanjing, 211106,P. R. China

[3]Faculty of Infrastructure Engineering, Dalian University of Technology, Linggong Road No.2, Dalian, 116024, P. R. China

Corresponding author's e-mail: ml.li@rioh.cn

Abstract. An ultra-thin layer wearing course with thickness between 15mm and 30mm are developed and applied for maintenance of asphalt pavement. The maximum size of the coarse aggregate is 10 mm. In this study, the high viscosity modified bitumen is used as binder for improving the anti-rutting and raveling performance of a porous mixture. Mixture designed study is carried out by combing the functional and structural performance measurements. The target air voids contents for those ultra-thin surfaces are between 15% and 20%. Important factors presenting the high temperature property, moisture feature, low temperature feature and anti-raveling performance are investigated in lab. Meanwhile, surface characteristic which reflect the sound absorption is also observed. A feasibility of using the ultra-thin wearing course mixture in China is proved. Trial sections are constructed in southeast part of China for studying the performance of ultra-thin waring course in maintenance engineering. The achievements from the study will provide technical support for the maintenance of asphalt pavement with moderate surface damage. It will also improve the skid resistance and the noise reducing function of the old pavement, which presents superior social and environmental benefits.

1. Introduction

A thin layer wearing course is commonly used as noise reducing pavement in European countries such as Germany and the Netherlands, and is used as open graded friction course (OGFC) in the United States. Due to definitions in different countries, it generally refers to the asphalt pavement with small size aggregate (the maximum particle size is less than 10 mm), thickness of 15-30 mm and designed air voids content greater than 14%. It can improve the safety and service life of road structure. In China, this surface type is called Ultra-Thin Wearing Course. A porous ultra-thin wearing course (PUC) has the air voids contents no less than 18%.

The most typical structural disease form of a porous asphalt pavement is the raveling caused by the strip of coarse aggregate. This disease is also particularly prominent for ultra-thin wearing courses. Because of the small nominal maximum particle size of aggregate and the low embedding force of coarse aggregate in PUC, it is easier to destabilize and disperse under the action of external loads and environmental factors. Therefore, asphalt binder is required to have strong encapsulation ability and high adhesion to coarse aggregate. For the binder of porous asphalt pavement and ultra-thin abrasive layer, there are many studies on rubber asphalt (AR), SBS, PG (performance-grade), matrix asphalt and epoxy asphalt at home and abroad. Woldekidan [1] analyzed the performance of asphalt mortar and mortar of porous asphalt pavement by dynamic rheological test, and simulated the fracture performance of asphalt mortar in porous asphalt pavement material system by means of mesoscale mechanics. Hagos [2] studied the aging mechanism and influencing factors of asphalt binder in the course of using porous asphalt pavement.

In the development of porous asphalt in China, in order to adapt to high temperature and heavy traffic conditions in summer, high-viscosity modified asphalt has been gradually used. The dynamic viscosity of this type of asphalt is greater than 20000 Pa·s at 60℃, which helps to enhance the bonding strength, improve the overall performance of pavement structure. Xu et al. [3] observed the high temperature performance of aged high viscosity modified asphalt and its adhesion to coarse aggregate by means of multi-stress creep recovery and ultraviolet spectroscopy. Qin et.al. investigated the influence of temperature and frequency on anti-rutting performance of high-viscosity modified asphalt [4], and Li et.al. analyzed the rheological properties of high viscosity modified asphalt made from low-grade asphalt [5].

For the performance of thin layer asphalt mixtures, Ahmed et al. [6] predicted and analyzed the temperature and reflection crack development of ultra-thin wear layer mixture by fracture energy test and finite element numerical simulation. Chen et al. [7] observed and evaluated the long-term performance of the ultra-thin wearing course.

The acoustic performance of asphalt pavement is mainly affected by the geometric structure of the pavement surface and the internal void condition of the surface layer. The acoustic properties of porous asphalt mixtures are generally analyzed from surface texture and noise absorption properties. Sanberg [8] studied the effects of noise texture and texture depth on porous asphalt pavement. Berengier et al. [9], Attenborough et al. [10] and other scholars have established phenomenological and microstructural models to simulate the acoustic impedance and sound absorption properties of porous asphalt mixture media. For the acoustic performance of thin layer wearing course, the main research is mainly based on experimental research. Li et al. [11] used laser scanning to analyze the apparent depth of the thin-layer overcoat structure and the sound pressure (P)-particle velocity (U).

Since the durability is lowered with the increasing air voids contents, the thin-layer course currently used has a voids content which is generally not high, and is mostly about 15% or less. In this study, the high-viscosity modified asphalt commonly used in China's porous asphalt was applied as the main binding material, and performances of PUC mixture with a maximum aggregate size of 10 mm was studied. The mixture and sound absorption performance were tested and investigated. With the analysis, a suitable PUC mixture that can be used in highway maintenance engineering is obtained. Application of this type of material was carried out in a test section in actual engineering. The research results provide a road surface structure with functional and structural advantages for highway maintenance.

2. Materials and Gradation Design

2.1. High viscosity modified asphalt

The binder used in the study is SBS modified asphalt, and the high and low temperature performances of the binder and mixture are improved by adding high viscosity modification additives (HVA). The test results of various performances of high viscosity modified asphalt with the ratio of SBS: HVA=92:8 are shown in Table 1. The test method is based on the Chinese standard “Standard Test Methods of Bitumen and Bituminous Mixtures for Highway Engineering” (JTG E20-2011).

Table 1. Features of high viscosity modified asphalt

Asphalt performance indicator	Unit	Test value	Standard value
Needle Penetration, 25°C，100g，5s	0.1mm	47.6	≮40
Softening point（$T_{R\&B}$）	°C	91.5	≮90
Ductility （5°C，5cm/min）	cm	32	≮30
Dynamic viscosity（60°C）	Pa·s	777548	≮400000
Brookfield viscosity（170°C）	Pa·s	1.069	≯3.0

2.2. Gradation design

The overall design principle of asphalt mixture grading is adjusted to meet the target air voids content by controlling the passing percentage of key mesh sizes (4.75mm, 2.36mm) and the sieve allowance below 0.075mm.

The size of fine aggregates of PUC-10 is from 0 to 3 mm and the coarse aggregate size is from 4.75 to 9.5 mm. Considering that the range of 4.75～9.5mm gear aggregate is too large, is was decided to increase a 7.5mm mesh for the mixture design. Among them, 4.75～7.5mm and 7.5～9.5mm are mainly used as the coarse skeleton structure to ensure the skeleton embedding, while the 0～3mm fine aggregate mainly plays the role of filling the gaps.

The test obtained the voids of coarse aggregate (VCA)of the coarse aggregate skeleton structure by comparing the different blending ratios of the 4.75-7.5 mm and 7.5-9.5 mm aggregates, as shown in Figure 1. The existing research shows that the smaller the skeleton gap ratio of the aggregate, the denser the aggregate and the stronger the intercalation between the aggregates. It can be seen from the figure that the VCA minimum is about 1:2, so the PUC-10 mixture skeleton structure selection (4.75-7.5mm): (7.5-9.5mm) is 1:2 ratio, and the mixture ratio is designed to The target void ratios of 20%, 18%, and 15% were met, and the three void ratios were determined as shown in the table.

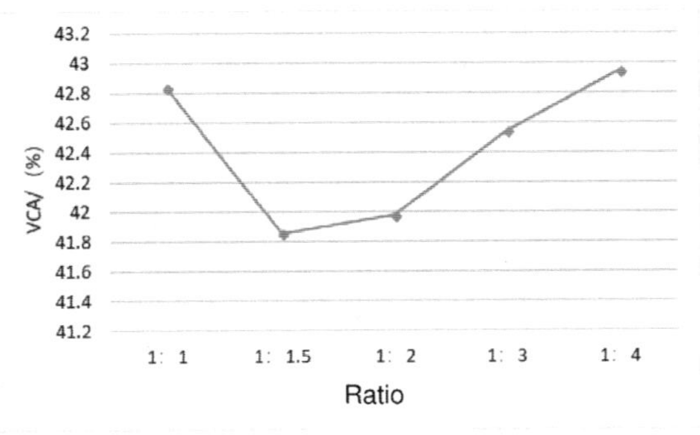

Figure 1. VCA of PUC-10 for different blending ratios of the 4.75-7.5 mm and 7.5-9.5 mm aggregates

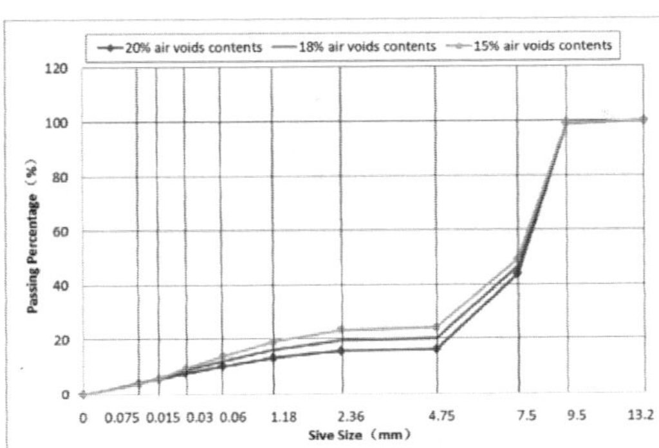

Figure 2. The gradation curve for PUC-10 mixture with different air voids content

2.3. Asphalt content

Based on the porous characteristics of the ultra-thin wearing course of large voids, the combination of the drainage and the raveling test is used to determine the optimum asphalt content according to the "inflection point" of the curve. In this study, for Gradation A, B and C, the optimum asphalt content are 4.7%, 5.0% and 5.1% respectively.

3. Sound absorption performance analysis

Aiming at the main influencing factors of road surface noise reduction, the sound absorption performance of porous thin layer pavement is analyzed. The influence of the air voids content changes on the sound absorption coefficient are to be compared with various thicknesses. Considering the thickness of ultra-thin wearing course, the common pavement surface layer and the sample thickness, three specimens with thicknesses of 25mm, 40mm and 63.5mm were respectively formed. The air voids content was designed to be 20%, 18%, 15%. The sound absorption coefficient test by impedance tube method on the PUC-10 mixture is shown in Figure 3.

As can be seen from the figure:

(1) The sound absorption coefficient-frequency curves of different air voids contents all showed similar trends. As the frequency increases, the sound absorption coefficient of the mixture gradually increases, and gradually decreases after reaching the peak value.

(2) Within a certain thickness range, the peak frequency of the same thickness mixture is in the range of 500~1400 Hz, and the peak of the sound absorption coefficient becomes "high and wide" with the increase of the air voids content, which is very obvious for the improvement of the sound absorption effect of the asphalt mixture. The peak value of the curve increases as the air voids content increases, and the corresponding frequency of the peak also gradually shifts toward the high frequency direction.

(3) In particular, when the thickness is increased to a certain range, even if the thickness of the mixture is kept constant, the peaks of the sound absorption coefficient of the mixture of different void ratios tend to gradually close together, so that the peak of the sound absorption coefficient corresponds to the tendency of the frequency to shift toward the high frequency.

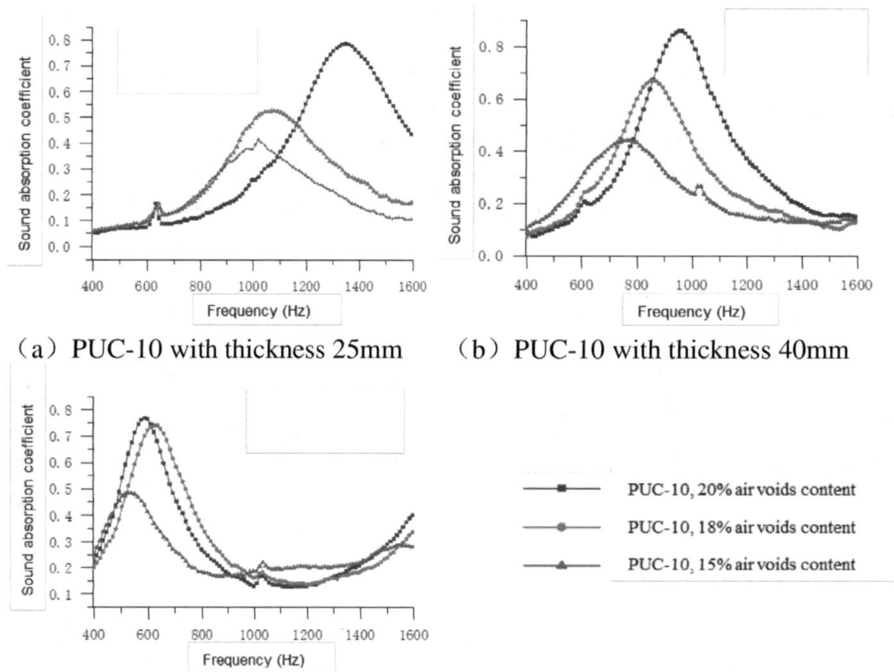

(a) PUC-10 with thickness 25mm (b) PUC-10 with thickness 40mm

(c) PUC-10 with thickness 63.5 mm

Figure 3. Sound absorption curves for PUC-10 mixtures

4. Mixture performance analysis

4.1. Freeze-thaw split tensile strength

The freeze-thaw split test is to determine the strength ratio of the asphalt mixture test sample before and after the water damage, so as to evaluate the water stability of the asphalt mixture. The test was carried out for two groups of specimens. The first group was stored at room temperature. The second group was vacuumed for 15 minutes and then immersed in the water for 0.5 h. The bag was placed in an incubator at -18 ° C for 16 h, and then placed in a constant temperature water bath at 60 ° C for

24h. The test results are shown in Figure 4.

It can be seen from Figure 4 that the 20% and 18% void ratio PUC-10 asphalt mixture freeze-thaw splitting tensile strength ratio is slightly lower than the conventional 80% requirement of China Highway, and the 15% void ratio asphalt mixture meets the specification requirements. In response to this situation, we conducted repeated experiments through the following three scenarios to improve the performance:

a. Replace asphalt for another SBS modified asphalt
b. Replace mineral filler
c. Replace the filler with 50% hydrated lime

The results of the freeze-thaw split test for the three schemes above are shown in Figure 5.

According to the test results:

(1) By replacing the test raw material asphalt and mineral filler, the influence on the performance is insignificant.

(2) Using 50% hydrated lime to replace the original filler can increase the freeze-thaw splitting tensile strength ratio, but this may bring about changes in other test indexes of porous asphalt mixture, which should be based on the actual engineering technology.

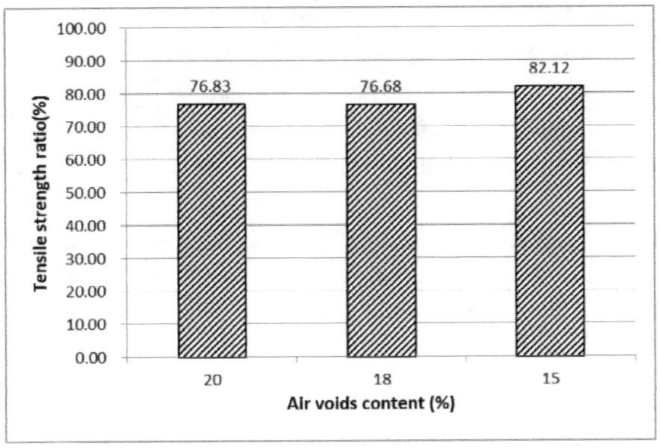

Figure 4. Residual split tensile strength for PUC-10 mixtures

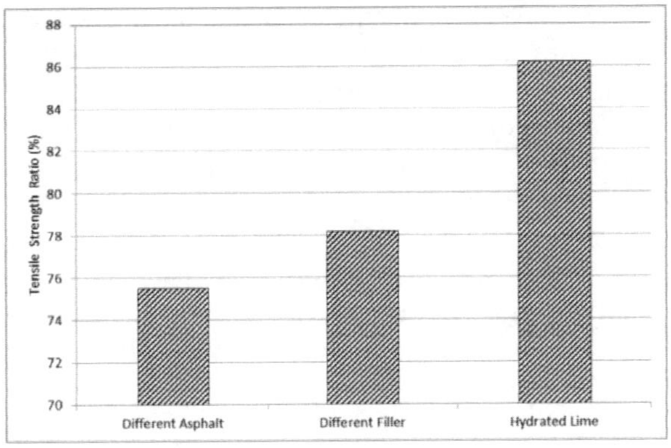

Figure 5. Residual split tensile strength for PUC-10 with material adjustment

4.2. Anti-raveling performance

Raveling is a significant disease of mixture with large porosity. The test was conducted by the Los Angeles Abrasion Tester. The test results of standard raveling and immersion raveling are shown in Figure 6. It shows that:

(1) The results of standard raveling test show that with the decrease of void ratio, the raveling loss of PUC-10 mixture decreases significantly, and the raveling loss percentage of asphalt mixture of 15%, 18% and 20% can meet the technical requirements of porous asphalt mixture in China, which is generally 15% for expressway.

(2) The results of water immersion raveling test show that the PUC-10 mixture with 15%, 18% and 20% air voids content has little water immersion raveling loss, and the correlation with void variation is not remarkable. Mixture with all the air voids content show good performance for anti-raveling.

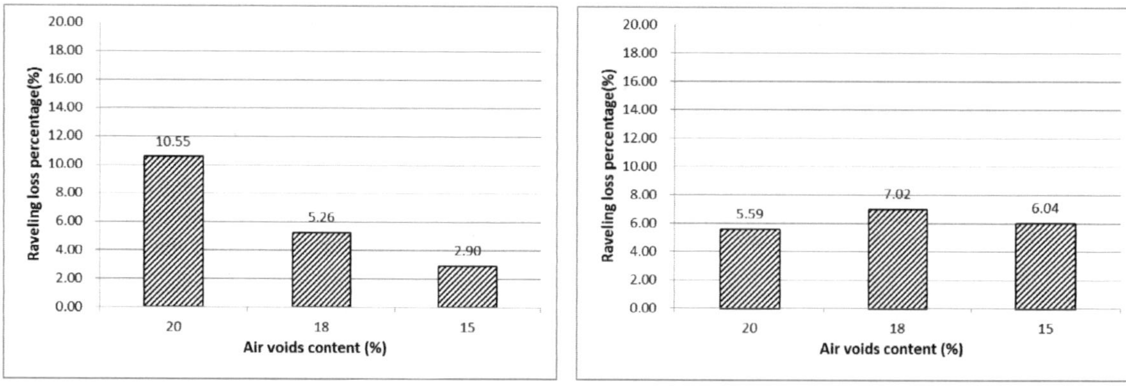

(a) Standard raveling measurement (b) Immersion raveling measurement

Figure 6. Raveling loss percentage for PUC-10 mixtures

4.3. Long-term aging performance

Asphalt mixture will gradually age in service process, and asphalt pavement damage often occurs in different periods. Therefore, it is necessary to conduct long-term aging test on porous asphalt mixture to analyze mixture performance.

In the test, PUC asphalt mixture was heated for 4 hours under the forced air condition of 135 ° C ± 3 ° C for short-term aging, and specimens were continuously heated in an oven at 85 ° C ± 3 ° C for 5 days. The test results are shown in Table 2.

It can be seen from Table 2 that after long-term aging, the immersion raveling rate increases by about 11%, and Marshall stability, residual stability, standard raveling loss, and freeze-thaw splitting residual strength were improved. All indicators meet the requirements of an engineering requirement in China, which means the current design of PUC-10 material can adapt to the environmental effects in long-term use.

Table 2. Test results of performances after long-term ageing

Performance	Test results		Requirements
	Before ageing	After ageing	
Marshall stability, kN	7.43	7.49	≥5.0
Residual Mashall stability percentage, %	94.30	100	≥85
Standard raveling loss percentage, %	10.55	7.3	≤15
Immersion raveling loss percentage, %	5.59	6.2	≤20
Residual Freeze-thaw split tensile strength ratio,%	86.5	88.6	≥80

4.4. Frost resistance

At low temperature, porous asphalt pavements may experience freezing and thawing process. Unlike ordinary pavement, water exists in the pores of a porous pavement during freezing and thawing. Therefore, after mixture was subjected to several freeze-thaw cycles in fully immersed state, the frost resistance of porous asphalt mixture was evaluated by attenuation degree of road performance.

4.4.1. Test conditions

The "4+2" cycle was adopted in the freeze-thaw test. The PUC-10 asphalt mixture with a voids content of 20% is firstly immersed in the water tank for 4 hours, and covered with plastic bag and placed in an incubator at -18 ° C for 4 hours. Then, it was taken out and placed in a water tank at room temperature for 2 hours. One cycle lasts for 6 hours and repeats continuously.

In particular, in the test process, it was difficult to achieve full saturated state after mixture specimen was covered with plastic bag. Most of them can only be partially infiltrated in the water, but this is also in line with the surface immersion of road as in rainy and snowy condition in practice. It is recommended to use a plastic box of appropriate specifications instead of a plastic bag to achieve full water saturation in the future, if it is necessary to simulate full saturated and frozen state of road surface.

4.4.2. Test results and analysis

In order to simulate the case where air voids content is relatively unfavorable, the test was carried out using PUC-10 specimen with 20% porosity. The changes in raveling loss rate and splitting tensile strength after 10, 20, 30, and 40 times freeze-thaw cycles of mixture are shown in Figure7.

According to the test results, following conclusions can be drawn:

(1) Asphalt mixture anti-raveling performance: As can be seen from Figure10, raveling loss of PUC-10 mixture with 20% void ratio increases with the number of freeze-thaw cycles. When number of cycles reaches 30 to 40 times, the speed of raveling was significantly accelerated, but after 40 times of "4+2" freeze-thaw cycles, the raveling rate is still below 15%, which means an acceptable performance for road engineering in China. The reason is considered that pressure generated by expansion of water in the voids of the specimens is bounded by adhesive force of asphalt binder; asphalt mixture has good toughness, and its resistance to deformation is greater than mixture's minor deformation caused by water frost.

(2) For splitting tensile strength, as shown in Figure7(b), it decreases with the number of freeze-thaw cycles; when within 30 times of freeze-thaw cycles, tensile strength is reduced by about 0.05MPa for every 10 times of freeze-thaw cycles; when the number of freeze-thaw cycles is more than 30 times, tensile strength is obviously decreased. Compared with results of unfrozen-thaw test in 4.2, the freeze-thaw splitting strength rate after 40 freeze-thaw cycles is still about 70%.

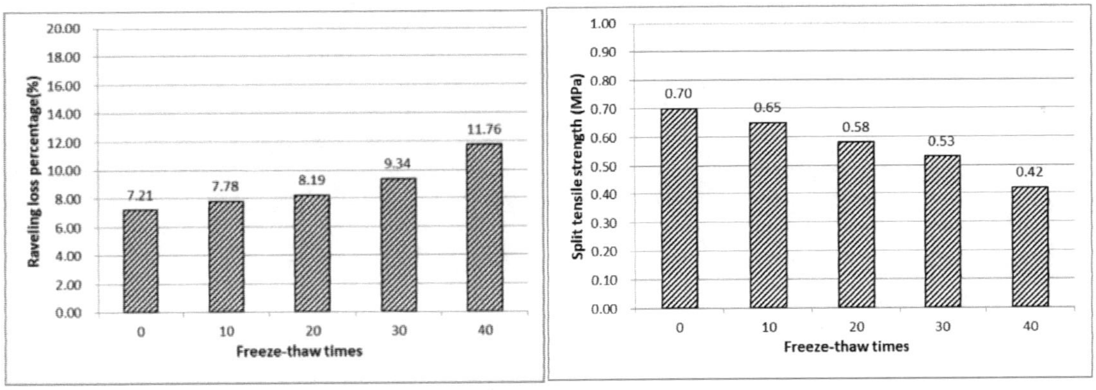

(a) Standard ravelling (b) Split tensile strength

Figure 7. Performance measurements for PUC-10 mixtures after freeze-thaw recycle

5. Trial section and road performance analysis

The porous asphalt pavement of coastal highways in Jiangsu province in China was completed and opened to traffic in 2005. It has been used in operation for more than 13 years. Field investigations showed that porous asphalt has been clogged to some extent. However, due to the rough texture of road surface helped to break the water film, it still has certain drainage function in rainy time. In order to improve safety performance in rainy days, it was planned to build PUC-10 porous ultra-thin wearing course for maintenance and repair of porous asphalt pavement. And during the service process, this technology will be verified and summarized. The specific maintenance section is K1125+200~K1124+190 in Nantong-Yancheng direction, and the total length is 1.0km. The thickness of the trial section is 25mm. The results of measurements on the mixture and pavement surface are shown in Tables 3 and Table 4 respectively. It can be seen from tables that both the performances of mixture and surface can meet requirements for road service. The long-term performance and function of pavement will continue to be investigated in subsequent studies.

Table 3. Measurement results for asphalt mixtures for PUC-10 trial section

Performance	Test results	Targets for the project
Theoretical maximum specific gravity	2.632	/
Air voids content, %	21.5	17~23
Marshall stability, kN	5.98	≮5
Drainage percentage, %	0.07	≯0.8
Standard raveling loss percentage, %	14.1	≯15
Immersion raveling loss percentage, %	19.0	≯20
Residual Mashall stability percentage, %	95.2	≮85
Residual Freeze-thaw split tensile strength ratio,%	96.1	≮80
Dynamic Stability, time/mm	4922	≮4000

Table 4. Measurement results on the newly paved PUC-10

Position	Lane	Thicknes s, mm	Air voids content, %	Compaction Degree, %	Permeability (ml/min)	British pendulum number (BPN)
K1125+ 000	Middle Lane	24.7	23.0	98.1	5556	57
K1124+ 700	Fast Lane	27.3	23.9	97.0	5310	61
	Middle Lane	27.2	22.7	98.5	5279	62
K1124+ 400	Fast Lane	26.5	23.3	97.7	5357	64
	Middle Lane	24.2	21.5	100.0	4700	65
	Heavy Lane	28.3	21.6	99.9	5187	60

6. Conclusions and recommendations

In this paper, the comprehensive evaluation on design, sound absorption property and mixture performance of 10mm PUC-10 mixture was carried out, and long-term performance observation is to be performed by constructing the test section. The main conclusions and recommendations are as follows:

(1) The laboratory evaluation and analysis on noise-reducing performance of porous ultra-thin wearing course mixture was performed. The study showed that air voids content affects mixture's

peak value and peak frequency of sound absorption coefficient, that is, when the air voids content is larger, peak value is higher, as well as a lower peak frequency correspondingly.

(2) According to the performance study on PUC mixture, for PUC type mixture with high viscosity modified asphalt as binder, when void ratio is 15%~20%, it can meet needs of high temperature, low temperature, water damage resistance, raveling resistance, long-term aging and frost resistance. The performance of water damage resistance can be improved by adding hydrated lime. For multiple freeze-thaw cycle test, comparison with dense graded ultra-thin wearing course is to be added in the future research work, and technical scheme for further strengthening fatigue resistance is also to be proposed.

(3) Based on the reconstruction project of Coastal Expressway in 2018, the preliminary application of PUC-10 porous ultra-thin wearing course technology was carried out. Mixtures produced in mixing plant and on-site inspections show that performance of mixture and road surface can both meet the requirements in using in China. Follow-up observations on trial section will be carried out, and construction techniques such as paving, rolling and temperature control will be improved as well.

References

[1] Woldekidan M F 2011 *Response modelling of bitumen, bituminous mastic and mortar* (Delft: Delft University of Technology)
[2] Hagos E T 2008 *The Effect of Aging on binder properties of porous asphalt concrete* (Delft: Delft University of Technology)
[3] Xu B, Chen J, Li M, Cao D, Ping S, Zhang Y and Wang W 2016 *Constr. Build. Mater.* **124** 681–9
[4] Qin X, Zhu S, He X and Jiang Y 2018 *Constr. Build. Mater.* **186** 476–83
[5] Li M, Zeng F, Xu R, Cao D and Li J 2019 *Materials* **12** 3776
[6] Ahmed S, Dave E, Buttlar E and Exline M 2013 *Int. J. Pavement Eng.* **14** 540–52
[7] Chen D and Scullion T 2015 *Constr. Build. Mater.* **95**:108-16
[8] Sandberg U 1987 *Appl. Acoust.* **21** 97-118
[9] Berengier M, Stinson M, Daigle G A and Hamet J F 1997 *J. Acoust. Soc. Am.* **101** 155–162
[10] Attenborough K, Bashir I and Taherzadeh S 2011 *J. Acoust. Soc. Am.* **129** 2806–19
[11] Li M, van Keulen W, van de Ven M, and Molenaar A 2015 *Appl. Acoust.* **88** 12–21

IOP Conf. Series: Materials Science and Engineering 1075 (2021) 012015 doi:10.1088/1757-899X/1075/1/012015

Effect of Polymer and Warm mix Modification on Asphalt Mixture Properties

Bhupendra Singh[1*], Nikhil Saboo[2] And Praveen Kumar[3]

[1]National Institute of Technology Patna, India.
[2] Indian Institute of Technology Varanasi, India.
[3]Indian Institute of Technology Roorkee, India,
*Corresponding author: bhupendrasingh@nitp.ac.in

Abstract. Among various kinds of asphalt binder modification methods, polymer modification and warm mix modification are two of the most popular ones. While polymer modification improves the lower and higher temperature performances of the asphalt mixture, warm mix modification reduces the mixing and compaction temperatures. So, the present study examines the effect of warm mix and polymer modification on the Marshall properties and moisture susceptibility of the asphalt mixes. In order to take advantage of both these popular modification methods, warm mix polymer modified bitumen (WMPMB) was prepared by mixing polymer and warm mix modifier in the conventional binder. The effect of the combined modification (warm mix and polymer) was studied with the help of experiments on the asphalt binders and mixtures. Based on the results it is concluded that using both the modifiers together we can avail the benefits of both the modification methods.

1. Introduction

The Increasing traffic, heavier axle loads, and extreme weathering conditions are imposing big challenges in front of the pavement researchers and practitioners. There are different methods which are being adopted to meet these challenges but one of the most common is the modification of asphalt binders. Even though asphalt binder makes 4-6% of the asphalt mixture but its properties affects the performance of the pavement considerably [1]. Among various modification methods, polymer modification and warm mix modification are two of the most widely used. Polymer modification is preferred over others as it has the advantage of improving low temperature performance as well as high temperature performance of the asphalt binders [2, 3, 4, 5]. On the other hand, Warm mix modification reduces the production temperature of the asphalt mixture by 20-40 °C. This reduced production temperature makes the mix economical by reducing the production cost and also making it environment friendly by reducing the fuel utilization in heating the materials [6, 7, 8].

One of the problems associated with polymer modified binder is that it has considerably higher production temperature as compare to the unmodified binder (reference). On the other hand, warm mix modification reduces the production temperature. Polymer modification and warm mix modification has been widely used by the researchers and found to be very beneficial in improving the properties of the asphalt binders and mix (reference) but the use of these two modification method is one area which has not been explored that much. Using these modification methods together may result in the production of mix which has superior performance along with the economic and environmental advantages. So, the current paper explores the prospects of using both the technologies together. In the study polymer modification and warm mix modification methods are used separately and together to produce polymer modified binder (PMB), warm mix asphalt (WMA) and Warm mix polymer modified binder (WMPMB). The effect of modification examined by using the different experimental procedures on the asphalt binder and mixture. The results obtained from different experiments are compared and based on that conclusions are given.).

2. Materials and experimental plan

In the present study viscosity graded VG-10 asphalt binder has been used as the base binder. The binder was modified using styrene butadiene styrene (SBS) polymer modifier and zycotherm a chemical warm mix modifier. By using these modifiers separately and together, three different modified binders were prepared. The optimum modifier content for SBS modification was determined with the help of separation test and fluorescence microscopy [9]. It was found to be 3% by the weight of base binder. The results of these test are not discussed here for brevity. For the preparation of WMA, optimum dosage of zycotherm i.e. 0.1% as suggested by the manufacturers was used. For WMPMB first PMB was prepared using 3% SBS and then 0.1% of zycotherm was added. The preparation details of these binders are given below in the table 1:

Table 1. Preparation details of the modified binders

Modified Binder	Base Binder	Modifier	Modifier Dosage	Blending Temp.	Blending Duration	Shear Rate
PMB	VG-10	SBS	3% by wt.	180°C	60 Minutes	1500 RPM
WMA	VG-10	Zycotherm	0.1% by wt.	150°C	15 Minutes	400 RPM
WMPMB	PMB	Zycotherm	0.1% by wt.	170°C	20 Minutes	1500 RPM

In order to characterise the materials first conventional tests were conducted on the asphalt binders and aggregates. Conventional test results for the asphalt binders is given in table 2:

Table 2. Conventional properties of asphalt binders

Tests	VG-10	WMA	PMB	WMPMB
Penetration Value, dmm	94	98	44	40
Softening Point, °C	46	42	65	67
Absolute Viscosity at 60 °C, Poise	980	1165	20800	22300
Ductility Value	>100	>100	75	74
Elastic Recovery (%)	-	-	68	66
High Temp. PG	PG 64-XX	PG 64-XX	PG 76-XX	PG 76-XX

From the above results it can be observed that polymer modification considerably reduced the penetration and ductility values whereas the softening point and viscosity values were increased. This can be credited to the continuous three-dimensional network which polymer additives forms in the base binder during the blending process. This polymer network, at the higher temperatures provides the required rigidity to the binder and on the other hand increases its elasticity at lower temperatures [10, 11]. In case of warm mix modification, the change in conventional properties were similar to that of polymer modification but the extent of changes was much smaller. This may not be the effect of warm mix modifier but these changes can be due to the fact that during the blending process binder is heated and that increases the stiffness of the binder. Polymer modification increased the performance grade (PG) of the binder whereas warm mix modification did not affect it.

The conventional properties of the aggregates are given below in the table 3. All the properties were found to be within the specified limits.

Table 3. Conventional Properties of Aggregates

Tests	MORTH* Specified Values	Obtained Values
Specific Gravity (Coarse Agg.)	-	2.70
Specific Gravity (Fine Agg.)	-	2.712
Water Absorption (IS:2386, Part III)	Maximum 2%	0.85%
Impact Value (IS:2386, Part IV)	Maximum 24%	15.53%
Los Angeles Abrasion (IS:2386, Part IV)	Maximum 30%	18.97%
Crushing Strength (IS:2386, Part IV)	-	16.55%
Combined Flakiness and Elongation Index (IS:2386, Part I)	Maximum 35%	22.45%

*Ministry of Road Transport and Highways, India.

In order to give the clearer understanding of the experimental procedure that has been followed in the study is given in figure 1:

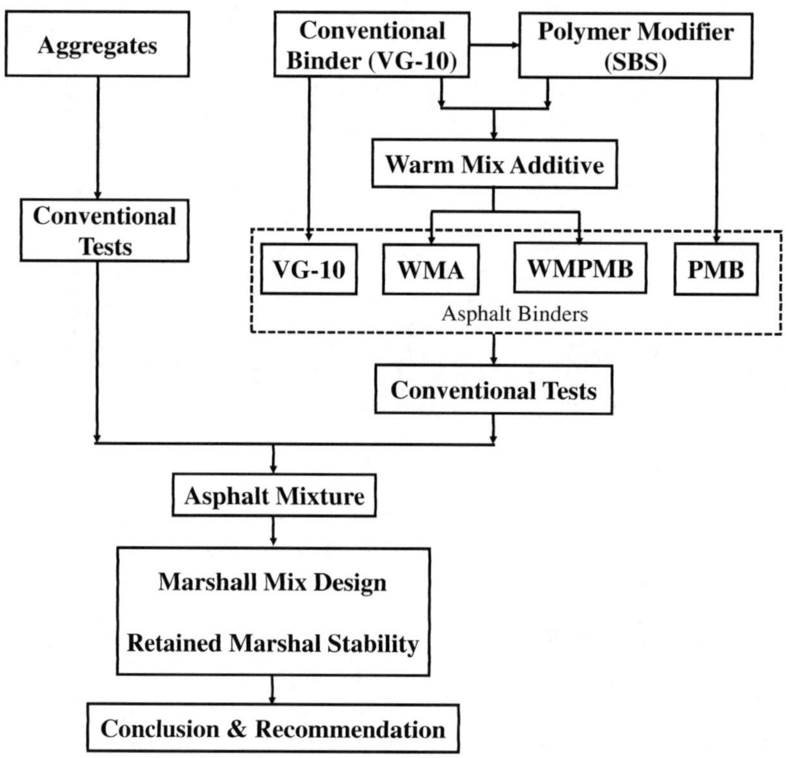

Figure 1. Experimental flowchart of the study

3. Results and discussion

In the study Marshall stability and retained marshall stability tests have been performed. The results of these tests are presented and discussed as follows.

3.1 Marshall Stability Test

The aggregate gradation adopted in the study has been given in the Table 4 and has been visually represented in the figure 2:

Table 4. Aggregate Gradation Adopted in the Study

Sieve Size (mm)	Cumulative % by weight of total aggregate passing	
	MORTH Specified limits	**Obtained Values**
26.5	100	100
19	79-100	89.5
13.2	59-79	69
9.5	52-72	62
4.75	35-55	45
2.36	28-44	36
1.18	20-34	27
0.6	15-27	21
0.3	10-20	15
0.15	5-13	9
0.075	2-8	5

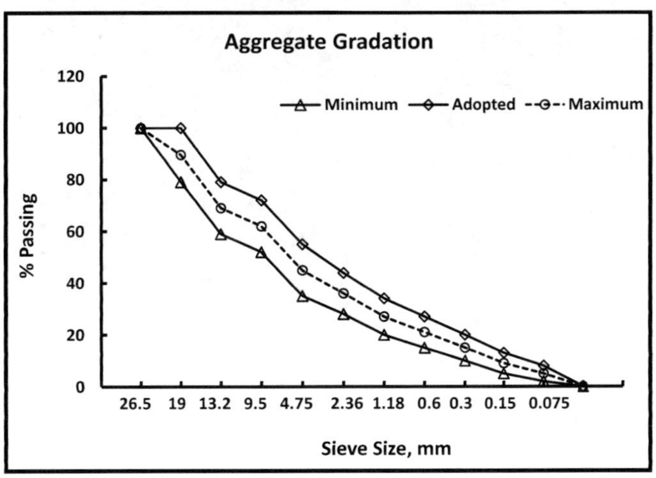

Figure 2. Aggregate gradation adopted in the study

After achieving the required gradation of the aggregates Marshall stability test was performed to evaluate the stability of the asphalt mix. Binder content was varied from 4.5% to 6.5% by weight of aggregate with an increment of 0.5%. About 1200 gram of aggregate (of the desired gradation) was taken and mixed with the different percentage of bitumen. The aggregate and bitumen was heated to the required temperature such that at no time the difference between their temperatures exceeds 14 °C. The mixing was done at a suitable predetermined temperature. The mixing and compaction temperature of WMA mixtures were adopted 20 °C lesser than their hot mix asphalt (HMA) counterparts as prescribed by the zycotherm manufacturers. The compaction and mixing temperatures of different binders is given below in table 5. After mixing the mixture was transferred to a pre-heated Marshall mould having a height of 63.5 mm and diameter of 102 mm. A mechanical hammer of standard weight is used to compact the sample at a suitable compaction temperature. The preheated hammer was placed in position and the mix was compacted by applying 75 blows on each face. Three identical samples at each binder content were prepared. Prepared Marshall samples are shown in figure 3. The compacted samples were allowed to cool at room temperature overnight. The samples were then transferred to a pre-heated water bath having a temperature of 60 °C for 30 to 40 minutes. Marshall stability test was performed on these samples. The sample were placed below the Marshall testing head. Compressive load is applied at a constant rate of 51 mm/minutes until the failure of the specimen. Marshall stability is the value of the maximum load at failure while the flow value the amount of deformation undergone by the sample as given by the reading of the flow meter.

Table 5. Mixing and compaction temperature of different binders

Binder	Mixing Temperature (°C)	Compaction Temperature (°C)
VG-10	150	130
WMA	130	110
PMB	172	155
WMPMB	152	135

Figure 3. Marshall test samples

Figure 4 presents the graphical representation of the Marshall stability and flow values obtained in the study. PMB was found to have 23% higher stability values than VG-10. Here warm mix modification was found to improve the stability values of the mix, for both the warm mix asphalt binders WMA and WMPMB, stability values were 8% and 10% (approx.) higher than the VG-10 and PMB respectively. Zycotherm, being a silane-based technology, is reported to be more effective than some other chemical warm mix additives [12]. This improved performance can be credited to the molecular level hydrophobic zone that it creates around aggregate surface. Organofunctional silanes present in the zycotherm forms a covalent bond between an organic and inorganic phase. The inorganic part of this couple forms hydrogen bond with the hydroxylated agent on surface of the stones while the organic part, in the presence of water condenses through the process of hydrolysis and forms hydrophobic siloxanes. So, when the temperature increases, hydrogen bond breaks and produces H_2O and covalently bonded metallosiloxane. This results in the formation of cross-linked siloxane (Si-O-Si) film structure over the aggregate surface. This resulted in better performance of WMA and WMPMB.

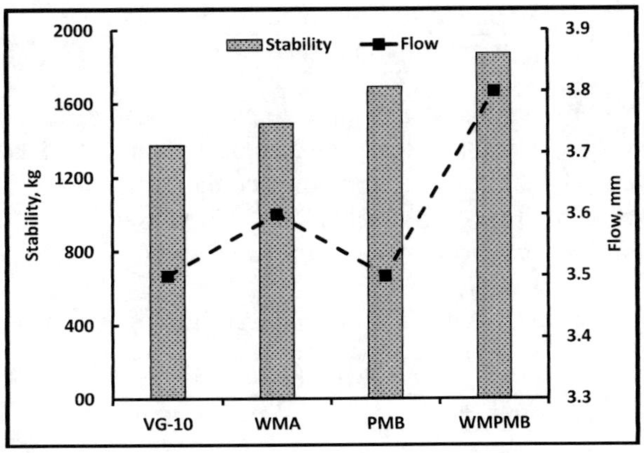

Figure 4. Graphical representation of Marshall Stability and Flow values

Marshall Quotient (MQ) is used as a measure of the permanent deformation; shear stress and rutting of the mix to be used in the road service. Marshall Quotient is estimated as the ratio of stability of the asphalt concrete mix to the flow value. A comparison of Marshall Quotient was done for asphalt concrete mix prepared using different asphalt binders at optimum content. It was observed that polymer modification improved MQ by 23% whereas in case of warm mix modification this improvement was marginal. Marshall quotient values of the different mixes has been given in figure 5.

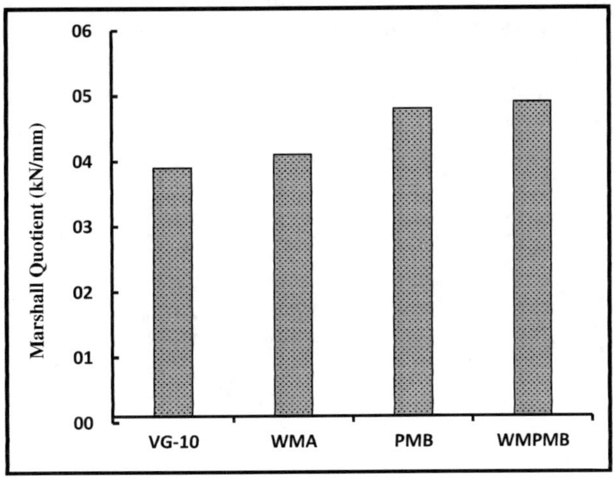

Figure 5. Marshall Quotient values for different asphalt mixtures

3.2 Retained Marshall Stability Test

Damage due to moisture is one of the biggest concerns in flexible pavements. The durability of an asphalt mixture can be indirectly assessed by its susceptibility to moisture. Retained Marshall stability test was carried out to examine the susceptibility of the asphalt mixes to moisture. For each type of mix, six different specimens were prepared and divided into two groups having three specimens each. Specimens from the first group were subjected to conditioning by immersing them in a water bath maintained at 60 °C for a period of 24 hours. On the other hand, specimens of group two were kept unconditioned following the normal immersing of specimens for 30 minutes at a temperature of 60 °C. All the samples were tested in a Marshall stability testing machine until failure by application of load at a constant deformation rate of 51 mm per minute. The average stability values for each group was calculated and the retained Marshall stability values were determined using the following equations.

$$Retained\ Stability = \frac{Marshall\ Stability\ of\ conditioned\ specimen}{Marshall\ Stability\ of\ standard\ specimen} \times 100 \tag{1}$$

Presence of moisture in a bituminous mix can lead to early failure of flexible pavements and can be a major hazard. The durability of asphalt mixture can also be evaluated by studying the susceptibility of asphalt mixtures to moisture. The loss of adhesion between the bitumen and aggregates was studied by utilising the retained Marshall stability test. A higher value of retained Marshall stability indicates lower moisture susceptibility and vice-versa. Table 6 presents the result of the retained Marshall stability.

Table 6. Retained Marshall Stability of Mixes

Binder	Stability (kg)		Retained Stability (%)
	Standard	Conditioned	
VG-10	1374.26	1081.64	78.71
WMA	1491.65	1245.61	83.51
PMB	1689.42	1481.25	87.68
WMPMB	1871.53	1711.14	91.43

Results clearly showed that mixes prepared using modified binders had higher retained Marshall stability than the conventional binder implying that the modified binders are less susceptible to moisture damage. Mix with WMPMB had the best retained Marshall stability. Figure 6 presents the retained Marshall stability values of the mix prepared using different binders. Here also it was observed that mixes containing binder with zycotherm i.e. WMA and WMPMB showed better moisture resistance than their corresponding base binders. This again can be credited to the siloxane (Si-O-Si) film structure over the aggregate surface which being a hydrophobic layer, improves the moisture susceptibility of the mixes [12]. Two major conclusions can be drawn from the results. First, mixtures prepared using modified binders have lower moisture susceptibility as compared to conventional binders and secondly addition of chemical warm mix additive zycotherm improves the moisture susceptibility of conventional as well as PMB.

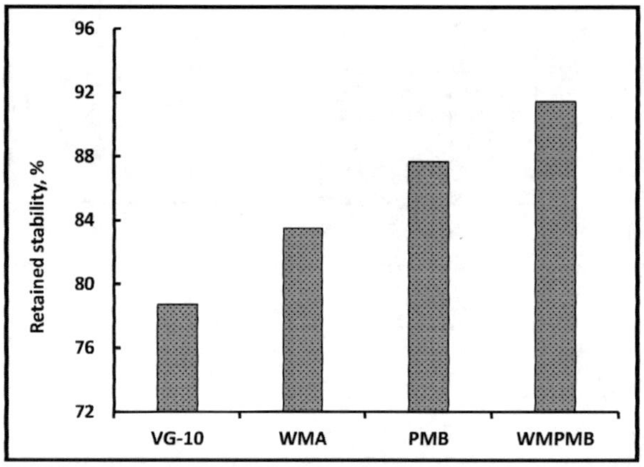

Figure 6. Retained Marshall Stability of the mix prepared using different binders

4. Conclusions

Present paper studied the use of both polymer modification and warm mix modification together. In order to check the stability of the asphalt mixtures, Marshall stability test was conducted. Retained Marshall stability test was conducted to assess the moisture susceptibility of the mixes. In the study it was found that polymer modification due to their three dimensional polymer network improved the Marshall Properties and moisture susceptibilities of the asphalt mixture considerably. Marshall stability values increased by 23%, Marshall Quotient was also increased by 23% whereas the retained marshal stability value increased by 10%. Warm mix modification also improved the properties of the mix. It improved Marshall stability by 8% and retained Marshall stability by approximately 5%. Using both the modification method proved helpful in utilising the advantage of both these modification methods. The WMPMB samples showed better performance with increased Marshall stability, retained Marshall stability and Marshall Quotient at 20 °C lower compaction and mixing temperature as compared to PMB. This makes it economic and environment friendly also. So the study concludes that polymer modification and warm mix modification can be used together to obtain better asphalt mixture structurally, economically and environmentally.

References

[1] Singh B, Kumar P 2015 Effect of modifiers on the ageing properties of bitumen: A review *proc. of 3rd Conference of Transportation Research Group of India, Kolkata, India.*

[2] Lu X, Ekblad UI 1999 Rheological properties of SEBS, EVA and EBA polymer modified bitumens *Mater. Struct.* **32** 131–139.

[3] Sengoz B, Isikyakar G 2008 Evaluation of the properties and microstructure of SBS and EVA polymer modified bitumen *Constr. Build. Mater.* **22** 1897–1905.

[4] Airey GD 2002 Rheological evaluation of ethylene vinyl acetate polymer modified bitumens *Constr. Build. Mater.* **16** 473–487.

[5] Isacsson ULF, Lu X 1999 Characterization of bitumens modified with SEBS, EVA EBA Polymers *Journal of Materials Science* **4** 3737–3745.

[6] Zaumanis M 2010 Warm mix asphalt Investigation *PhD Thesis, Riga Technical University, Kgs. Lyngby, Denmark.*

[7] Rubio MC, Martínez G, Baena L, Moreno F 2012 Warm Mix Asphalt: An overview *Journal of Cleaner Production* **24** 76–84.

[8] Banerjee A, de Fortier S, Prozzi JA 2012 The effect of long-term aging on the rheology of warm mix asphalt binders, *Fuel* **97** 603-611.

[9] Singh B, Kumar P 2019 Effect of polymer modification on the ageing properties of asphalt binders: Chemical and morphological investigation *Constr. Build. Mater.* **205** 633-641.

[10] Airey GD 1997 Rheological Characteristics of Polymer Modified and Aged Bitumens *PhD Thesis, University of Nottingham.*

[11] Singh B, Saboo N, Kumar P 2017 Effect of short-term aging on creep and recovery response of asphalt binders *Journal of Transportation Engineering, Part B: Pavements* **143(4)**, 04017017.

[12] Mirzababaei P 2016 Effect of zycotherm on moisture susceptibility of Warm Mix Asphalt mixtures prepared with different aggregate types and gradations *Constr. Build. Mater.* **116** 403-412.

IOP Conf. Series: Materials Science and Engineering 1075 (2021) 012016 doi:10.1088/1757-899X/1075/1/012016

Blending of RAPM samples with virgin binder: a study using FTIR Spectroscopy

Bhaswati Bora[1*] **and Animesh Das**[1*]

[1]Department of Civil Engineering, IIT Kanpur, 208016, Kanpur, UP, India
*Corresponding author: adas@iitk.ac.in

Abstract. In the bitumen recycling process, aged binder from reclaimed asphalt pavement material (RAPM) and virgin binder (which is added externally) get blended. The extent of blending depends on several factors. In this study, RAPM sample is kept in a pool of virgin binder and the extent of blending is studied (by taking samples from the blended binder) with the variation of temperature and RAPM size. Subsequently, the extent of blending is studied for two different RAPM samples of different ageing levels. Appropriate level of temperature and adequate mixing (by stirring) is maintained so as to create a favourable condition for blending between the aged and virgin binder to occur. After the test, the blended binder is retrieved and its chemical composition is analysed using FTIR spectroscopy, and the results are compared.

1. Introduction

In the bitumen recycling process, reclaimed asphalt pavement material (RAPM) is mixed with virgin aggregates (VA) and virgin binder (VGB). The binder present in RAPM (termed as, RAPM-binder in this paper) eventually blends with the virgin binder [1–4].

Three stages of blending may be considered, namely (a) beginning of blending (at time, $t = 0$) (b) continuation of blending due to mixing (due to stirring) and diffusion process (at any time t) (c) completion of blending (at time, $t \rightarrow \infty$). These stages are schematically presented in figure 1. In this figure, the greyscale shade of the blended binder schematically represents the degree of blending. The concentrations of RAPM binder away from the surface of the RAPM for these three stages are schematically presented in figure 2.

Diffusion is a natural process which occurs when the aged binder attached to RAPM comes in contact with virgin binder added externally [5, 6]. The extent of diffusion (between the RAPM binder and virgin binder) in the recycled mix depends on time and temperature, and theoretically full blending is achieved over infinite time (as explained through figures 1 and 2). A number of factors affect the extent of blending, for example, (i) extent of stirring done (during mixing the ingredients at the time of recycling), (ii) temperature during the process of recycling, (iii) surface area available for diffusion to take place, (iv) age of the RAPM binder and so on [2, 5, 6].

The diffusion rate is expected to increase with more stirring, elevated temperature and availability of more surface area of the RAPM. On the other hand, if the RAPM is significantly aged (that is, RAPM binder is more oxidized) the rate of diffusion may be low. In that case, even after considerable passage of time ($t \gg 0$), the concentration of binder may still appear to be the one as shown in figure 2(a), and not as shown in figure 2(c) [7]. This idea is the basis of the present study.

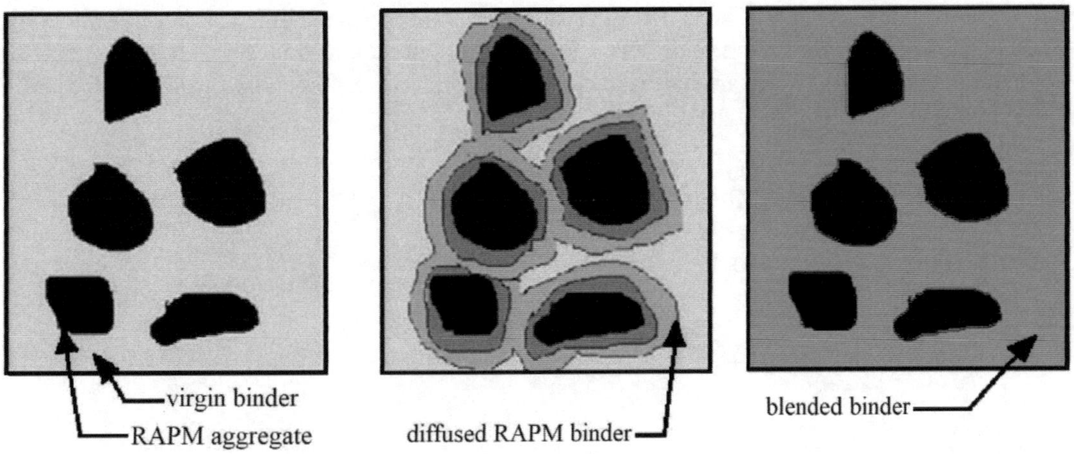

(a) Stage I: blending yet to happen (b) Stage II: blending happening due to diffusion (c) Stage III: complete blending is achieved

Figure 1. Various stages of blending between virgin binder and RAPM binder.

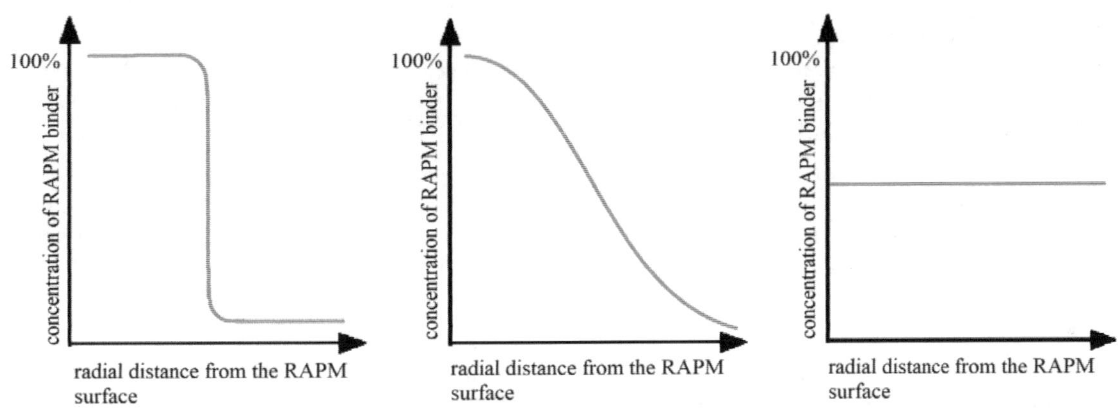

(a) Stage I: blending yet to happen (b) Stage II: blending happening due to diffusion (c) Stage III: complete blending is achieved

Figure 2. Change in concentration of RAPM binder from the surface of the RAPM.

In the present study RAPM, is kept in the pool of virgin binder. Temperature is kept elevated and sufficient stirring is done – that is, favourable condition is created for RAPM binder to get blended with virgin binder. Subsequently, at the end of the experiment, a sample of the blended binder is collected and its composition is chemically analysed. Fourier Transform Infrared (FTIR) Spectroscopy is used in the present study for the chemical analysis. For different experiments, different levels of temperature and size of RAPM are used. Two different RAPM samples (of different ageing levels) are used in the present study for comparison purpose.

2. Background

FTIR Spectroscopy is used to recognize different organic and inorganic compounds from the vibration of their bonds [8, 9, 10]. The presence of a bond representing a specific compound gives a peak in the spectra. This is represented in a plot between absorbance intensity versus wavenumber. The presence of a compound is estimated either from the height of the peak or, from the area under the peak [8, 9, 11].

Ageing in binder leads to the oxidation of certain compounds. The ageing can be determined from the oxidation of the carbonyl, sulfoxide, or aromatic compounds and therefore, these are called the oxidation indicators [12–17]. The aromatic index increases with oxidation under certain conditions. The oxidation indices used in this study are as follows [18–20]:

$$\text{Carbonyl index} = \frac{A_{1700} \times 100}{\sum A} \tag{1}$$

$$\text{Sulfoxide index} = \frac{A_{1030} \times 100}{\sum A} \tag{2}$$

$$\text{Aromatic index} = \frac{A_{1600} \times 100}{\sum A} \tag{3}$$

$\sum A = A_{1700} + A_{1600} + A_{1460} + A_{1376} + A_{1030} + A_{864} + A_{814} + A_{743} + A_{724} + A_{2953,2923,2851}$.
Here, A_w denotes any peak at the wavenumber, w.

Thus, extent of blending between RAPM binder and virgin binder can be recognized from the quantity of oxidation indicators present in the blend after the test [12].

3. Experimental studies

The RAPM samples (two in number) used for this study are taken from the stored material available in the Transportation Engineering Laboratory of IIT Kanpur. These two samples are identified as $RAPM_1$ and $RAPM_2$ in the present study.

The RAPM is separated from each other with the help of manual hammering. The material is washed in water to remove dirt. Then, it is sun-dried and finally kept at room temperature for 48 hours to remove any moisture that may be present on its surface. This drying scheme is as per the procedure outlined in Taha et al. [21]. The material is sieved and two sizes are considered in this study - these are 9.5 mm passing and 4.75 mm retained (or in short, 9.5P–4.75R) and 2.36 mm passing and 1.18 mm retained (or in short, 2.36P–1.18R).

The properties of the $RAPM_1$ binder, $RAPM_2$ binder (extracted following the procedure outlined in ASTM D2172 (2005) and ASTM D1856–09 (2015)) [22, 23], and the virgin binder are presented in table 1.

Table 1: Properties of the binders.

Binder properties	$RAPM_1$ binder	$RAPM_2$ binder	Virgin binder
Softening point	84.5 °C	82.5 °C	49.5 °C
Ductility	3cm	3cm	88cm
Specific gravity	1.205	1.149	1.056
Absolute viscosity at 60°C	-	-	2897 Poise
Kinematic viscosity at 135°C[#]	8948 cSt	17124 cSt	458 cSt
Flash point	270°C	266°C	250°C
Carbonyl index	3.374	2.366[a]	0.067
Sulfoxide index	0.162	0.745[a]	0.326
Aromatic index	4.503	2.639	1.329

Notes [a] data from tests conducted earlier, see [12]; [b] results for one test only

Virgin binder is placed in a beaker of 250 ml and the RAPM samples are poured into it (refer to figure 3). The mix is placed inside a glycerine bath, and the bath is placed over a hotplate. A thermometer capable of measuring temperature up to 360°C is used to measure the temperature of the glycerine bath. A mechanical stirrer with a speed of 8 rpm is used to stir the mix of RAPM and virgin binder. The setup is as shown schematically in figure 4. A photograph of the setup is shown in figure 5.

Figure 3. Mix of RAPM and virgin binder.

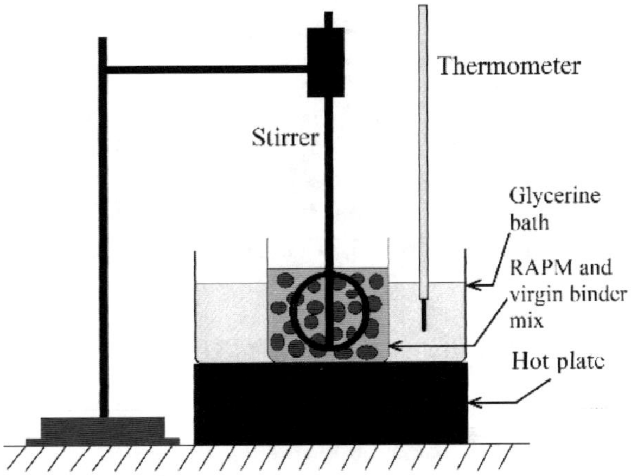

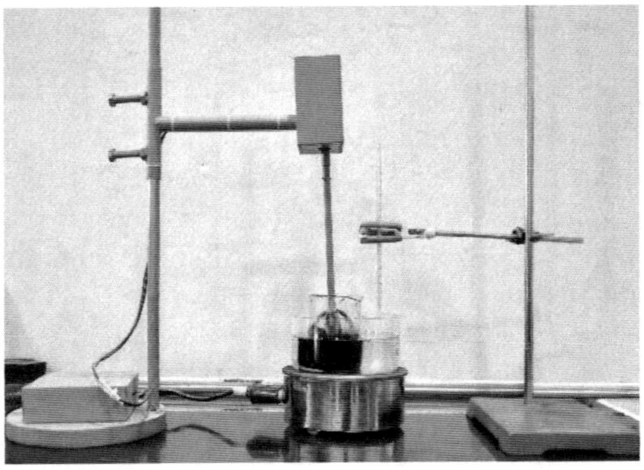

Figure 4. Schematic diagram of the test setup. **Figure 5.** Photograph of the test setup.

Tests are carried out at two different temperatures, namely 130°C and at 150°C. Duration of stirring is kept fixed as 30 minutes.

For $RAPM_1$ samples, tests are conducted at two temperatures (130°C and 150°C), two sizes (P2.36-R1.18 and P9.5-R4.75) and three proportions between virgin binder and $RAPM_1$ (1:2, 1:2 and 1:3 by weight). For each test, three (3) replicate samples are prepared. Hence, total number of samples tested for $RAPM_1$ are $2\times2\times3\times3 = 36$. The detailed description of the mixes prepared for this study are tabulated in table 2. A sample calculation for estimation of RAPM binder percentage with respect to the total binder (considering both RAPM binder and VGB) for sample # VI is presented in the appendix.

Table 2: Experimental parametric scheme for blending of $RAPM_1$ and virgin binder.

Sample identification	$RAPM_1$ size	Temperature	VGB: $RAPM_1$ (by weight)	RAPM binder %	No of samples
I	P2.36-R1.18	150°C	1:1	5.5	3
II	P2.36-R1.18	150°C	1:2	10.4	3
III	P2.36-R1.18	150°C	1:3	15.5	3
IV	P9.5- R4.75	150°C	1:1	3.5	3
V	P9.5- R4.75	150°C	1:2	6.1	3
VI	P9.5- R4.75	150°C	1:3	12.1	3
VII	P9.5- R4.75	130°C	1:1	3.5	3
VIII	P9.5- R4.75	130°C	1:2	6.1	3
IX	P9.5- R4.75	130°C	1:3	12.1	3
X	P2.36-R1.18	130°C	1:1	5.5	3
XI	P2.36-R1.18	130°C	1:2	10.4	3
XII	P2.36-R1.18	130°C	1:3	15.5	3

Since it is expected that the oxidation indices would be higher at a higher temperature and for smaller RAPM size (because, of larger surface area), for the $RAPM_2$ samples, tests are conducted at 150°C and for size P2.36-R1.18. Three proportions between virgin binder and $RAPM_2$ (1:2, 1:2 and 1:3) are taken. For each test, three (3) replicate samples are prepared. Hence, total number of samples tested for $RAPM_2$ are $3\times3 = 9$. Experimental parametric scheme for the mix of $RAPM_2$ and virgin binder is tabulated in table 3.

Table 3: Experimental parametric scheme for blending of $RAPM_2$ and virgin binder.

Sample identification	$RAPM_2$ size	Temperature	Virgin binder: $RAPM_2$	RAPM binder %	No of samples
XIII	P2.36-R1.18	150°C	1:1	8.2	3
XIV	P2.36-R1.18	150°C	1:2	12.2	3
XV	P2.36-R1.18	150°C	1:3	13.4	3

1.5gm of the blended binder from the mix is taken for FTIR analysis. The samples are completely dissolved in carbon disulphide (CS_2). The concentration of binder in the solution is maintained as 16.67wt/vol %. KBr pellet method is used to make the FTIR samples, following the same process elaborated in [12]. The samples are prepared in 15ml glass bottles with an aluminum foil cover as shown in figure 6. A micro pipette of one micro litre capacity is used to maintain equal volume of solution sample on the KBr pellet.

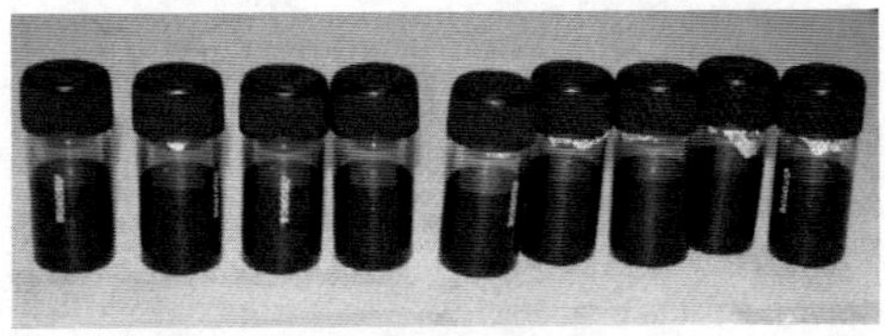

Figure 6. FTIR sample solutions.

4. Results and Discussions

The carbonyl, sulfoxide and the aromatic indices for $RAPM_1$ samples (refer to table 2) are presented in figures 7, 8 and 9, respectively. Lines are drawn to show the trends, by connecting the mean values of the data-points with smooth curves. This is done for all the graphs presented in this paper.

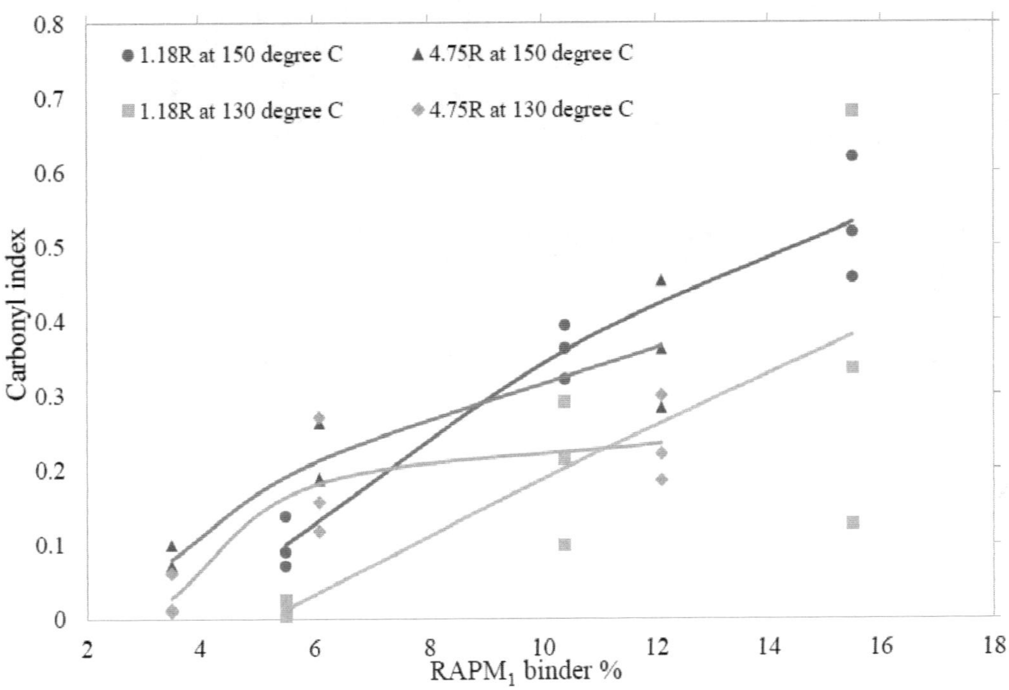

Figure 7. Carbonyl index of $RAPM_1$ samples.

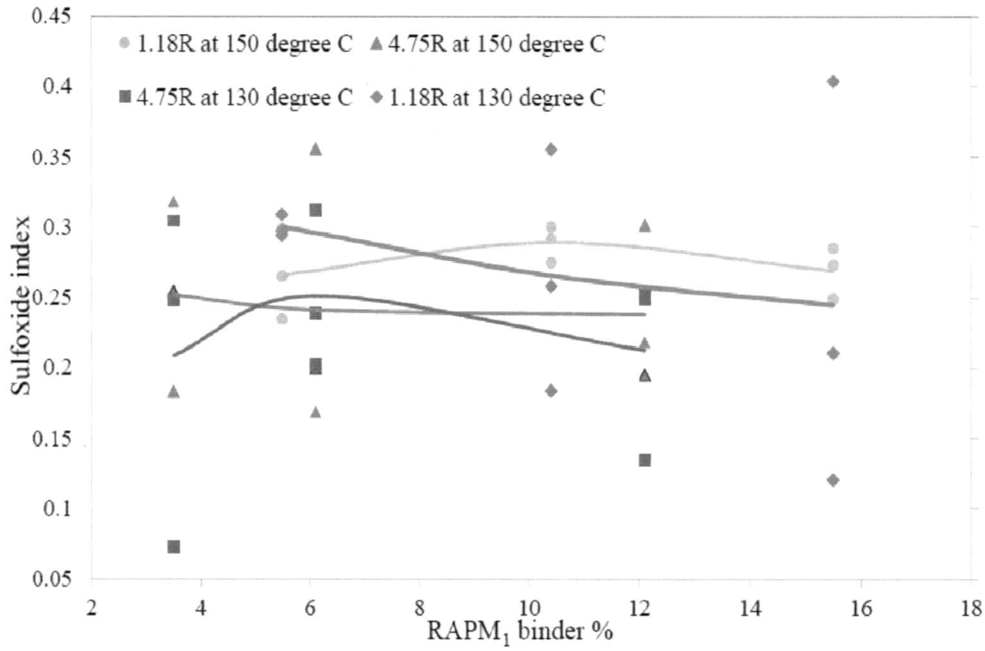

Figure 8. Sulfoxide index of RAPM$_1$ samples.

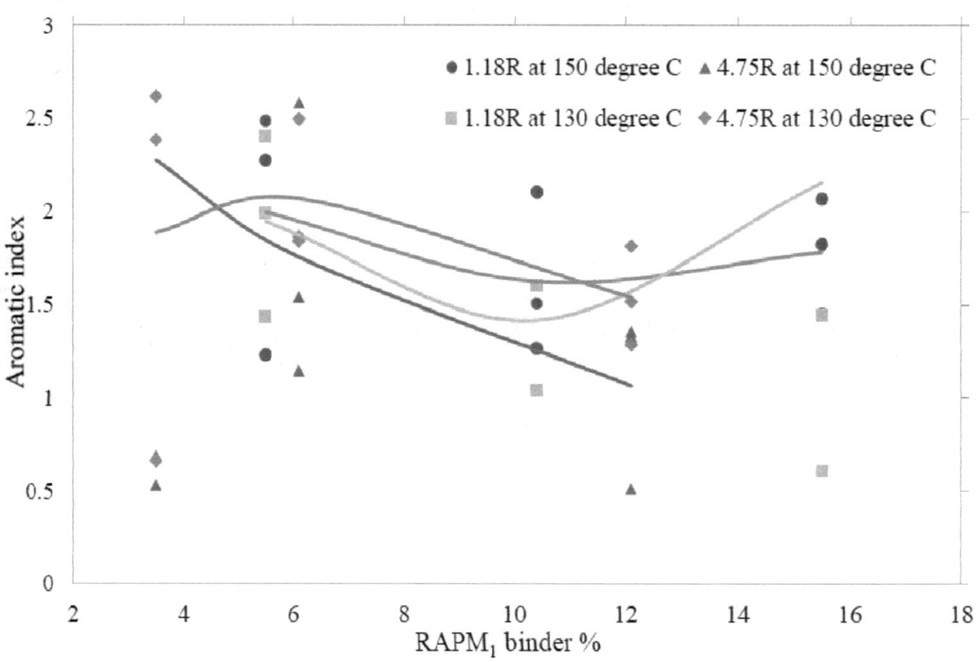

Figure 9. Aromatic index of RAPM$_1$ samples.

In figure 7 an increasing trend of carbonyl index with the increase in percentage RAPM$_1$ binder is observed. In figure 8 no specific trend is observed in the sulfoxide index. This may be due to the decomposition of sulfoxides at a higher temperature as also reported by other researchers [14, 20, 24]. In figure 9 a somewhat decreasing trend is observed of the aromatic index. Formation of ketones, carboxylic acids and anhydrides at higher temperature may have caused this decreasing trend [20, 24]. In general (although there is some experimental scatter) more oxidation is observed to happen at higher temperature and for smaller RAPM size. Smaller size of RAPM samples, because of higher specific surface area, are expected to contain higher quantities of aged binder, as well as larger available surface area for diffusion to occur.

Finally, the carbonyl index is compared for $RAPM_1$ and $RAPM_2$ samples in figure 10. Since, $RAPM_1$ samples are more aged (refer carbonyl index value presented in Table 2), $RAPM_1$ plot is showing higher values in figure 10 for all percentages of RAPM binder. It may be noted that this comparison is made for P2.36-R1.18 size and test temperature 150°C for both $RAPM_1$ and $RAPM_2$ samples.

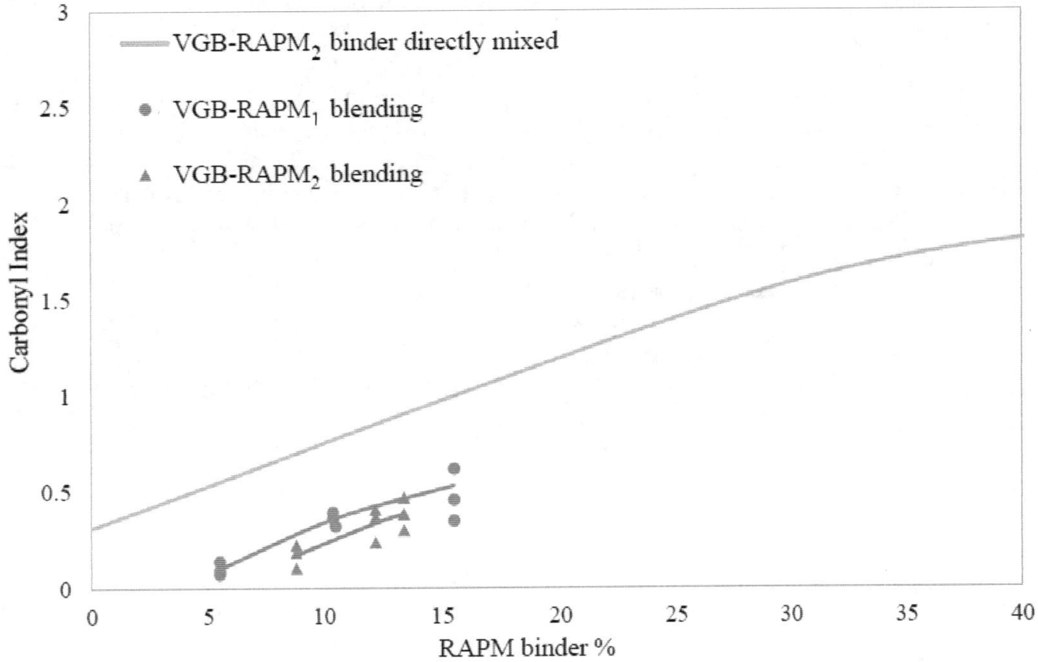

Figure 10. Carbonyl Index of $RAPM_1$ and $RAPM_2$.

However, in figure 10 the carbonyl values of all the blended binders are lower than that of the corresponding RAPM (refer to Table 1). That is, complete blending is not occurring in any of the RAPM binders irrespective of their age.

In a separate study, the $RAPM_2$ binder and VGB was physically mixed (that is, two binders were mixed directly and therefore it did not involve any RAPM diffusion process) [12], this can be termed as 'full blending'. It is interesting to compare (in figure10) the relevant data [12] with the present observation. It is seen that that the values of carbonyl indices observed in the present study (where a favourable condition is created for blending to happen) is consistently lower than the full blending of binders.

5. Closure

In the present work, extent of blending is studied in an experimental setup developed in-house for different temperatures, RAPM size and different ageing levels. FTIR is used for characterizing the extent of blending. It is observed that complete blending of the RAPM and virgin binder may not necessarily occur even if the maximum favourable conditions are provided for blending to occur.

Acknowledgements
The first author wishes to acknowledge Dr. Veena Singh, Department of Chemistry, IIT Kanpur, for her help in operating the FTIR machine and in preparation of the KBr pellets.

References
[1] Cavalli M C, Partl M N and Poulikakos L D 2017 Measuring the binder film residues on black rock in mixtures with high amounts of reclaimed asphalt *Journal of Cleaner Production* **149** pp 665-672
[2] Ekblad J and Lundström R 2017Soft bitumen asphalt produced using RAP *Materials and Structures* **50** 13
[3] Soleymani H R, Anderson M, McDaniel R and Abdelrahman M 2000 Investigation of the black rock issue for recycled asphalt mixtures *Association of Asphalt Paving Technologists Proc.* **69** pp 366–390
[4] Al-Qadi I L, Carpenter S H, Roberts G, Ozer H and Aurangzeb Q 2009 Determination of usable residual asphalt binder in RAP(Illinois: University of Illinois at Urbana-Champaign Department of Civil and Environmental Urbana)*Federal Highway Administration Report FHWA-ICT-09-031*
[5] Kriz P, Grant D L, Veloza B A, Gale M J and Blahey A G 2014 Blending and diffusion of reclaimed asphalt pavement and virgin asphalt binders *Road Materials and Pavement Design* **15** S1 pp 78–112
[6] Rad F Y, Sefidmazgi N R, and Bahia H 2014 Application of diffusion mechanism degree of blending between fresh and recycled asphalt pavement binder in dynamic shear rheometer *Transportation Research Record* **2444** pp 71–77
[7] Huang B, Li G, Vukosavljevic D, Shu X, and Egan B K 2005 Laboratory investigation of mixing hot-mix asphalt with reclaimed asphalt pavement *Transportation Research Record* **1929** pp 37–45
[8] Smith BC 2011 *Fundamentals of Fourier Transform Infrared Spectroscopy 2nd Edition* (Boca Raton:CRC Press, Taylor and Francis Group)
[9] Coates J 2006 *Interpretation of Infrared spectra, a practical approach, Encyclopedia of Analytical Chemistry,* ed RA Meyers (Chichester: Wiley) pp 10815–10837
[10] Stuart B 2004 *Infrared spectroscopy: fundamentals and applications 1st Edition*(New Jersey:Wiley)
[11] Campbell PG and Wright JR 1963 Infrared spectra of asphalts: some aspects of the changes caused by photooxidation *Journal of Research of the National Bureau of Standards*-C, *Engineering and Instrumentation* **68 C**2 pp 115–123
[12] Bora B and Das A 2019Estimation of binder quantity in binary mixture of asphalt binders using FTIR *15thWorld Conf. on Transport Research – WCTR* (Mumbai) 26–31May 2019
[13] Beitchman BD 1959 Infrared spectra of asphalts *Journal of Research of the National Bureau of Standards-A. Physics and Chemistry* **63A**2 pp 189–199
[14] Nivitha M R Prasad E and Krishnan J M 2016 Ageing in modified bitumen using FTIR spectroscopy *International Journal of Pavement Engineering* **17**7 pp 565–577
[15] Karlsson R and Isacsson U 2003Application of FTIR-ATR to characterization of bitumen rejuvenator diffusion *Journal of Materials in Civil Engineering* **15**2 pp157–165
[16] Ahmed F 2016 *Analysis of bitumen rejuvenation using FTIR- ATR M Sc thesis* (Helsinki: Department of Chemical Technology, Aalto University)
[17] Nasrazadani S Mielke D Springfield Tand Ramasamy N 2010 Practical applications of FTIR to characterize paving materials (Texas:University of North Texas)*Federal Highway Administration Report FHWA/TX-11/0-5608–1*
[18] Lamontagne J Dumas P Mouillet Vand Kister J 2001Comparison by Fourier transform infrared (FTIR) spectroscopy of different ageing techniques: application to road bitumens*Fuel* **80**4 pp 483–488

[19] Graca A L F 2016 Predictors of aging and potential of FTIR to detect protein aggregates (Averio:Universidade de Averio, Instituto de Biomedicina (iBMED))http://hdl.handle.net/10773/17157
[20] Petersen J C1984 Chemical composition of asphalt as related to asphalt durability: state of the art *Transportation Research Record* **999** pp 13–30
[21] Taha R Al-Harthy A Al-Shamsi K and Al-Zubeidi M 2002 Cement stabilization of reclaimed asphalt pavement aggregate for road bases and subbases *Journal of Materials in Civil Engineering* **14**3 pp 239–245
[22] ASTM D1856-09 2015 Standard test method for recovery of asphalt from solution by Abson Method(West Conshohocken, PA:ASTM International)
[23] ASTM D2172-052005 Standard test methods for quantitative extraction of bitumen from bituminous paving mixtures(West Conshohocken, PA:ASTM International)
[24] Robertson R E 1991 Chemical properties of asphalts and their relationship to pavement performance(Washington DC: National Research Council)*Strategic Highway Research Program SHRP-A/UWP-91-510*

Appendix

A sample calculation for RAPM binder percentage with respect to total binder
(for sample # VI of Table 2)

$$\begin{aligned}
\text{Weight of virgin binder} &= 40\text{gm} \\
\text{Weight of RAPM} &= 120\text{gm} \\
\text{Weight of washed RAPM}_1\text{ aggregates} &= 114.47\text{gm} \\
\text{Weight of RAPM}_1\text{ binder} &= (120 - 114.47)\text{gm} = 5.53\text{gm} \\
\text{Percentage of RAPM}_1\text{ binder} &= \frac{\text{Weight of RAPM}_1\text{Binder}}{\text{Total weight of binder}} \times 100 \\
&= \frac{5.53}{40 + 5.53} \times 100 \\
&= 12.1\ \%
\end{aligned}$$

IOP Conf. Series: Materials Science and Engineering 1075 (2021) 012017 doi:10.1088/1757-899X/1075/1/012017

Characterization of pavement surface undulations using side view images

Subham Jain[1] and Animesh Das[1]*

Department of Civil Engineering, IIT Kanpur, Kanpur, UP-208016, India
*Corresponding author: adas@iitk.ac.in

Abstract. Pavement surface undulations affect friction, ride comfort, water drainage, noise etc. Various indices have been proposed in the literature to characterize the pavement surface undulations. The present paper compares a few pavement surface undulation indices in the spatial domain and the frequency domain. For this study, core samples of pavement surface were collected from freshly constructed pavements, which had not been opened to the traffic yet. The core samples were sliced using a high-speed diamond saw. Images of the side view of the slices were captured using a digital camera. Suitable image processing scheme was employed to detect the pavement surface profiles. Various indices were subsequently estimated and compared.

1. Introduction

Pavement surface undulations affect various engineering properties, arising from tyre-pavement interaction, namely friction [1], ride comfort [2], water drainage [3], noise [4], etc. [5]. Some of the traditional methods of measuring the pavement surface undulations are sand patch test [6], outflow meter test [7], etc. and some of the emerging methods of measuring the pavement surface undulations are laser-based triangulation [8,9], close-range photogrammetry [10,11] etc.

Various spatial domain based indices have been proposed in the mechanical engineering literature (for example, the Average Maximum Profile Peak Height, Maximum Profile Peak Height, Maximum Profile Valley Depth, Mean Width of Profile Elements, Arithmetic Average Height and so on [12,13]) and in the pavement engineering literature (for example, the Mean Profile Depth (MPD) [14,15]). Various frequency domain based indices have been used in the mechanical engineering literature (for example, the power spectral density (PSD) [16]) and moment of PSD [17]) and in the pavement engineering literature (for example, the PSD [18,19]). In this study, a few spatial domain and frequency domain based indices are estimated (for various pavement surface profiles) and compared. The indices compared in this study are the Average Maximum Profile Peak Height (R_{pm}) [13],Arithmetic Average Height (R_a) [13] and the power of the surface undulations (P) in two different frequency bands ($P_1$and P_2)[20].

2. Background

In this section, a brief discussion on the indices used in this study (R_{pm}, R_a and P) is presented.

2.1. Average Maximum Profile Peak Height

Let a point measured in the profile have its height Y_i and be located at a distance X_i , as illustrated in figure 1. The level of the 'mean height', as shown in figure 1, is the average of the measured height of all the points in the profile. Let this profile of arbitrary length 'L' be divided into 'n' segments of equal lengths 'L/n'. Let, the maximum height of the profile in i^{th} segment be R_{pi}, for example, the maximum height of the first segment is R_{p1}, as illustrated in figure 1.

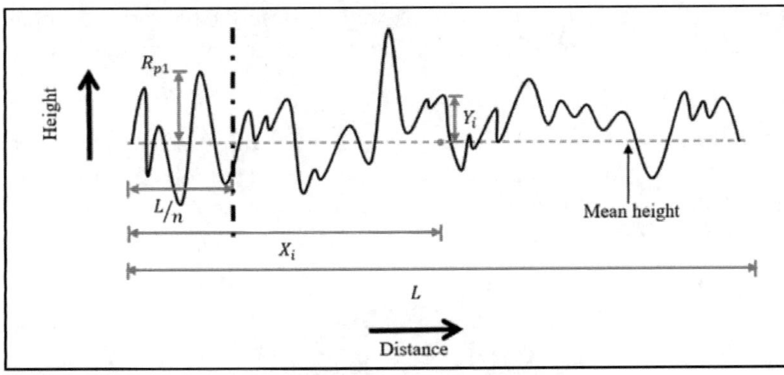

Figure 1. Schematic diagram explaining the concept of the Average Maximum Profile Peak Height.

Then, the Arithmetic Maximum Profile Peak Height (R_{pm}) is given by equation (1)[13]:

$$R_{pm} = \frac{\sum_{i=1}^{n} R_{pi}}{n} \tag{1}$$

2.2. Arithmetic Average Height
The Arithmetic Average Height (R_a) is the sum of the absolute value of the heights of each point in the profile, divided by the length of the profile [13]. Let the number of such points measured in that profile shown in figure 1 be *N*, then the $R_a$is given by equation (2)[13]:

$$R_a = \frac{\sum_{i=1}^{N} |Y_i|}{N} \tag{2}$$

2.3. Power of the surface undulations
Power of a signal, in electrical engineering terminology, is defined as the sum of the squares of the amplitude of the sinusoids (obtained by Fourier Transform of the signal) divided by N^2[20]. Assuming the sampling frequency of the signal as F_s, the frequency and the amplitude of the ith sinusoid as f_i and A_i respectively, then the power of the surface undulations is given by equation (3) [20]:

$$P = \frac{2}{N^2} \sum_{\forall i} |A_i|^2 \tag{3}$$

where, $f_i = i \times F_s/N$. In the present work, the frequency band is divided into two parts: any frequency ranging between 0.1 cycles/mm to 8 cycles/mm and any frequency ranging between 8 cycles/mm to 20 cycles/mm. Thus the indices $P_1$and P_2 are expressed as given in equation (4) and equation (5):

$$P_1 = \frac{2}{N^2} \sum_{\forall i} |A_i|^2 \text{ for } 0.1 \le f_i \le 8 \tag{4}$$

$$P_2 = \frac{2}{N^2} \sum_{\forall i} |A_i|^2 \text{for } 8 \le f_i \le 20 \tag{5}$$

3. Experimental studies
In the present study, one hundred twenty-five (125) pavement slices were obtained from 17 pavement cores acquired from different parts of India. These cores were extracted from freshly constructed

pavements on which no vehicle had passed, as the purpose of the study was to characterise pavement surface which had not undergone polishing and/ or wearing due to the passage of tyres on them. The gradation of all the core samples were Bituminous Concrete Grading 2, specified in MoRT&H guidelines, Government of India [21]. The following sections describe the sample preparation, the image acquisition and subsequently, the profile detection and the estimation of the indices.

3.1. Sample preparation

Various stages of sample preparation are shown in figure 2. Figure 2(a) shows a representative pavement core sample. To protect aggregates on top of the core from being dislodged during the cutting process, the pavement core surfaces were covered with an epoxy resin mixture. Pavement core samples were cut using a high-speed diamond saw to obtain slices (figure 2(b)).

Figure 2(a). The original pavement core sample.

Figure 2(b). The pavement slice was cut from the core.

3.2. Image acquisition

Images of the side view of these slices were captured (see figure 3) using a digital camera (aperture f/6.3, shutter speed 1/60th of a second and focal length 24mm). A graph paper (with a known dimension of the grid boxes) was placed on the top of the slices for calibration. A white sheet was kept as background (with pixel values close to 255), and care was taken so that the presence of dirt on the white paper (with pixel values close to 0) did not interfere with the profile detection scheme employed subsequently.

3.3. Profile detection and estimation of indices

The following steps were taken to detect the profile:

- Thresholding [22] was applied to the images acquired. The thresholding operation converted all the pixels covered by pavement core into black (pixel value 0) and all the pixels covered by the resin and the white paper background into white (pixel value 255) making it a binary image. Figure 4 shows a representative binary image of the pavement slice after thresholding.

Figure 3. A typical image acquired of the pavement slice.

Figure 4. A typical image after thresholding.

- It was envisaged that by enhancing the contrast between the pavement core and the resin layer, thresholding would work more efficiently and surface profile detection will be unambiguous. Thus, a backlit light source was placed, which illuminated the resin layer and the pavement core. A schematic diagram of the setup is shown in figure 5. With this illumination, the image of the pavement core became darker than it originally was, and the image of resin layer become lighter than it originally was. Hence, the boundary demarcation between the resin layer and the pavement core became sharper than the former boundary.
- The change of the pixel value from 255 (white) to 0 (black) was treated as the boundary (see figure 6). This boundary indicates the pavement profile.
- The actual dimension of the graph paper grid was used for scaling the profiles to their actual lengths (individually for all the 125 slices). The length of each slice was kept uniform (75 mm) for further analysis.

Subsequently, the indices (R_{pm},R_a, $P_1$and P_2) of the 125 profiles were obtained. The definition of these indices has already been discussed in Section 2, titled 'Background'. For calculation of R_{pm} the profile was divided into five equal halves of 15 mm each. For the calculation of $P_1$and P_2, Discrete Fourier Transform was performed on MATLAB®[23].

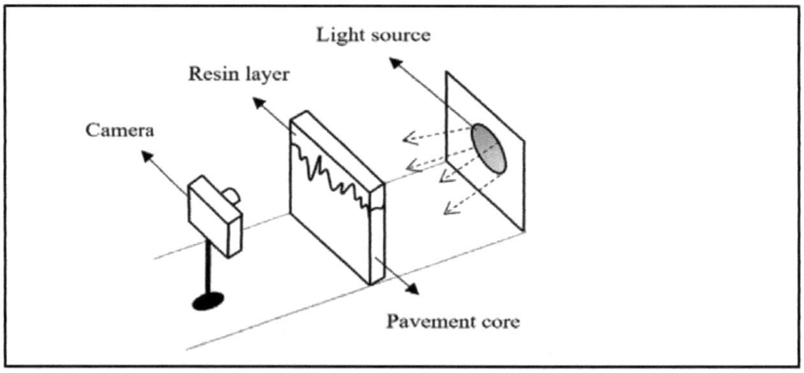

Figure 5. A setup shown for enhancement of contrast between the resin and the pavement core.

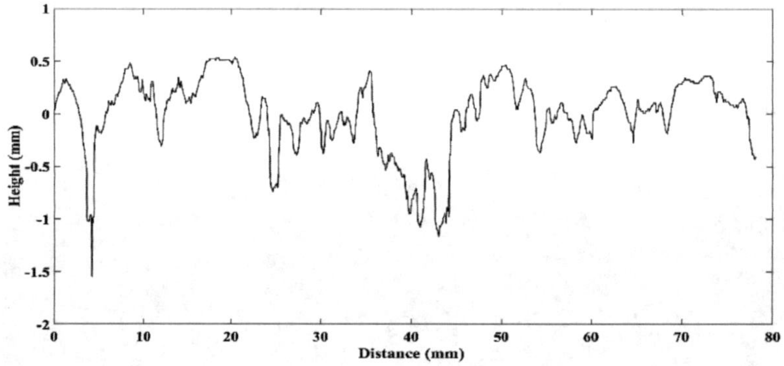

Figure 6. A typical pavement surface profile obtained after image processing.

4. Results

The R_{pm}, R_a, $P_1$and P_2 values are plotted in figures 7-10. As seen in figures 7-10, across the various profiles, the R_{pm} and R_a follow a similar pattern, while $P_1$and P_2 do not follow the pattern of R_{pm} and R_a. The possible reason is explained in the following:

- The actual dimension of the graph paper grid was used for scaling the profiles to their actual lengths (individually for all the 125 slices). The length of each slice was kept uniform (75 mm) for further analysis. Thus, it may be possible that two different pavement surface profiles, having disparate amplitude-frequency distribution of the sinusoidal waves, can have comparable values for the peak heights present in both profiles. Thus, the R_{pm} values may turn out to be same, although the two profiles may be characteristically different.
- Similar reasons can be said for R_a.

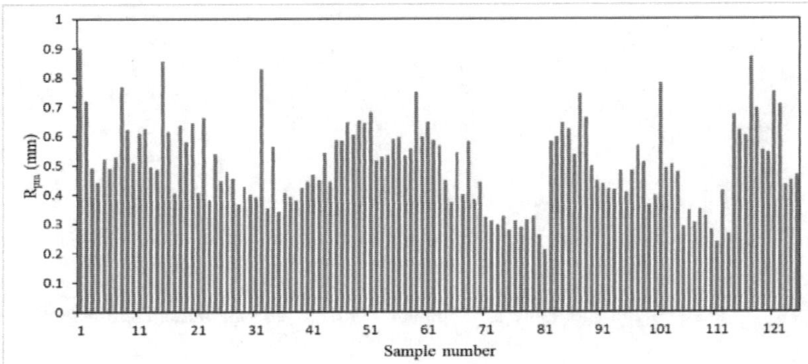

Figure 7. R_{pm} obtained for the various pavement surface profiles.

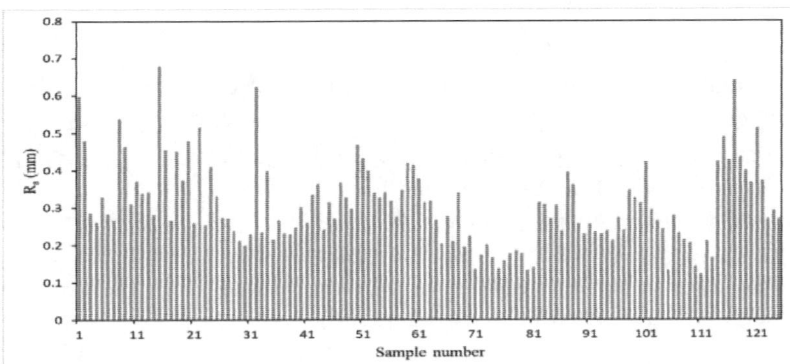

Figure 8. R_a obtained for the various pavement surface profiles.

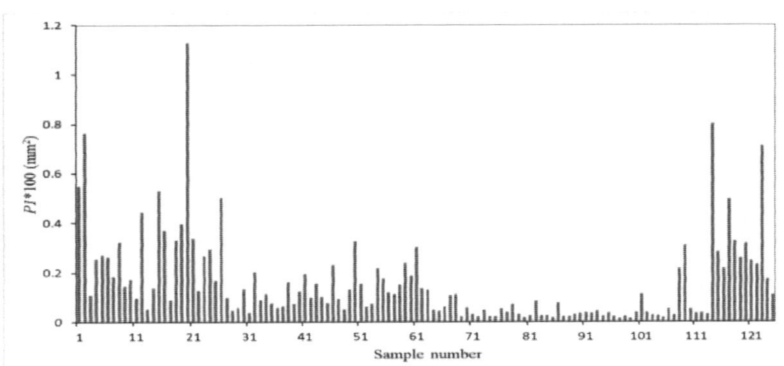

Figure 9. P_1 obtained for the various pavement surface profiles.

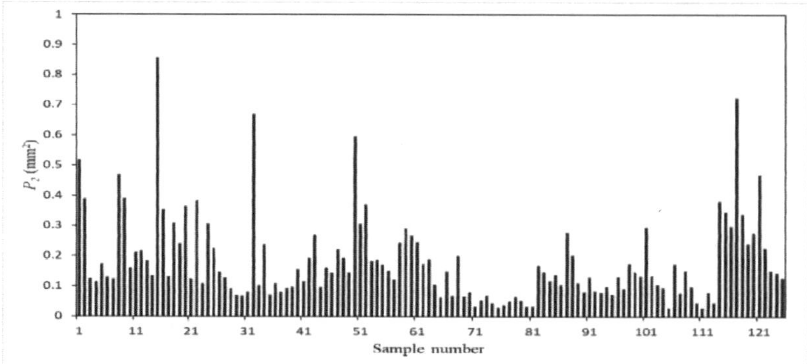

Figure 10. P_2 obtained for the various pavement surface profiles.

5. Discussions

In this study, two spatial domain and two frequency domain based indices have been compared for characterizing pavement surface undulations. The indices compared are the R_{pm}, R_a and the power of the surface undulations in two different frequency bands. It is observed that R_{pm} and R_a vary in a similar pattern, while P_1 and P_2 do not follow their pattern.

In the present method, illumination scheme discussed in Section 3.3 had been adopted. This helped in enhancing the contrast. However, since backlit light-source was used, the surface undulations detected were the highest of elevation value of all possible elevation values of the points from one edge (A-B) of the slice to the other (C-D), as schematically explained through figure 11. Making the slices thinner would reduce this problem. Thus, efforts were made to reduce the thickness of the slices as far as practicable.

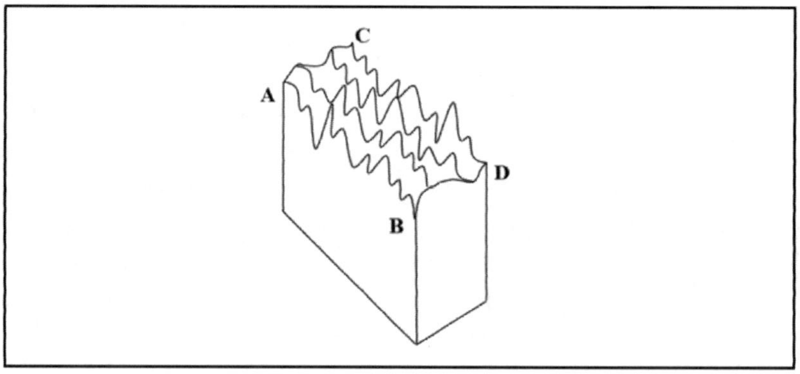

Figure 11. Schematic diagram of the sliced portion.

Acknowledgements

The authors sincerely thank Dr. K S Venkatesh, Professor, Department of Electrical Engineering, IIT Kanpur for the useful discussions. The authors also thank Mr. Shailendra Kumar Tiwary from Oriental Structural Engineers Pvt. Ltd., Mr. Bidur Jha and Mr. Shahu J Patil from LEA Associates South Asia Pvt. Ltd., Mr. Anuj Narula from M/s Teckkonnect, Mr. Sunil Yadav from H.G. Infra Engineering Limited, and Ms. Atasi Das from G R Infraprojects Limited for providing the pavement cores used in this study. The authors wish to thank Mr. Rahul Kumar, Experimental Rock Deformation Laboratory, Department of Earth Sciences, IIT Kanpur for his help in cutting the pavement cores.

References

[1] Hall J W, Smith K L, Titus-Glover L, Wambold J C, Yager T J and Rado Z 2009 *Guide for Pavement Friction* (Washington, DC: National Cooperative Highway Research Program)
[2] Sayers M W and Karamihas S M 1998 *The little book of profiling* (Ann Arbor: University of Michigan)
[3] Tan S A, Fwa T F and Chai K C 2004 Drainage considerations for porous asphalt surface course design *Transp. Res. Rec.* **1868** 142–9
[4] Huschek S 1996 Characterization of pavement surface texture and its influence on tire/road noise *Proc. of the 3rd Int. Symp. on Pavement Surface Characteristics* (Christchurch, New Zealand) pp 301–18
[5] Jain S and Das A 2018 Effect of undulations on pavement friction – a literature review *Int. Conf. on Pavements and Computational Approaches (ICOPAC 2018)* (New Delhi: CSIR-CRRI) pp 189–94
[6] ASTM E965 - 96 2006 Standard test method for measuring pavement macrotexture depth using a volumetric technique (West Conshohocken, Pennsylvania: ASTM International)
[7] ASTM E2380 2009 Standard test method for measuring pavement texture drainage using an outflow meter (West Conshohocken, Pennsylvania: ASTM International)
[8] Bitelli G, Simone A, Girardi F and Lantieri C 2012 Laser scanning on road pavements: a new approach for characterizing surface texture *Sensors* **12** 9110–28
[9] Cigada A, Mancosu F, Manzoni S and Zappa E 2010 Laser-triangulation device for in-line measurement of road texture at medium and high speed *Mech. Syst. Signal Process.* **24** 2225–34
[10] Ahmed M, Haas C T and Haas R 2011 Toward low-cost 3D automatic pavement distress surveying: the close range photogrammetry approach *Can. J. Civ. Eng.* **38** 1301–13
[11] Kogbara R B, Masad E A, Woodward D and Millar P 2018 Relating surface texture parameters from close range photogrammetry to grip-tester pavement friction measurements *Constr. Build. Mater.* **166** 227–40
[12] Smith G T 2002 *Industrial Metrology: Surfaces and Roundness* (London: Springer Verlag)
[13] ASME B46 2009 Surface Texture (Surface Roughness, Waviness, and Lay) (New York: American Society of Mechanical Engineers)
[14] ISO 13473-1 1997 Characterization of pavement texture by use of surface profiles–Part 1: Determination of mean profile depth (Geneva: International Organization for Standardization)
[15] ASTM E1845 2015 Standard practice for calculating pavement macrotexture mean profile depth (West Conshohocken, Pennsylvania: ASTM International)
[16] Gujrati A, Khanal S R, Pastewka L and Jacobs T D B 2018 Combining TEM, AFM, and profilometry for quantitative topography characterization across all scales *ACS Appl. Mater. Interfaces* **10** 29169–78
[17] Nayak P R 1971 Random process model of rough surfaces *J. Lubr. Technol.* **93** 398–407
[18] Hartikainen L, Petry F and Westermann S 2014 Frequency-wise correlation of the power spectral density of asphalt surface roughness and tire wet friction *Wear* **317** 111–9
[19] Mahboob Kanafi M and Juhani Tuononen A 2016 Top topography surface roughness power spectrum for pavement friction evaluation *Tribiology Int.* **107** 240–9
[20] Smith S W 2002 *Digital Signal Processing: A Practical Guide for Engineers and Scientists* (Burlington, Massachusetts: Newnes)
[21] MoRTH 2001 *Specifications for road and bridge works* (New Delhi: The Indian Roads Congress)
[22] MathWorks Image Thresholding, https://in.mathworks.com/discovery/image-thresholding.html, last accessed March 04, 2019.
[23] MathWorks MATLAB , http://www.mathworks.com/products/matlab, last accessed March 16, 2019.

IOP Conf. Series: Materials Science and Engineering 1075 (2021) 012018 doi:10.1088/1757-899X/1075/1/012018

Innovative bitumen technology to reduce impact on air quality

V Guwe[1], J Read[2], R Taylor[2], J Vasudevan[1] and R Hanumanthgari[3]

[1]Shell Malaysia Trading Sdn Bhd., Level 23, Menara Shell, 211, Jalan Tun Sambanthan, 50470, Kuala Lumpur, Malaysia
[2]Shell International Petroleum Company, Shell Centre, York Road, London, SE1 7NA. UK
[3]Shell India Markets Private Limited, Plot No-7, Bangalore Hardware Park, Devanahalli Industrial park, Mahadeva-Kodigehalli, Bangalore - 562 149, India
Corresponding author: Vincent.V.Guwe@shell.com

Abstract. With increasing environmental awareness, air quality is a topic of intense discussion and focus with many environment authorities/agencies setting targets aiming to improve air quality. Reducing the impact on air quality from road construction and maintenance activities plays a part in a holistic approach to improving air quality, particularly in the urban environment. This paper describes an innovative bitumen technology which can help reduce the impact on local air quality from bitumen and asphalt mixtures during asphalt manufacture and pavement installation. It presents laboratory experiments designed to replicate the end-to-end aspects of the bitumen supply chain; starting from refineries and storage depots to paving at road construction sites. The laboratory results have shown substantial reductions in a range of air quality indicators, such as SOx, NOx, particulate emissions, volatile organic content emission as well as in other potential nuisances such as H2S and odour. Full scale field evaluations carried out in major cities have reflected the results obtained in the laboratory and show a reduction in the emission of specific gases and particulates at various points of the asphalt preparation and pavement laying process. This bitumen technology plays a role as part of the holistic solution in improving air quality in the urban environment.

1. Introduction

'Air quality' refers to the condition of the air within our surroundings [1,2,3]. Good air quality is the degree to which the air is clean, clear and free from pollutants such as smoke, dust and smog among other gaseous impurities.

Air quality is determined by assessing a variety of indicators. An Air Quality Index (AQI) is a number used by many government agencies [1,2,3,4,5] to communicate to the public the current quality of the air or how the air quality might change in the future, a little like a weather forecast. Different jurisdictions [2,3,4,5] have their own air quality indices, corresponding to particular air quality standards, but all tend to follow similar principles.

It is known generally that there are a mixture of different compounds of gases i.e. nitrogen oxides, sulphur dioxide, carbon monoxide, particulates, volatile organic compounds (VOCs) etc. which can be released from bitumen and asphalt mixtures at elevated temperatures that may impact air quality in the immediate vicinity of an operating asphalt plant or during paving applications. Generally, the gases from bitumen and its application are normally present at low levels in outdoor air [6,7,8].

Specific gases and particulates from hot mix asphalt (HMA) plants which impact air quality are categorised under two types of emissions: ducted sources and fugitive sources.

The primary sources of specific gases and particulates associated with asphalt production are the dryers, hot bins, and mixers, which emit particulate matter and a variety of gaseous emissions during production [9,10,11,12,13].

Other sources of specific gases and particulates found at asphalt plants and paving applications are shown in figure 1.

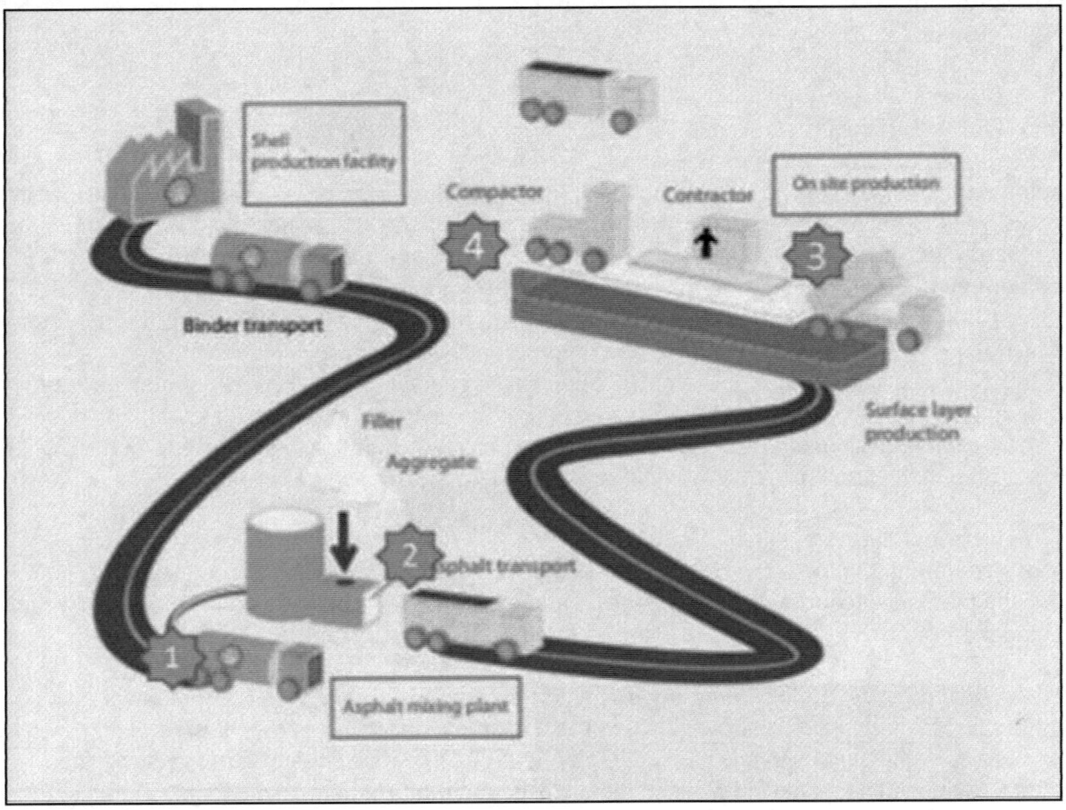

Figure 1. Schematic diagram to indicate the locations to measure gases and particulates:
1: Hot liquid bitumen storage tanks.
2:

- Asphalt storage silos which temporarily store the asphalt;
- Truck load-out operations, in which the loose asphalt mixture is loaded onto trucks for hauling to the job site;
- Hot oil heaters, which are used to heat the asphalt storage tanks;
- Yard emissions, which consist primarily of fugitive emissions
- Most of the other potential specific gases and particulates, such as the dust generated during the drying of aggregate, are captured by baghouse filters or similar controls and not released to the environment.

3: Asphalt in truck beds and the discharge of loose asphalt mixture from truck to paver tipper
4: Asphalt laying and compaction at paving sites.

There are also other sources comprising vehicular traffic generating CO, NO and fugitive dust on paved and unpaved roads while aggregate material handling, and other aggregate processing and drying operations.

Shell investigated the impact of its innovative bitumen technology which helps reduce emissions of specific gases and particulates from bitumen and asphalt mixtures during asphalt manufacture and pavement installation.

2. Materials and characterisation

Shell Bitumen FreshAir consists of an innovative active additive bitumen technology and was developed over a number of years. The mixture of inhibiting components of Shell Bitumen FreshAir act directly with selective compounds which are the source of gases, particulates and odour releasing molecules in bitumen.

2.1. Laboratory testing

Shell's global R&D team conducted an extensive laboratory study which examined a range of twenty bitumens from different geographical regions spanning a wide range of grades from very hard (10/20 penetration grade) to very soft bitumens (160/220 penetration grade) from different refineries. The laboratory experiments were designed to simulate the end-to-end application from bitumen storage, transportation, asphalt mixing (and compaction) and early pavement life to investigate the impact of Shell Bitumen FreshAir in relation to different air quality indicators.

The studies included analysis of specific gases emitted from bitumen during bitumen storage at elevated temperatures (140˚C and180˚C) for a one-month duration.

Specific gases and particulates were measured from asphalt mixing at multiple temperature 140˚C, 160˚C and 180˚C, and early life of the pavement at room temperature and at 60˚C.

Additionally, any potential impact upon bitumen properties and asphalt mixture performance as a result of the Shell Bitumen FreshAir addition was assessed.

2.2. Testing and Analysis

Various testing and analytical methods were used, developed or modified to measure the number of gases and particulates during the study to correspond to various stages of the bitumen supply chain and its application:

2.2.1. Bitumen testing. In the laboratory, to study the bitumen storage effect, bitumen was stored in an aluminium toothpaste tube and placed in a vertical static position in the oven at multiple temperatures over a period of time and specific gases (NO, NH_3, SO_2, CO) were measured in vapour space i.e. 30% headspace (the air left above the content in the closed aluminium toothpaste tube).

Gaseous emissions were measured using different Dräger sampling tubes and systems enabling the identification and measurement of different gases (NO, NH_3, SO_2, CO, NH_3). These tubes are glass vials filled with a chemical reagent that reacts to a specific chemical or family of chemicals. A calibrated 100 ml sample of air was drawn through the tube with a Draeger accuro® pump from the aluminium toothpaste tube. Multiple gaseous tubes with varying detection ranges were used.

Volatile organic content and semi-volatile organic content were measured using an analytical technique that combined separation properties by gas chromatography (GC) with the detection feature of Mass Spectrometry (MS) and Flame Ionization Detector (FID) to identify different substances within a test sample in the headspace i.e. the air left above the contents in a sealed container.

2.2.2. Asphalt preparation and testing. A 30 L asphalt mixer was used to study the emissions from hot asphalt mixtures. The measurements were carried out at three temperatures, viz. 140˚C, 160˚C and 180˚C. These temperatures were selected to cover the normal range of temperatures for asphalt mixtures. For gases like NO_2, NH_3, CO, O_3 and the TVOC, $PM_{2.5}$ and PM_{10} it was not possible to directly measure the emissions since the detectors did not have a sampling pump to withdraw the air sample for analysis. To withdraw samples from the 30 L asphalt mixer the detectors were kept in an enclosure through which a continuous stream from the headspace in the mixer was withdrawn using an air pump, at a flow rate of 2.0 L/min. Specific gases (NO, NO_2, NH_3, SO_2,) and particulate matter

($PM_{2.5}$, PM_{10}) were measured in a 30 L mixing unit with special modification to the mixer in order to control the dilution of emissions while mixing in the mixing unit. The measurements were carried out at multiple mixing temperatures (140˚C, 160˚C and 180˚C) over a period of 20 minutes of mixing.

Specific gases (NO, NO_2, NH_3, SO_2) and particulate matter ($PM_{2.5}$, PM_{10}) were monitored at room temperature and elevated temperatures on the prepared slab at 60˚C. The asphalt slab was placed in a plastic enclosure containing both gaseous emission and PM detectors. The sample within the plastic enclosure as shown in figure 2 was then kept for monitoring in an environmental chamber for a duration of one month.

Table 1. Devices used to monitor specific gases and particulates in bitumen and asphalt while mixing and monitoring in environmental chamber

Specific gases and particulates	Detector used
SO_2, NO, CO, NH_3 & H_2S in Bitumen	Dräger sampling tubes and system
VOC	Agilent GC/MS-FID
SO_2	Honeywell, BW GasAlertMicro 5
NO_2	Honeywell, BW GasAlert NO_2 EXTREME
NO	Honeywell, BW GasAlert NO EXTREME
NH_3	Honeywell, BW GasAlert NH_3 EXTREME
$PM_{2.5}$ and PM_{10}	TSI Inc. DustTrak DRX Aerosol Monitor 8534

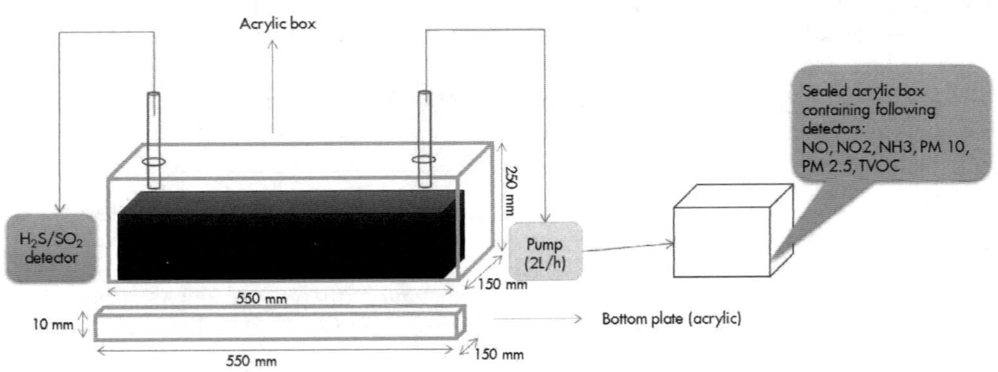

Figure 2. Schematic of experimental setup used to measure the emissions from an installed asphalt pavement immediately after compaction

2.3. Field Trials

To validate the findings in the laboratory study, a number of externally monitored field trials were carried out to assess the efficacy of Shell Bitumen FreshAir during real life asphalt production and paving and to provide a statistical validation of the laboratory results.

To minimise field variations as much as possible all field measurements were monitored by an independent environmental consultancy specialising in dust and air quality monitoring and consultancy.

Trials were carried out using Shell Bitumen FreshAir in cities around the world – London (United Kingdom), Saraburi (Thailand), Lelystad (Netherlands), and Nantes (France). The results were analysed to better understand how this solution can help road industry work towards cleaner asphalt production and paving.

2.3.1. Field trials and measurements. The field trial assessments and monitoring guidelines were selected and modified from the US Environmental Protection Agency (US EPA), European Environmental Agency (EEA), and UK Environmental agency to suit the trials that were carried out.

The external agency's equipment (calibrated and certified portable devices) and test methods selected were used at all trial sites to minimise experimental set up variations. The measurements were carried out both upwind and downwind of the various measurement locations. All weather conditions were recorded using a portable weather station for all the trials. The asphalt mixtures used at the different trials had different mixture designs, aggregates and contained varying quantities (0-40%) of recycled asphalt pavement (RAP).

At the asphalt plant: measurements were taken at the mixing unit and truck loading silos and at paving site measurements were taken during unloading of the asphalt mixture into the hopper of the paver, at the screed of the paver and behind the paver, during and post compaction. The results provided are an average of these measurements.

3. Results

The studies performed in the laboratory indicate that the amount of gaseous and particulates generated during the bitumen usage from end-to-end application varies, depending on the source of the bitumen, the grade of bitumen and the processes used to make the bitumen.

3.1. Effect of Shell Bitumen FreshAir on gaseous emissions from bitumen

The monitoring of Shell Bitumen FreshAir samples resulted in a reduction in the concentration of SO_2 and H_2S measured in the headspace of the toothpaste tubes after 7 and 15 days storage at different temperatures (140˚C - 180°C). The studies performed in the laboratory indicate that the amount of gases generated during the bitumen storage varies, depending on the source of the bitumen, the grade of bitumen and the processes used to make the bitumen. The data of the 20 bitumens tested was given to statistical analysis experts and the conclusions are provided in this paper. The reduction in the SO_2 and H_2S concentrations average a reduction of 25% and were found to be independent of the application temperature (140˚C - 180°C), see Tables 2 and 3.

Table 2. Effect of Shell Bitumen FreshAir on the concentration of SO_2 in the headspace of the toothpaste tube – 95% confidence intervals

Time of storage (days)	Mean reduction measured (%)	Lower bound of reduction measured (%)	Upper bound of reduction measured (%)
7	25.8	11.1	40.5
15	26.9	15.9	37.9

Table 3. Effect of Shell Bitumen FreshAir on the concentration of H_2S in the headspace of the toothpaste tube – 95% confidence interval

Time of storage (days)	Mean reduction measured (%)	Lower bound of reduction measured (%)	Upper bound of reduction measured (%)
7	26.9	14.8	39.0
15	30.3	21.4	39.2

NH_3, CO and NO were detected at very low levels for a few of the bitumens and some were below detection levels (BDL) during the measurements.

3.2. Effect of Shell Bitumen FreshAir on emissions from hot asphalt mixture during laboratory mixing

It was observed that NH_3, NO_2, CO and O_3 were not detected in significant concentrations in the 30 L mixer over a time of 20 minutes and at all temperatures tested. Air quality indicators, SO_2 and NO were detected in the 30 L mixer while mixing the asphalt mixture. Apart from this H_2S was also detected, and whilst this is important it is not an air quality indicator. Significant reductions were seen with respect to the concentrations of SO_2, NO and H_2S, irrespective of temperature. Without a rigorous statistical analysis of the results it could be said that the Shell Bitumen FreshAir resulted in a decrease in emissions during mixing compared with conventional bitumen, which could improve air quality indicators.

The analysis of the data with respect to SO_2 showed a minimum mean reduction of 62% and NO of 28%, see Table 4. The temperature of the asphalt mixture was seen to have no statistical influence on the reduction in SO_2 and NO observed.

Data collected on the $PM_{2.5}$ and PM_{10} was not used as stable data could not be obtained, potentially as a consequence of the experimental design. It is recommended that field data be used to for any the effectiveness of Shell Bitumen FreshAir the reduction in $PM_{2.5}$ and PM_{10} levels.

Table 4. Reduction of SO_2 and NO observed from asphalt mixtures – 95% confidence interval

Gaseous Emission	Mean reduction measured (%)	Lower bound of reduction measured (%)	Upper bound of reduction measured (%)
SO_2	90.0	62.2	117.8
NO	57.5	28.5	86.3

3.3. Effect of Shell Bitumen FreshAir on emissions from hot asphalt pavements and during early life

Measurements of the emissions from the hot asphalt slabs showed that the emissions, if any, were below the detection limit of the detectors used. The detectors used typically had a lower detection limit of 1 ppm. This would indicate that emissions, if any, were extremely small. Emissions from slabs when maintained at 60℃ showed a similar trend, i.e. that emissions, if any, were too low to result in their detection during the course of the experiments.

After the first four asphalts were tested at all three mixing temperatures with no detection of emissions, further testing was discontinued.

3.4. Field Measurements

The reduction of specific gases and particulates are presented in table 5 which are given on an average basis over time, although minimum and maximum reductions were also recorded during the field measurements. The measurements were taken at multiple points at each paving site (unloading of asphalt mixture into paver hopper, at the screed, behind paver and before and post compaction. The results presented are an average of all measurement points and compared to conventional bitumen.

In the asphalt plant, during production i.e. near the mixing unit and during loading of loose asphalt mixture into trucks, Shell Bitumen FreshAir has shown, on average reduction of gases by 40% along with potential nuisance vapours such as H_2S and odour causing compounds associated with bitumen and asphalt mixtures.

In the field gaseous emissions (SO_x, NO, NO_2, VOC, CO) and PM were reduced on average by 40% compared to conventional bitumen during paving application. It was observed that the emissions reduce very quickly and disperse in the atmosphere within the first hour of paving application.

Field measurements for particulate matter in the Netherlands at the paving site were anomalous and need further research as well as in London where the measurements were impacted by contamination of some of the emissions (NO and VOC) at the paving site, during which different pavers, rollers/compacter were operated by the crew on conventional bitumen and Shell Bitumen FreshAir. During the field measurements there are many external factors which can impact the outcome of the results.

Table 5. Overall % reduction of specific gaseous and particulate emissions from field measurements

Location	SO_2	NO	NO_2	CO	VOC	$PM_{2.5}$	PM_{10}
Nantes, France	-50	-54	-41	-81	-33	-28	-28
London, UK	-99	70	-67	-90	33	-88	-90
Saraburi, Thailand	-19	-84	BDL*	-67	-41	-47	-56
Lelystad, Netherland	-50	-25	BDL*	-50	-58	NA	NA

*BDL: Below Detection Level

4. Conclusions

The outcome of the laboratory findings corresponded well with field measurements outcomes in terms of measuring a reduction in specific gases and particulates when mixing and paving asphalt.

There is a significant, 40% on average, reduction between the concentrations of emissions in terms of SO_2, NO, NO_2, CO, PM and total volatile organic compounds measured from loose mixture (e.g., in a truck, when discharged into the paver hopper) and the road surface, immediately during and after compaction, when using Shell Bitumen FreshAir.

It was observed the emissions reduce very quickly and disperse in the atmosphere within the first hour of paving application.

With the field measurements, it is clear that considerable care needs to be taken to account for external parameters which have the propensity to alter the outcome of the field results e.g. type and kind of fuel used, paving equipment, handling of paving instruments, traffic, vehicle movement, location and surroundings of the paving application, weather, particularly wind, along with mixture designs, asphalt type and temperatures.

From laboratory data and field measurements, Shell Bitumen FreshAir showed significant reduction of specific gases (NO_x, SO_x, CO, VOC) and particulates as well as in other potential nuisance vapours such as H_2S and odour causing compounds associated with bitumen storage, transport, asphalt mixing and compaction.

Shell Bitumen FreshAir is an innovative bitumen technology which can help reduces the impact to local air quality from bitumen and asphalt mixtures during asphalt manufacture and pavement installation and plays a role as part of the holistic solution in improving air quality in the urban environment.

References

[1] World Health Organisation, http://www.who.int/airpollution/en/
[2] United States Environmental Protection Agency, https://www.epa.gov/environmental-topics/air-topics
[3] European Environment Agency, https://www.eea.europa.eu/themes/air
[4] Environnent Agency, UK. https://www.gov.uk/government/organisations/environment-agency
[5] Environmental Protection Department, The Government of the Hongkong. https://www.epd.gov.hk/epd/english/environmentinhk/air/air_quality_objectives/air_quality_objectives.html
[6] Kuklinska K, Wolska L and Namiesnik J 2015 Air quality policy in the U.S. and the EU – a review *Atmospheric Pollution Research* **6** 129-137
[7] United States Environmental Protection Agency 2002 National emission standards for hazardous air pollutants: Revision of source category list under Section 112 of the Clean Air Act *Federal Register* **67** 6521–36 http://www.gpo.gov/fdsys/pkg/FR-2002-02-12/pdf/02-3348.pdf
[8] Jin Y, Andersson H and Zhang S 2016 Air pollution control policies in China: A retrospective and prospects *Int. J Environ. Res. Public Health* **13** 1219
[9] Connolly U 2001 Clearing the air *Hot Mix Asphalt Technology* **6** 21-2

http://www.flexiblepavements.org/sites/www.flexiblepavements.org/files/clean_air_2_pg_article.pdf

[10] United States Environmental Protection Agency 2000 Hot mix asphalt plants - emission assessment report EPA 454/R-00-019 https://www3.epa.gov/ttn/chief/ap42/ch11/related/ea-report.pdf

[11] United States Environmental Protection Agency 2004 Emission Factor Documentation for AP-42 Section 11.1 Hot Mix Asphalt Plants Final Report RTI Contract No. AGMT DTD 10/31/02

[12] AQEG 2005 Particulate matter in the UK: Summary Defra London https://uk-air.defra.gov.uk/assets/documents/reports/aqeg/pm-summary.pdf

[13] Hodan W M and Barnard W R 2004 Evaluating the contribution of PM2.5 precursor gases and re-entrained road emissions to mobile source PM2.5 particulate matter emissions https://www3.epa.gov/ttnchie1/conference/ei13/mobile/hodan.pdf

IOP Conf. Series: Materials Science and Engineering 1075 (2021) 012019 doi:10.1088/1757-899X/1075/1/012019

Application of asphalt rubber cape seal as asphalt pavement surface course for rehabilitated trunk highway in east Gansu

L Han[1,2], H Han[3], H Guo[1] and J Zhu[4]

[1] Highway and Railway Engineering School, Shaanxi College of Communication Technology, Xi'an, Shaanxi, China, 710018;
[2] Key Laboratory for Special Area Highway Engineering of Ministry of Education, Chang'an University, Xi'an, Shaanxi, China, 710064;
[3] HuaBang Construction & Investment Group Co., Ltd., No. 408 East Yanjiang Road, Guangzhou, Guangdong, China, 510030;
[4] Gansu Road and Bridge Construction Group Co., LTD, Lanzhou, Gansu, China, 730000
[*] Corresponding author: chdhanlili@yahoo.com

Abstract. Although been frequently used as a preventative maintenance measure for asphalt pavements, the Cape seal process has not been widely adopted as a possible alternative for rehabilitated pavement. This research evaluated the feasibility of applying asphalt rubber Cape seal structure into the surface course of reconstructed highway in west China. The Cape seal is composed of 11-16 mm chips spread over asphalt rubber membrane sprayed in advance, by which followed a micro surfacing layer with SBR latex modified asphalt emulsion. Based on a test section in the Banxiao highway project, the construction of asphalt rubber Cape seal and micro-surfacing were investigated experimentally. In addition, to guide the engineers, major issues during construction of the Cape seal were discussed and corresponding countermeasures were suggested.

1. Introduction

Cape seal is a kind of anti-skid and water proof compound sealing technology which is composed of chip seal and slurry seal or micro-surfacing[1,2,3]. Originally invented in South Africa in the 1950s, preliminary cape seal technology was realized by spreading stone chippings on coarse aggregate chips which was used mainly in newly built roads with day traffic volume less than 300 vehicles[4]. The currently prevalent Cape seal technology was improved in 1957 and widely used in Australia and North America thereafter[3-4].

Nowadays, with the rapid development of asphalt rubber and emulsified asphalt technology, Cape seal has evolved into the combination of asphalt rubber chip seal and micro-surfacing using modified asphalt emulsion[3-4]. Such combination can, on the one hand, improve the anti reflective property of the entire sealing, and on the other, ensure adequate anti skid performance and surface evenness.

Firstly introduced in China in the year 2004, Cape seal was applied into pavements of several urban streets and rural roads. For example, in 2005 one city street in Chongqing adopted 25mm thick Cape seal as pavement surface course over semi rigid base. Results indicated this kind of pavement was comparable to dense grade asphalt concrete pavement[3].

In 2011, cape seal structure using asphalt rubber chip seal and micro-surfacing mixture with SBR latex modified asphalt emulsion was built in a city street in Xi'an[3]. Additionally, based on local climate and traffic condition, researchers in Xinjiang also investigated the impact of gradation on chip seal performance, and mixture design method of micro-surfacing. However, the performance of cape seal is highly related to its construction quality. Poor chip seal application and a lack of inter layer bonding may lead to pavement distresses such as unevenness, bleeding, shoving and insufficient skid resistance. Up to date, application of this technology in rehabilitated highways was barely reported. Therefore, more investigations are needed to better improve this compound sealing technology.

In this view, this research aims to evaluate the feasibility of applying multi-layer Cape seal structure into asphalt pavement of rehabilitated highways. Used as the surface course, a three layer Cape seal system was developed and one test section located in a reconstructed highway project in East Gansu was built.

2. Raw materials and mixture design

2.1. Raw materials

2.1.1. Asphalt binder Cape seal

Asphalt binder Cape seal is a compound seal structure that is primarily composed of one underlying chip seal layer and an above micro-surfacing layer. Asphalt binders in a Cape seal system work in two ways. One is to be used as chip seal binder. Such binder is frequently sprayed by special asphalt distributor prior to application of chips to provide reliable bottom adhesion for the chips. Commonly used binders include asphalt rubber, SBS modified binder as well as asphalt emulsion. The other way is to take effect in the micro-surfacing mixture as a mixture binder.

In this research, the high viscosity asphalt rubber supplied by a factory in Gansu was selected as binder for the chip seal layer. Based on the climate in Qingyang district, the SK 70$^{\#}$ paving grade virgin asphalt and 20% （by weight of virgin asphalt） 40 mesh crumb rubber powder from Hubei were chosen in preparing asphalt rubber. To ensure sufficient swelling of crumb rubber powder in virgin asphalt, constant stirring to the rubber-asphalt blend at 180℃ was highlighted, which should be lasted more than 2 hours. The technical properties of asphalt rubber were tested and are shown in table 1.

Table 1. Properties of the asphalt rubber used in chip seal.

Technical property	Unit	Test value	Standard value[5]	Test method[6]
penetration（25℃,100g,5s）	dmm	46	40-60	T0604-2011
5℃ ductility	cm	11	≥10	T0605-2011
Softening point	℃	68	≥58	T0606-2011
Elastic recovery（25℃,10cm）	%	76	≥60	T0662-2011
180℃ Brookfield viscosity	Pa.s	3.510	⩾3.5	T0625-2011

The SBR latex modified asphalt emulsion was selected as the binder for micro surfacing slurry mixture, which was in situ produced. Its technical indicators are shown in table 2.

Table 2. Properties of the SBR latex modified asphalt emulsion.

Technical property		Unit	Test value	Standard value [7]
Percentage retained, 1.18mm		%	0.04	≤0.1
Engler viscosity E25		/	5.10	3-30
Storage stability，1d		%	0.8	≤1
Residue after evaporation	content	%	65.2	≥60.0
	penetration	0.1mm	68	40-100
	5℃ductility	cm	35	≥20.0
	Softening point	℃	66	≥53.0

2.1.2. Mineral aggregates.

First is chips commonly adopted chip size covers 4～6,6～10, and 10～14 mm, which could be carefully selected according to the traffic volume pavement confronted and requirements on pavement anti-skid performance. To adapt the traffic of Qingyang area as one of the most important crude oil transport corridors in China, the 11～16 mm single size chips quarried locally were chosen. Its gradation is shown in figure 1. Uniform single size chips have adequate void, which leave enough leeway for asphalt rubber retention to ensure reliable adhesion between chips and asphalt membrane. Aside from that, single size chips could guarantee considerable surface texture which is indispensable for good anti skidding performance. In addition to gradation, other physical properties of the chips were also tested. e.g. bulk density 2700 kg/m^3, crushed stone value 19.5%, elongated and flaky particle content 4.9%, adhesion grade to asphalt Ⅳ.

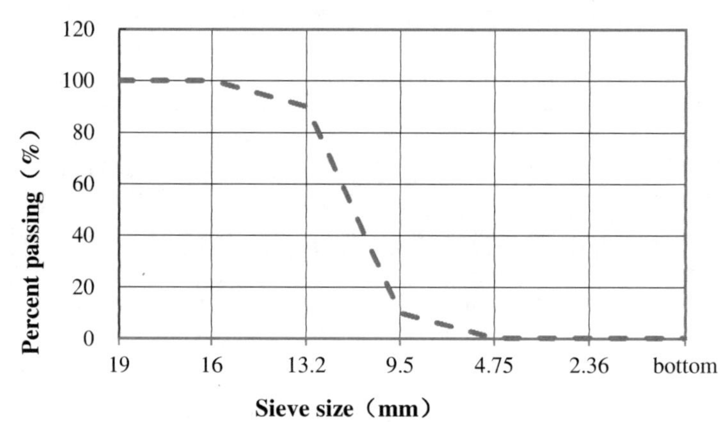

Figure 1. Gradation of the chips used.

Secondly is aggregates for micro surfacing mixture. The frequently used MS-3 type gradation was chosen for micro surfacing mixture, which was obtained by blending three different single size aggregates together (0～3mm, 3～5mm and 5～10mm) as shown in figure 2. According to the laboratory test, the ultimate blending proportion was 0～3mm：3～5mm：5～10mm=47:27:26. Since sand equivalent of particles finer than 4.75mm plays an important role in mixture performance, it should be carefully tested and strictly controlled. In this research, this index for 0～3mm machine made sand was tested as 67%, greater than the minimum value 65% in the code.

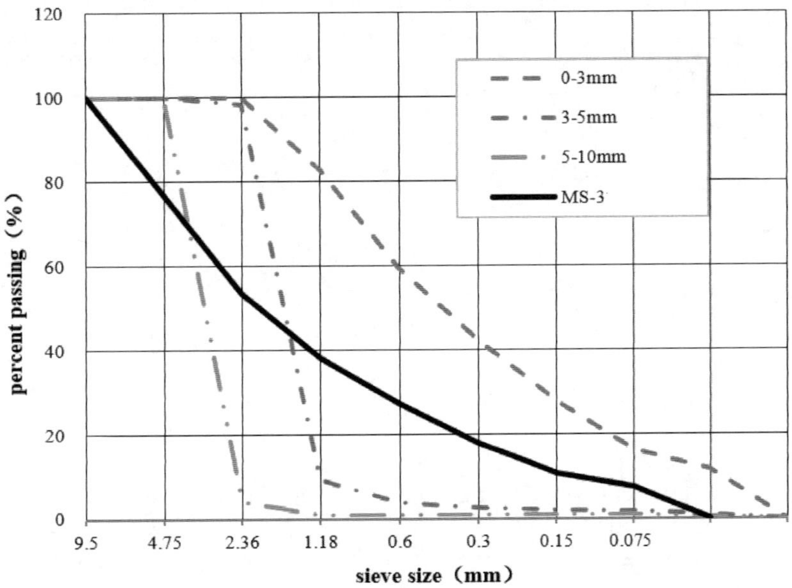

Figure 2. Gradation of aggregates for micro surfacing mixture.

2.1.3. Other materials.
The P.O. 42.5 grade cement with an application rate of 1% by aggregate weight was adopted as the filler for micro surfacing mixture to adjust the mixing time, consistency as well as productivity. Additionally, water free from soluble salts and contaminants was transported from local rivers to meet the mixing procedure. The water content was determined by the mixing time test.

2.2. Proportion design and properties of micro surfacing mixture
In the technical guide, micro surfacing mixture can be divided into two types MS-2 and MS-3 according to its nominal maximum particle size. Featuring a nominal maximum particle size 9.5 mm, the latter is applicable to rutting filling, usually 8～10 mm thick and with the typical asphalt-aggregate ratio and cement content of 6.0%～8.5% and 0～3% respectively.

The test section is located in Qingcheng county of east Gansu which is one of the main producing areas of Changqing oilfield. The test section was built in the Banqiao to Xiaoyuanzi second class trunk highway. To meet the demand for heavy traffic passage such as crude oil transport and oil production equipment transferring, the much stricter MS-3 mixture gradation was selected for the cape seal test section in this research.

2.2.1. MS-3 gradation.
The gradation curve of MS-3 used was designed as shown in figure 2, which meets the code requirement.

2.2.2. Mixture properties.
Based on the chosen aggregate gradation, the content of asphalt emulsion, cement and water were preliminarily selected. Then mixtures were prepared and the mixing time test, wheel track abrasion test as well as loading wheel test were performed on various mixture samples to determine the final material proportion. Finally, the material proportion was determined as: content of SBR modified asphalt emulsion, cement and water was 9.5%, 1% and 4% respectively. Their technical indices were tested (as shown in table 3) and verified to ensure they meet the code standard.

Table 3. Technical properties of the micro surfacing mixture.

Properties		**Unit**	**Test value**	**Standard value** [7]
Mixing time，25℃		s	126,moderate consistency	≥120
Wet track abrasion test	water soaking 1h	g/m^2	341	≤540
	water soaking 6d		775	≤800
Load wheel test,25℃		g/m^2	388	≤450
Stability and resistance to compaction test,25℃		%	4.4	≤5

3. Multi-layer Cape seal structure

Traditional two-layer Cape seal includes the chip seal bottom layer that could absorb pavement stress a great deal and hence mitigate its impact on surface layer, and slurry seal or micro-surfacing layer above which ensures good evenness, anti-skid performance and imperviousness of pavement. The depth of two-layer Cape seal ranges from 2.5～3cm, depending on the size of chips.

In this research, the three layer Cape seal (asphalt rubber chip seal + asphalt rubber chip seal + micro-surfacing layer) structure was developed and used as the surface course of asphalt pavement in the test section. Construction of the three-layer cape seal was undertaken after the binder course was finished. According to the design document, original pavement structure was two–layer system, involving 4cm thick dense grade asphalt concrete AC -13C and 6cm dense grade asphalt concrete AC -20C. It should be noted that the three-layer Cape seal structure is expected to better resolve the issue of pavement thickness transition from test section to normal sections. Moreover, such structure reserved greater stress-absorbing capability for pavement confronted with overloading vehicles.

4. Cape seal construction technique

Performance of Cape seal is highly dependent on its construction quality. Main steps for building Cape seal pavement include pre-construction preparation, application of asphalt rubber chip seal as well as construction of micro-surfacing mixture. Based on the Banxiao project, one 100m long multi-layer Cape seal test section was built to investigate the building technique. The pavement width of the test section was 8.5m.

Prior to the application of chip seal, binder course that had been accomplished should be cleaned thoroughly. To ensure reliable intercourse bonding, asphalt emulsion tact coat should be sprayed with an application rate of 0.9～1.1kg/m^2.

4.1. Application of asphalt rubber chip seal

4.1.1. Spraying asphalt rubber. During the spraying process, asphalt temperature should be strictly controlled within 190℃～200℃. The application rate of asphalt rubber was 1.5～1.6kg/m^2 and the moving speed of asphalt distributor was controlled within 5～10km/h to ensure smooth and even spraying.

4.1.2. Spreading chips. Asphalt spray was followed closely by the spreading of chips to avoid impairing the chip-asphalt bonding. The chips were pre-heated to above 50℃ and the spreading rate was determined as 18～19 kg/m^2, resulting in a coverage rate of no less than 80%.

Compaction to the chip seal layer started from the very beginning of the chip spreading work. Primary compaction was implemented 3 times using two pneumatic tired rollers with the travelling speed of 2～3km/h. Subsequently, the final compaction was accomplished using a 5～8t steel wheel roller. 1 to 2 hours after compaction, the chip seal layer was swept and cleaned to remove swimming stone.

Right after the spreading, compaction and sweeping for the first asphalt rubber chip seal, the second chip seal layer was carried out as per similar application craft as shown in figure 3 a）.

4.2. Laying micro-surfacing mixture

The micro-surfacing layer must be laid at least 2 days after the bottom chip seal was finished to guarantee that most of the chips be firmly bonded to the asphalt rubber membrane(as shown in figure 3 b）). Once the laboratory proportion of micro-surfacing mixture was determined, calibration to the slurry seal machine was necessary. Several machine parameters such as opening degree of aggregate bin, asphalt pumpage et. c could be fixed accordingly.

Prior to the laying process, the lower layer must be fully cleaned and wetted. Unevenness at the longitudinal joints should be no more than 6mm. Besides, evident scratching resulted from oversize particles should be controlled.

(a) (b)

Figure 3. Construction of Cape seal test section in Banxiao project: (a) chip seal layer; (b) micro-surfacing layer.

4.3. Quality control and inspection

Asphalt rubber application rate for chips seal, consistency and asphalt-aggregate ratio of the micro-surfacing mixture are three key points in the quality control of cape seal construction. In case of too high asphalt rubber application rate, bleeding would be prone to occur during hot days. Whereas in terms of too low application rate, the chip-asphalt bonding would be weakened and flying stone could frequently be observed.

To control the application rate of asphalt rubber, the authors adopted both the mass-based-control method and field measurement method. While in terms of the consistency control for micro-surfacing mixture, we used the scratching method combined with the sunlight illumination test. During the laying of the mixture, the asphalt -aggregate ratio was monitored and adjusted by shifting the opening degree of aggregate bin, asphalt emulsion pumpage as well as water pump pressure with its relative error less than ±0.3%.

5. Major issues and countermeasures

Upon the completion of asphalt rubber chip seal and micro-surfacing, inspections were performed on various layers regarding appearance quality, surface smoothness and so on. The inspection indicated that the test section was well constructed in general. However, subject to relative cold climate when the test section was built and other factors such as lack of machinery, technical staff et c, some major issues still need to be addressed as listed below.

IOP Conf. Series: Materials Science and Engineering 1075 (2021) 012019 doi:10.1088/1757-899X/1075/1/012019

(1) Uneven spreading of chips was identified during the application of asphalt rubber chip seal, especially at horizontally curved section. So manual re-spreading and levelling was indispensable in such case.
(2) The alignment of Cape seal border was not smooth enough, hence the Cape seal border controlling stakes are recommended to be densified to better guide the moving of synchronous chip sealer and slurry seal machine.
(3) Partial spalling was observed after one week curing to the micro-surfacing layer, which maybe resulted from poor intercourse bonding between chip seal and micro-surfacing layer, low asphalt–aggregate ratio of micro-surfacing mixture. This issue could be better addressed by appropriately increasing tact coat application rate and asphalt–aggregate ratio.

References

[1] Mansour S and Thomas W 1998 *Evaluation of the cape seal process as a pavement rehabilitation alternative* Center For Transportation Research Bureau of Engineering Research, University of Texas At Austin.
[2] Li C, He Z and Yan D 2007 Application of cape seal in rural highways Highway **12** 129-31.
[3] Chen F 2009 *The research on application of cape seal at county road in Chongqing* Chongqing Jiaotong University master thesis.
[4] Jack V K and Shu M 2012 *Application of recycled tyre rubber modified asphalt cape seal technology* Proceeding for 2012 China Highway Maintenance Congress 55-61.
[5] Ouyang H and Han J 2010 *Technical guide for chip seal construction in Shaanxi Province* Beijing.
[6] Ministry of Transport, China 2011 *Standard test methods of bitumen and bituminous mixtures for highway engineering (JTG E20-2011)* Beijing.
[7] Highway Research Institute, Ministry of Transport, China 2005 Technical guide for micro-*surfacing and slurry seal (JTG/T F40-02-2005)* Beijing.

Assessment of RCA with RAP materials for pavement applications

S Jayakody[1, *], C Gallage[2] and J Ramanujam[2]

[1]Department of Civil Engineering, South Eastern University of Sri Lanka, Oluvil, Sri Lanka
[2]School of Earth, Environment and Biological Sciences, Science and Engineering Faculty, Queensland University of Technology, Brisbane, Australia
[*] Corresponding author: jshiran@gmail.com

Abstract. Demolished concretes are being recycled to produce recycled concrete aggregates (RCA) in different scales around the world. Therefore, applications of RCA as a construction material are being popular by replacing conventional materials. Many research studies have confirmed the potential use of RCA to be applied as an unbound pavement material. However, presence of constituents such as reclaimed asphalt pavement (RAP) in RCA effects on its native properties. Therefore, this study evaluates performance characteristics of RCA with RAP under dynamic loading condition. A series of "repeated load tri-axial (RLT)" test was conducted on RAP blended RCA samples to evaluate the elastic and plastic deformation characteristics with increase of load cycles. The elastic deformation was characterised by resilient modulus and slightly dropped with increase of RAP from 0, 5, 10 to 15% in RCA. Further, it observed a trend of small increase of the plastic deformation of the RCA with the increase of RAP portion. However, presence of RAP up to about 15% in RCA did not significantly affect on the accumulation of permanent strain.

1. Introduction

Recycling and reusing of C&D waste creates many economic and environmental benefits while enhance the sustainability in construction material industry. Concrete is the major component of total among the various types of C&D waste which is greater than 80% in Australia [1]. Therefore, recycling the concrete to produce the crushed aggregates is one of the sustainable outcomes of waste management. Utilization of recycled concrete aggregates in granular layers in flexible pavements is a viable alternative since road are typically demanded a huge volume of crushed aggregates. The reclaimed asphalt pavements (RAP) is one of the major constituents in RCA in Australia and its influence on the properties of RCA has not been completely investigated. The bitumen in the RAP is sensitive on the temperature, thus could be made an impact on physical and mechanical properties of RCA. Therefore, Road authorities do not rely on consistent properties of RCA due to insufficient technical information of the effect of RAP on RCA.

Investigations on the strength properties of RCA have revealed feasibility of RCA to be used as unbound pavement materials under different conditions [2, 3]. However, Investigations on the RCA and RAP together are very limited. A comprehensive study was conducted by [4] to investigate the effect of RAP on RCA to be used as unbound pavement material in subbase layers. The results introduced to allow only up to 15% of RAP to be mixed in RCA to assure the required performance. The study was further emphasised not to use 100% of RAP as subbase course material. The results of this study were confirmed by the findings of [5]. A detailed study on RCA and RAP was conducted by [6] to investigate the confined strength and unconfined strength of RAP blended RCA specimens and revealed a gradual decrease of strength in RCA when increase the RAP portion.

Performance characteristics of pavement materials which are the elastic and plastic deformation need to be investigated under the dynamic load conditions prior to their applications. RCA has been subjected to such experiments and revealed comparable performance characteristics. The research findings have revealed that the physical properties of RCA such as particle size distribution, degree of compaction, degree of saturation and presence of constituents have a greater influence on the elastic and plastic deformation of RCA [3, 7-9].

RCA itself has recorded inconsistent properties as a construction material due to the presence of cement fines, different qualities of parent aggregates, presence of minor constituents such as bricks, glass etc., in different regions [9-11]. In addition, RAP also presence in RCA as a major constituent which mixed with RCA at the recycling process. Therefore, investigations are needed to determine the effect of RAP as a constituent on the properties of RCA and, limited past studies are recorded [4, 12, 13]. This study was focused to investigate the influence of RAP on the performance characteristics of RCA as a granular pavement material. The study was planned to conduct a series of repeated load tri-axial (RLT) test to analyse the elastic and plastic deformation characteristics under dynamic load condition. The elastic response of the specimens was analysed by the behaviour of resilient modulus and plastic response was characterised by the accumulation of vertical strain of the RAP mixed RCA specimens with the number of load cycles.

2. Materials

Recycled concrete aggregates (RCA) and reclaimed asphalt pavements (RAP) were obtained from a concrete recycling plant in Queensland, Australia. Figure 1 (a) and (b) show pictures of the samples of RCA and RAP respectively.

(a) (b)

Figure 1. (a) Recycled concrete aggregates-RCA (Scale 1:5) (b) Reclaimed asphalt pavement-RAP (Scale 1:3.2)

Physical properties of the RCA and RAP were extracted from the [6] since same materials were subjected to performance test in this study. Figure 2 shows the Particle size distribution curves of RAP and RCA materials with upper and lower limit of the standard base layer materials. Physical properties of RCA and RAP are shown in Table 1 and Table 2 respectively and the values present as reference for the purpose of comparison in similar studies in future. The descriptions of these properties have been presented in the previous publication of [6].

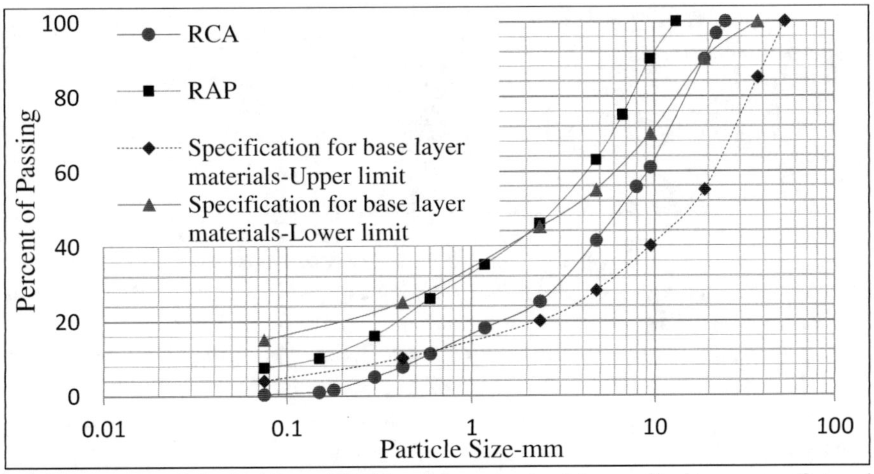

Figure 2. PSD curves of the RAP and RCA with upper and lower limit curves of standard base layer materials

Table 1. Physical properties of RCA [6]

Property	Values	Standard granular pavement material
Liquid Limit (LL) [%]	21.00	Maximum 25 [14]
Plasticity Index (PI)	5.40	Maximum 6 [14]
Linear Shrinkage (LS) [%]	1.00	Maximum 3.5 [14]
Water absorption (particles > 4.25 mm) [%]	5.35	<10 [9]
Water absorption (particles < 4.25 mm) [%]	6.50	<10 [9]
Specific gravity (Gs)	2.64	2.85 [15]
Maximum dry density (MDD) [g/cm^3]	1.75	>1.79 [9]
Optimum moisture content (OMC) [%]	13.20	8-15 [9]
California Bearing Ratio (4 days soaked) [%]	90-95	> 80% [14]

Table 2. Physical properties of RAP [6]

Property	Values	Standard test method
Binder (bitumen) content [%]		
Solvent extraction method	3.87	[16]
Ignition method	4.81	[17]
Specific gravity	2.59	[18]
Water absorption (particles < 4.25 mm) [%]	2.85	[18]
Water absorption (particles > 4.25 mm) [%]	2.81	[19]

3. Methodology

Test programme was designed to conduct Repeated load tri-axial (RLT) test to investigate the performance characteristics of RAP mixed RCA samples. RAP was mixed in RCA in different ratios by weight and shown in Table 3 with new sample names.

RLT tests were conducted under two test series as shown in Table 4 to determine the resilient and residual properties of the samples shown in Table 3. Tests were conducted in 2 different water contents. Water content-1 was selected as the corresponding optimum moisture contents to evaluate the performance at high moisture level. Water content-2 was the corresponding moisture levels at the degree of saturation of 60% of all the specimens. 60% of DoS was to represent the water content which is applied to compact the granular materials when the unbound granular layers are constructed [20].

Table 3. Sample names with the mix ratio of RCA and RAP

Blended Sample Name	RCA (% by weight)	RAP (% by weight)
RM1-100/RAP0	100	0
RM1-95/RAP5	95	5
RM1-90/RAP10	90	10
RM1-85/RAP15	85	15
RM1-80/RAP20	80	20

Materials were dried in oven overnight and cooled to room temperature before mix with water. The water mixed materials were kept in sealed containers for about three hours prior to the compaction and compacted samples were again kept in sealed containers for four days as it was revealed as the required curing time for the compacted RCA specimens [21]. Samples were compacted at their corresponding maximum dry densities (MDDs). RLT test were performed as the standards of QDTMR (Q137-Permanent deformation and resilient modulus of granular materials [22]) and Each test was run for 50,000 load cycles.

The obtained results of elastic deformation behaviour of the samples then evaluated with resilient modulus values. It was compared the design specifications of resilient modulus of base layer materials which use in pavement designing and assessed the feasible applications.

The plastic strain values were evaluated under limited shakedown concept to determine the behaviour of accumulation of plastic deformation. Results used to rank the samples and to assess the effect of RAP on RCA on the formation of permanent deformation.

Table 4. Repeated load tri-axial test programme

Sample Name	Confining pressure-kPa	Vertical stress-kPa	Test Series for Water content- 1 (OMC)	Test Series for Water content-2 (DoS-60%)	MDD-t/m^3	No: of load cycles
RM1-100/RAP0	125	750	13.2 %	11.62 %	1.748	
RM1-95/RAP5	125	750	12.2 %	10.81 %	1.785	
RM1-90/RAP10	125	750	12.7 %	10.74 %	1.785	50000
RM1-85/RAP15	125	750	12.3 %	10.58 %	1.789	
RM1-80/RAP20	125	750	12.9 %	10.55 %	1.789	

4. Results and discussions

4.1 Effect of RAP and moisture content on plastic properties of RCA

Minimize the formation of rut depth in pavement structure is the key aspect in pavement designing. Prediction of rutting in flexible pavement is extremely complex since it basically depends on the quality of the base and subbase materials as well as the environmental conditions (e.g. Freezing/thawing, flooding, natural drainage system etc.). Therefore, examine the plastic properties of the pavement materials under dynamic load condition is the foremost factor to understand and assess the behaviour of materials in pavement rutting [23].

Figures 3 and 4 illustrate the accumulation of plastic deformation of the five RCA samples at OMC and 60% of DoS level within 50,000 of load cycles. Rapid plastic deformations of the five samples were observed due to the rapid settlements at the initial cyclic loads. The settlements were occurred due to the re-compaction and densification at both moisture levels. Size of the initial settlement was high with increasing the RAP portion. Effect of RAP on the densification of the compacted samples significantly noticed with 20% of RAP at the both moisture levels and presence of RAP up to about 15% in RCA did not show a significant influence on the initial stabilization. The densification effect was high on deformation of the specimens at the initial load cycles due to the reorientation of particles [24]and this can be minimized by applying high effort on compaction of RCA.

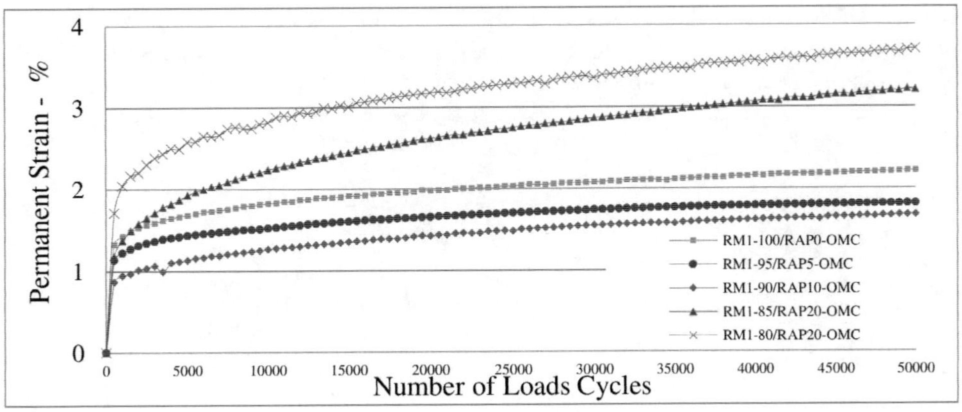

Figure 3. Permanent strains of RAP mixed RCA samples at OMC within 50,000 load cycles

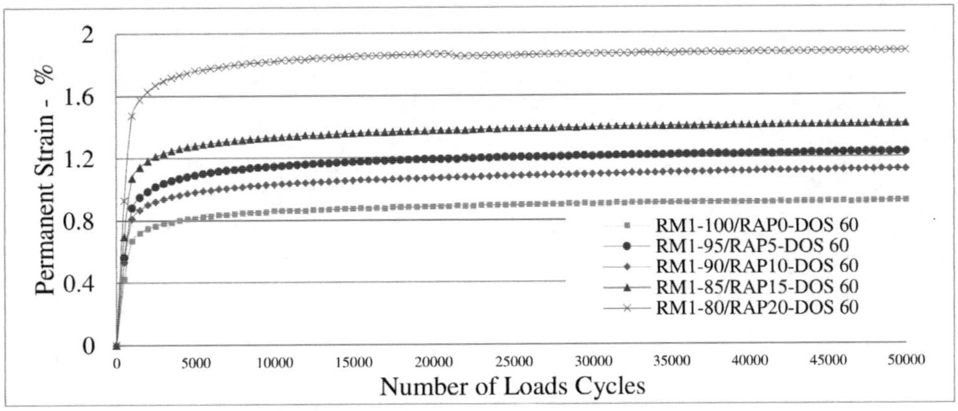

Figure 4. Permanent strains of RAP mixed RCA samples at 60% of DoS within 50,000 load cycles

According to the deformation curves in Figure 3 the accumulated plastic strains of the samples of RM1-100/RAP0, RM1-95/RAP5, and RM1-90/RAP10 fluctuated within a small range of 1.5-2% indicating a continuous decrease of the rate of deformation consequently, highlights a negligible effect of RAP when the quantity was small in RCA. The specimens of RM1-85/RAP15 and RM1-80/RAP20 exhibited comparatively higher deformation and the rates of strain accumulation were being still increased at the 50,000 load cycles. Figure 5 (a) further illustrates the rates of deformation with the accumulation of vertical depth after initial settlement of the specimens at OMC. The specimens had high deformation rates and greater initial settlements at the beginning of load cycles. A rapid drop of the rate of deformations within a small range of rut depth were exhibited in all the specimens at their OMC values. After rapid settlements, the rates of deformation approached zero value indicating a promising sign of stiffening the compacted RCA with more load cycles.

According to the figure 4, the specimens mixed with 20% of RAP (RM1-80/RAP20) still showed the highest deformation at the 60% of DoS level. In contrast, it has significantly noticed that RCA mixed with 15% of RAP (RM1-85/RAP15) has considerably decreased its plastic deformation at 60% of DoS level than OMC (DoS=68%). However, five samples have given appreciable results in accumulating their plastic deformation at the 60% of DoS level. Ultimately, all the specimens exhibited below 2% of permanent strains within 50,000 load cycles. Figure 5(b) further illustrates the rate of deformation of the specimens with formation of permanent deformation at the 60% of DoS level. Unlike the Figure 5(a) the curves do not exhibit rapid drop of deformation rates at the beginning of load cycles. All specimens show a gradual decrease of deformation rate up to a certain level and there after a rapid drop were indicated. After accumulation of permanent deformations which is less than 4.00 mm, all specimens lean towards to stabilize with gradual decrease of deformation rates which approached zero. Therefore, it revealed the hardening of the accumulation of plastic strain within a small range of permanent deformation at the 50,000 load cycles.

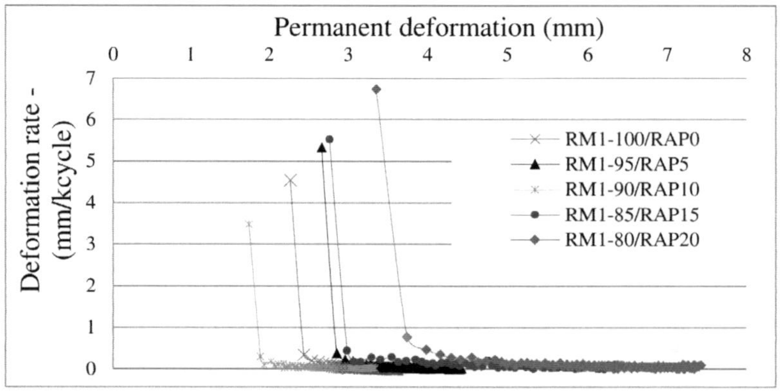

(a)

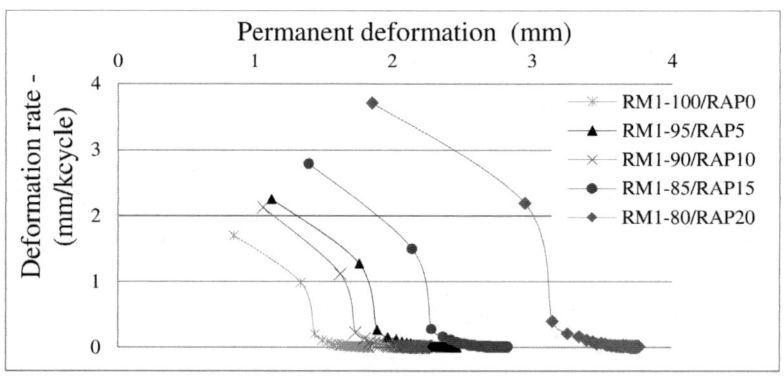

(b)

Figure 5. Rate of permanent deformation versus permanent axial deformation at (a) optimum moisture contents and (b) DoS = 60% of the RAP mixed RCA samples

Fives RCA samples illustrated great performance in terms of permanent deformation at the 60% of DoS level. More water at the corresponding OMC values has adversely affected on gaining of permanent strain as well as on the rate of strain accumulation at the end of 50,000 load cycles. More water in the pores of compacted specimens causes excess pore water pressure and lowers the shear resistance and decrease the stiffness. Water at 60% of DoS seems to be the adequate moisture contents to positively impact on gaining high stiffness to reduce the plastic deformation under dynamic wheel load. According to the Vuong and Hazell's [25] interpretation on plastic deformation behaviour, first three specimens (RM1-100/RAP0, RM1-95/RAP5, and RM1-90/RAP10) could be defined as "Stable" and the other two specimens (RM1-85/RAP15, RM1-80/RAP20) as "Unstable" at the optimum moisture contents. In contrast, all the samples were in the "Stable" state at the 60% of DoS level which satisfied the requirement for base layer material for the pavements subjected to heavy traffics (10^7 ESA).

It is important to highlight that the net accumulation of permanent axial deformation has become sensitive on the increase of RAP portion. Although, it is not much significance at the little potion of RAP up to about 15% and it can be observed a trend of high deformation when the RAP portion over 15% in RCA.

4.2 Effect of RAP and moisture content on elastic properties of RCA

Figure 6 shows the resilient modulus (Mr) results of the five samples at the corresponding OMCs and 60% of DoS levels at the end of 50,000 load cycles. The samples exhibited a gradual decrease of Mr but, within a small range at both moisture levels with increase of RAP. However, it illustrates the trend of decreasing the Mr with high portion of RAP in RCA. More RAP materials supply more medium size particles in RCA samples which the gradation adversely effects on compaction and there by lower the stiffness. Further, Thom and Brown [26] explained lubricating effect on particles which makes an adverse impact on resilient behaviour due to the lower inter-particle contact. The similar effect can be expected in bitumen coated RAP particles with hydrophobic surface. These particles' surface made such a lubricating effect at the wet condition and more RAP particles contribute more surface area with bitumen coat and lubricating effect increases in the compacted specimens.

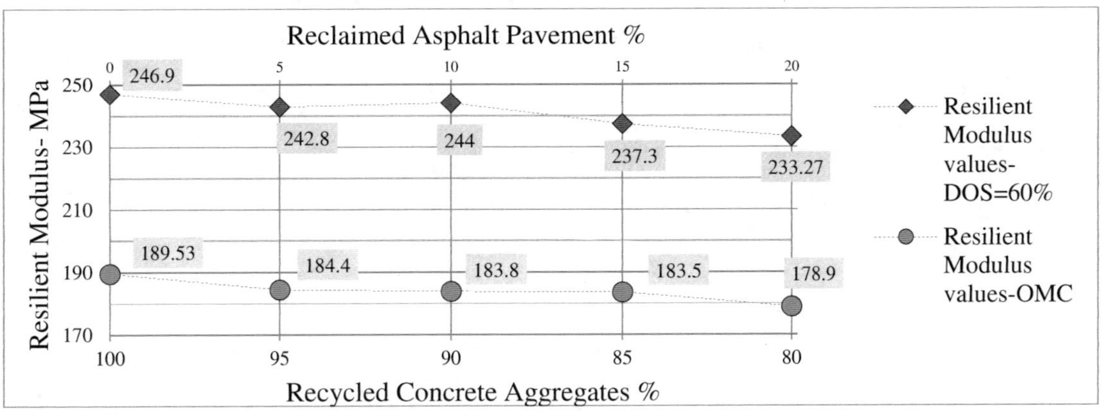

Figure 6. Resilient modulus values of the RAP mixed RCA samples at OMC and 60% of DoS level at the end of 50,000 load cycles

Poor performance observed at the high moisture contents at OMCs compared to the low moisture content (60% of DoS) since excess pore water pressure generates with more moisture and decline the shear resistance. Stiffness of the compacted specimens has reduced while the shear resistance was weak and deform more elastically resulting low resilient modulus. Furthermore, samples exhibited increasing of Mr at the low moisture content (60% Of DoS) as shown in figure 7 while the Mr curves at OMCs steadily behaved as shown in figure 8. Continuous increase of Mr depicts the decreasing of elastic deformation of the RCA specimens at 60% of DoS. Therefore, elastic deformation of the five samples interpreted "Stable" state and further high stiffness and consequently high resilient modulus can be predicted with more load cycles beyond 50,000.

Some of the materials attribute their resilient behaviour such that become to steady state with a decreasing rate of their Mr while some of the pavement materials become to steady state with increasing rate of Mr. Although, the tested RCA samples exhibited low Mr values (range of 233-247 MPa at 60% of DoS) than some of high-quality base layer materials, the shapes of the Mr curves are promising sign of reducing the elastic deformation with more load cycles. Thus, it can be predicted gaining of high Mr values with more load cycles beyond the tested value of 50,000 load cycles.

Australia has introduced specific standard values of resilient modulus for normal standard and high standard base layer materials for structural design of pavements [27]. The standard guidelines of pavement structural designs have recommended minimum Mr as 150 MPa for the normal standard base layer materials and 210 MPa for the high standard base layer materials [27]. The obtained Mr of the five RCA samples exhibited greater Mr than the required minimum value of normal standard base materials at the 68% of DoS (OMCs) while the Mr values were greater than the minimum value of high standard base layer materials at the 60% of DoS. However, the presented Mr values of the RCA samples were obtained at the 50,000 load cycles. The shapes of curves of Mr depict an increasing rate of Mr at the end of the 50,000 load cycles and can be predicted greater Mr values with more load cycles beyond 50,000.

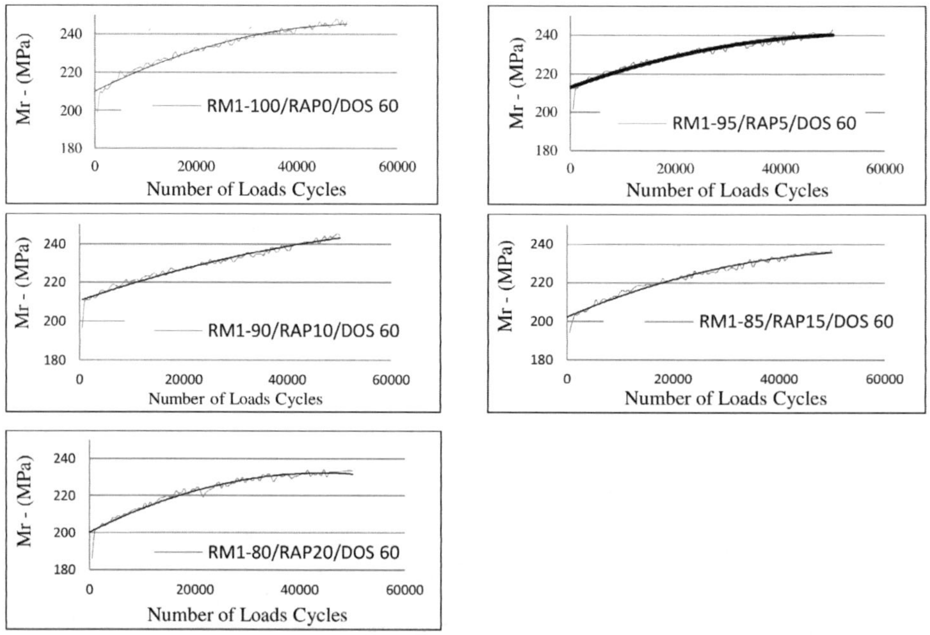

Figure 7. Resilient behaviour of the RAP mixed RCA samples at 60% of DoS level

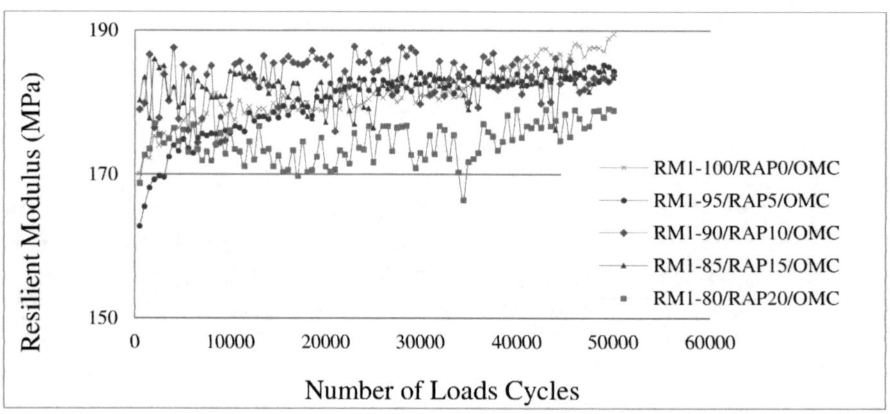

Figure 8. Resilient behaviour of the RAP mixed RCA samples at OMC levels

5. Conclusions

This study was carried out to investigate the performance characteristics of RCA with and without RAP material at two different water contents. The conclusions are drawn below:

- Accumulations of permanent deformations were higher at the OMC than 60% of DoS in all RAP mixed RCA samples. However, RCA exhibited well performance even at the OMC when the RAP portion was about 10.
- Five samples illustrated high resistance to plastically deform at 60% of DoS level. The results confirmed the presence of RAP up to about 15% in RCA specimens did not significantly affect on the accumulation of permanent strain. However, the accumulation of permanent axial strain increased with the increase of RAP portion in RCA.

- Corresponding moisture contents at 60% of DoS level of the samples exhibited best resilient moduli. Greater resilient modulus can be predicted with RAP up to 20% in RCA since the resilient modulus curves exhibited continuous increasing rate at the end of 50,000 load cycles. In contrast, resilient moduli were low at the OMCs. However, gaining of resilient moduli was steadily behaved within 50,000 of load cycles.
- Resilient modulus has varied within a small range with increasing the RAP from 0 to 20% in RCA at 60% of DoS which reflects insignificant impact on the resilient modulus when presence of RAP up to 20% in RCA.
- According to the analysed results of accumulation of permanent strain and resilient modulus, it can be concluded that presence of RAP in small quantities as a constituent in RCA does not significantly affect to degrade the quality and performance as unbound pavement material.

Acknowledgments
Authors acknowledge the Queensland University of Technology (QUT), Brisbane, Australia. The authors gratefully acknowledge Alex Fraser Queensland for providing Recycled Concrete Aggregates (RCA) for this experimental program. Queensland Department of Transport and Main Roads (QDTMR) and Brisbane City Council (BCC) are greatly acknowledged for providing some of testing facilities.

References
[1] Tam VWY 2009 Comparing the implementation of concrete recycling in the Australian and Japanese construction industries *J. Cleaner Production* 17, 688-702.
[2] Jayakody S, Gallage C and Kumar A 2014 Assessment of recycled concrete aggregates as a pavement material *Geomechanics and Eng.* 6, 235-248.
[3] Arulrajah A, Piratheepan J, Ali M and Bo M 2012 Geotechnical Properties of Recycled Concrete Aggregate in Pavement Sub-Base Applications *Geotechnical Testing J.* 35, 743-751.
[4] Arulrajah A, Piratheepan J and Disfani MM 2014 Reclaimed Asphalt Pavement and Recycled Concrete Aggregate Blends in Pavement Subbases: Laboratory and Field Evaluation *J. Mat. in Civil Eng.* 26, 349-357.
[5] Kazmee H, Tutumluer E and Beshears S 2017 Using Accelerated Pavement Testing to Evaluate Reclaimed Asphalt Pavement Materials for Pavement Unbound Granular Layers *Journal of Mat. in Civil Eng.* 29, 04016205.
[6] Jayakody S, Gallage C and Ramanujam J 2019 Effects of reclaimed asphalt materials on geotechnical characteristics of recycled concrete aggregates as a pavement material *Road Mat. and Pavement Design* 20, 754-772.
[7] Nataatmadja A and Tan YL 2001 Resilient Response of Recycled Concrete Road Aggregates *J. Tran. Eng.* 127, 450-453.
[8] Leite FDC, Motta RDS, Vasconcelos KL and Bernucci L 2011 Laboratory evaluation of recycled construction and demolition waste for pavements *Cons.and Building Mat.* 25, 2972-2979.
[9] Arulrajah A, Piratheepan J, Disfani MM and Bo MW 2013 Geotechnical and Geoenvironmental Properties of Recycled Construction and Demolition Materials in Pavement Subbase Applications *J. Mat. in Civil Eng.* 25, 1077-1088.
[10] Padmini AK, Ramamurthy K and Mathews MS 2009 Influence of parent concrete on the properties of recycled aggregate concrete *Cons.and Building Mat.* 23, 829-836.
[11] Gobieanandh V and Jayakody S 2016 *Evaluate the strength of cement treated recycled construction and demolition aggregates as a pavement material*. The 7th Int. Conf. on Sustainable Built Environment, Kandy, Sri Lanka.

[12] Arshad M and Ahmed MF 2017 Potential use of reclaimed asphalt pavement and recycled concrete aggregate in base/subbase layers of flexible pavements *Cons. and Building Mat.* 151, 83-97.

[13] Gabr AR and Cameron DA 2012 Properties of Recycled Concrete Aggregate for Unbound Pavement Construction *J. Mat. in Civil Eng.* 24, 754-764.

[14] Roads DTM 2010 *Main roads technical standards: MRTS05 - unbound pavements*, Department of Transport and Main Roads, Brisbane, Australia.

[15] Vegas I, Ibañez JA, San José JT and Urzelai A 2008 Construction demolition wastes, Waelz slag and MSWI bottom ash: A comparative technical analysis as material for road construction *Waste Management* 28, 565-574.

[16] Roads DTM 2011 *Test method Q308A Binder content and aggregate grading of asphalt (refluxing)* Department of Transport and Main Roads, Brisbane, Australia.

[17] Roads DTM 2100 *Test method Q308D-2002, Binder content and aggregate grading of asphalt (Ignition oven) Queensland Government*, Department of Transport and Main Roads, Brisbane, Australia.

[18] Roads DTM 2011 *Test method No Q109A-1996, Apparent particle density of soil (Fine fraction)*, Department of Transport and Main Roads, Brisbane, Australia.

[19] Roads DTM 2011 *Test method Q214B-1996, Particle density and water absorption of aggregates (Coarse fraction)*, Department of Transport and Main Roads, Brisbane, Australia.

[20] Bodin D, Moffatt M, Jameson G and Group A 2013 *Development of a wheel-tracking test for rut resistance characterisation of unbound granular materials*, Austroads technical report Austroads Ltd, Sydney NSW, Australia.

[21] Jayakody S, Gallage C and Kumar A 2012 *Assessment of recycled concrete aggregate for road base and sub-base*. Proceedings of the Second Int. Conf. on Geotechnique, Cons. Mat. and Environment, Kuala Lumpur, Malaysia.

[22] Roads DTM 2013 *Test method Q 137, Permanent deformation and resilient modulus of granular unbound materials*, Department of Transport and Main Roads, Brisbane, Australia.

[23] Lekarp F, Isacsson U and Dawson A 2000 State of the Art. II: Permanent Strain Response of Unbound Aggregates *J. Trans. Eng.* 126, 76-83.

[24] Allen JJ 1973 *The Effects of Non-constant Lateral Pressure on the Resilient Response of Granular Materials*, University of Illinois at Urbana-Champaign.

[25] Vuong BT and Hazell D 2003 Development of performance-based specifications for unbound granular materials: Issues and recommendations *Road & Transport Research* 12.

[26] Thom NH and Brown SF 1988 The effect of grading and density on the mechanical properties of a crushed dolomitic limestone, In *Australian Road Research Board (ARRB) Conference, 14th*, Canberra, Australia.

[27] Jameson G and Group A 2008 *Guide to Pavement Technology Part 2: Pavement Structural Design*, Austroads Incorporated, Sydney NSW Australia.

IOP Conf. Series: Materials Science and Engineering 1075 (2021) 012021 doi:10.1088/1757-899X/1075/1/012021

Structural Design of Asphalt Pavements in Germany

B Hristov[1,*], F Wellner[2], G C Falla[2] and S Leischner[2]

[1]Department of Engineering Sciences, Chair in Transportation, University of Applied Sciences (HTW Berlin), Wilhelminenhofstraße 75A, 12459 Berlin, Germany
[2]Department of Pavement Engineering, Technical University of Dresden (TU Dresden), Georg-Schumann-Straße 7, 01187 Dresden, Germany
[*] Corresponding author: Borislav.Hristov@HTW-Berlin.de

Abstract. The motivation for this research comes from the inadequacy of the methods currently used for sizing and optimization of building materials used in pavement construction based on increasing traffic loading and changing climatic conditions. Using examples, the paper highlights how deficits can be identified by a detailed investigation of the relationships in order to form a foundation for solving these deficits.

1. Introduction

Pavement design in Germany has so far been standardized according to the experience based on layer thickness specifications instead of being dimensioned on the basis of calculations, as it is in the rest of civil engineering. The assessment of the quality of the building materials itself is empirical and based on past experiences. The quality of materials like asphalt is evaluated by assessing the mix constituents. Typical examples for that are Needle penetration test, Ring and Ball softening point (for bitumen) or grain size distribution (for aggregates) and mix composition requirements (binder content, void content, etc.). An examination of the materials behavior under use-relevant stresses is not considered or it is only very simplified to derive any meaningful results.

Rapid globalization of economies and improving trade relations between countries has promoted a huge demand for mobility in the world. This combined with the increasing industrialization (road as "storage area" of the industry) has resulted in increased traffic load during the past few decades. Road freight traffic in Germany quadrupled from less than 100 billion tonne-kilometre (tkm) in 1980 to 500 billion tkm in 2010. The forecast for the next 30 years shows a similar trend with a road freight traffic increase between 200-300%. In addition to the changes in climatic and loading conditions, financial pressure is also supposed to increase as absence of cheap resources increases the cost of pavement construction.

All factors described above lead to an urgent need for a scientific and mechanical approach towards pavement structure to make it sustainable. These factors also highlight a need for complete optimization strategy that must be employed on the building materials and the entire structure in order to achieve high load bearing pavement structures.

2. State of the art

2.1 Design according to current guidelines

An example of an asphalt pavement structure is shown in Figure 1.

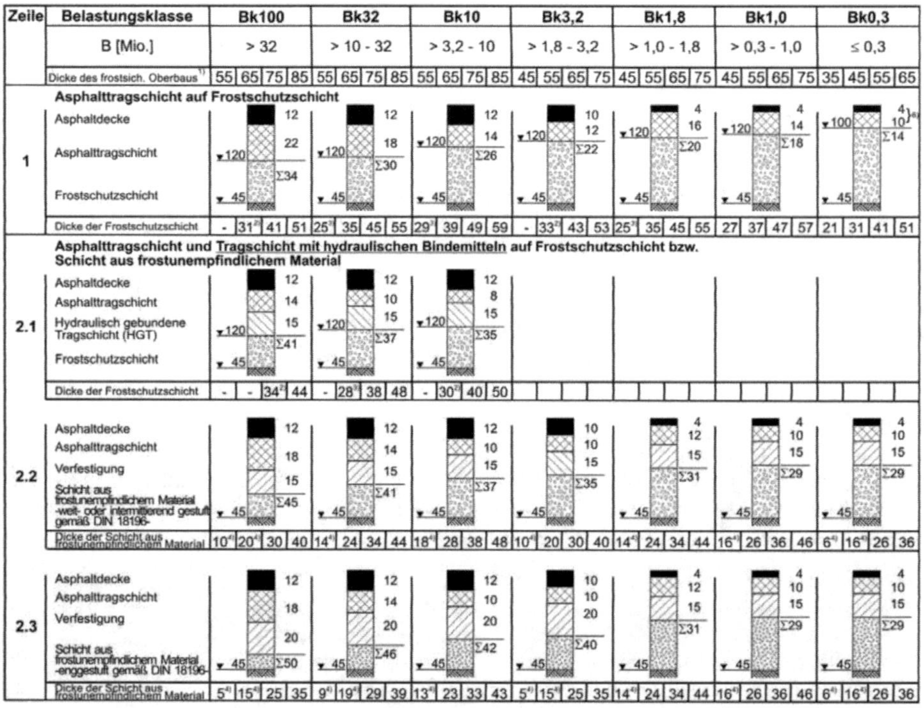

Figure 1. Pavement structure [1]

Asphalts are conventionally designed based on very extensive set of rules according to the guidelines. These regulations apply not only to the composition of asphalt but also to simple properties of layers in the built-in state (void content, degree of compaction) [2, 3]. The design of pavement structure based on the requirements of building materials can be done in two ways. The first relatively simple way is the standardization according to [1]. In this method the layer thicknesses are specified for certain load, climatic and substrate conditions. Secondly, mechanistic design based on various evidence of the resistance of the structure against fatigue cracking [4]. To ensure frost resistance, the thickness of the overall structure depends on the composition of the upcoming layers and the layers of the substructure. The layer thicknesses as well as the total thickness of the pavement structures were determined empirically based on past experience. This standardized method is almost exclusively used today.

The mathematical procedure described below is only used in exceptional cases (side offers, PPP projects, functional construction contracts). The guidelines introduced for this purpose [4], currently contain a procedure, which includes the computational verification of the individual layers according to the criteria of fatigue and deformation resistance. For building materials, a linear-elastic behavior along with a temperature or a frequency dependency is assumed. The material properties required for the verification in the asphalt layers are described by the temperature-stiffness modulus function and the fatigue function, which are determined using test procedures as standardized in [5]. To calculate the total axle load and temperature spectrum over a year, simplified assumptions are made for the load. A rough classification of the axle loads at the corresponding geographical location (11 axle load classes of 0-2 t, 2-4 t … > 22 t) with a traffic condition similar to the frequency distributions of the classes and only a few temperature profiles over the asphalt layer (13 gradients for the temperature classes from -15°C to -10°C, -10°C to -5°C and +45°C to +50°C) is done. However, this results in at least 143 load states, which must be taken into account in the design/forecast calculations. The evaluation of the individual load cycles with individual load conditions with respect to 'consumption' of service life then takes place according to Miner's hypothesis. The model-related deviations of the calculated and observed results are currently covered with the help of a safety factor. The required safety factors (total safety factors) were determined by calibration calculations on the currently standardized thickness of the pavement structure, on the assumption that these pavements reach a service life of 30 years for certain calibration conditions (assumptions for material properties, loads and temperatures).

2.2 Evaluation and further procedure for the regulations

An important question is, whether the method described for the design (catalogue design according to [1]) and the corresponding additional rules for determining the mixture composition or requirements [3, 6] provide a sufficiently reliable procedure for achieving the durability, particularly with respect to the deformation and fatigue resistance of the constituent materials, and the whole pavement is possible. Regulations, for example: the deformation or fatigue behavior of the asphalt are not yet defined despite the newly established rules for computational design [4].

Due to the use of bitumen as a "binder" and crushed rock as aggregate, the asphalt mix shows stress-dependent elastic and viscous plastic material behavior. Due to the strong temperature-dependent viscosity of the binder, the same loads on the same pavement produce distinctly different stress states and load reactions at different temperatures. Nevertheless, initially linear-elastic, temperature and frequency dependent material behavior is assumed for asphalt. Also, for the unbounded layers, linear-elastic behavior is initially assumed, although the non-linear behavior of this is long known. To evaluate the damage caused by individual load cycles under different loading conditions, the use of Miner's hypothesis remains in discussion. Although, studies conducted at Institute of Urban and Pavement Engineering at Technische Universität Dresden have shown that acceptable results can be calculated with this hypothesis [7], however there is a need for further in-depth research.

In order to bring it closer to reality, the axle load classes described in Chapter 2.1 should be modified. Now, 22 axle load classes (from 1 t to 22 t with increment of 1 t per class) instead of 11 axle load classes and assigned frequencies of occurrence should be taken into account during the pavement design process. The temperatures which occur in the asphalt layers during the year and during the day should be described in order to improve the reliability. As a next step, at least 200 different temperature profiles at defined surface temperatures and frequencies of occurrence are created. Combining these with the axle load conditions result in 4400 loading conditions. Nevertheless, the computational time and effort must be kept within practice-relevant limits.

Usually, during the construction process of the asphalt, layers are paved as "hot on cold". In the pavement structure shown in Figure 1, this results in layer boundaries between the asphalt surface course and the asphalt binder course, between the asphalt binder course and the asphalt base course as well as in the middle of the asphalt base course (when the total thickness exceeds 18 cm) since it is paved in two separate steps to ensure a proper compaction of the asphalt base course.

The bond between the layers is currently created by spraying the surface of the lower layer with bitumen emulsion (tack coat) prior to the paving of the top layer. The process ensures a largely rigid and permanent bonding between the layers. However, the bitumen emulsion used for gluing is a liquid with temperature and frequency dependent viscosity. Road surface temperatures of 50 °C to 60 °C can be reached during summer time at intense sunlight. In approximately 4 cm depth below the surface layer, the temperatures between the surface and binder course are not significantly different from those on the surface. Does the bitumen act as an adhesive or as lubricant in these conditions? Assuming the latter, what impact does it have on the stresses and on the deformation and fatigue resistance of the asphalt pavement? In order to achieve answers to these questions, investigations have been carried out at the Institute of Urban and Pavement Engineering at Technische Universität Dresden, Germany that will be discussed in the following parts of the paper.

3. State of research

3.1 Preliminary remarks

Ten years ago an intensive research activities have been carried out both internationally and in Germany to develop improved pavement design methods and assumptions related to traffic and climatic loading in order to predict accurately the structural response of the pavement structure. The research has already led to better understanding of material behavior and its impact on interlayer stresses and therefore ultimately on the service life of the pavement.

3.2 Assessment with the help of previous simplified approaches – fatigue criteria

Using the procedure mentioned above, tests were carried out for three asphalt samples produced in the laboratory under the following assumptions: same traffic load, same pavement structure (same layer thickness), same temperature conditions and same asphalt composition according to the regulations. The aggregates consisted of only one quarry stone material with particle size distribution produced in the lab by sieving. The only difference between these asphalts was the different origin of the bitumen 50/70 (according to the same classification regulation). Test specimens were prepared based on these asphalts, the stiffness modulus as a function of temperature and fatigue was determined in accordance with the valid instructions for the performance of splitting tensile tests [5] and based on the calculations. Figure 2 shows the result of the calculations. Although the bitumen belonged to the same grade (50/70), subjected to identical test conditions, differences in the service life from 10 to 30 years were observed. Taking into account the fact that the common practice of mixing plants nowadays is to change the bitumen supplier mainly for financial reasons without checking or changing the asphalt composition, it is clear that a desired or predetermined service life cannot be achieved with empirical-based dimensioning methods in combination with the requirements defined in the set of rules.

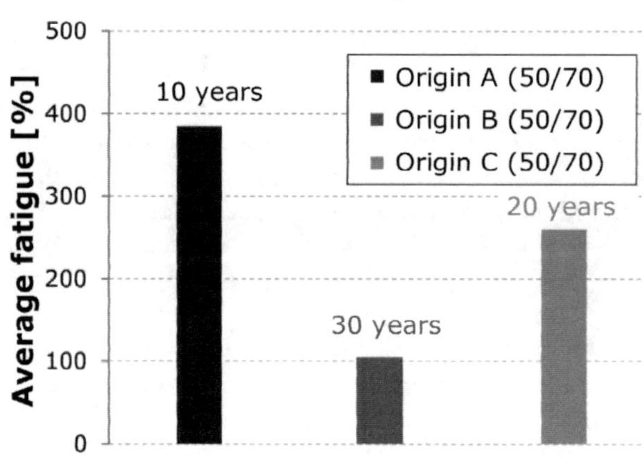

Figure 2. Calculated service life of asphalt pavements with binders of different proveniences by identical other mixture parameters

Starting from this first example, the aim of further investigation at the Institute of Urban and Pavement Design [10] was to determine the influence of the asphalt mix composition on its performance and durability of the pavement structure based to fatigue criterion. Based on the fatigue and stiffness modulus, temperature functions were obtained by trial and error method, the calculations were then made on the selected pavement variants. A total of 18 asphalt base course variants of the AC 22 T S specimen were produced. The variations of the particle size distribution, the binder content, the binder viscosity and the binder manufacturer were carried out. In order to be able to analyze the influence of the particle size distribution of an asphalt base course mixtures, the passage of the rock aggregate at 0.063 mm, 2.0 mm and 16.0 mm grain size was determined so that it was exactly in the middle of the required tolerance range of [3] (mean grain size distribution = mKGV), secondly on the lower limit (coarse grain size distribution = gKGV) and thirdly on the upper limit of the grain size range (fine particle size distribution = fKGV) of an AC 22 TS. The required minimum binder content (Bmin = 3.5 M.-%) was calculated according to TL asphalt on the basis of the bulk density of the aggregate mixture (amphibolite) used for all grain size distributions. Then the binder contents were increased in 0.5 M.-% or in 1.0 M.-% steps to max. 6.5 M.-%. The binders used were paving grade bitumen 50/70 from two binder manufacturers (A and B) and bitumen 30/45 (A) from manufacturer A. Table 1 shows an overview of the tested asphalt base course variants.

Table 1. Tested asphalt mixtures

Grain size distribution	Binder type	Binder manufacturer	Binder content
medium	50/70	A	3,5 M.-%; 4,0 M.-%; 4,5 M.-%; 5,0 M.-% 5,5 M.-%
coarse			3,5 M.-%; 4,5 M.-%; 5,5 M.-%
fine			3,5 M.-%; 4,5 M.-%; 5,5 M.-%; 6,5 M.-%
	30/45		3,5 M.-%; 4,5 M.-%; 5,5 M.-%
	50/70	B	3,5 M.-%; 4,5 M.-%; 5,5 M.-%

From this table the designation of the respective mixture variant results in the following way: mKGV 3,5 M-% 50/70 A means: Average particle size distribution, binder content 3,5 M-%, bitumen 50/70 of the manufacturer A. These designations / Abbreviations can be found in the following figures. The stiffness module-temperature functions determined for these mixture variants are shown in Figure 3, the fatigue functions in Figure 4.

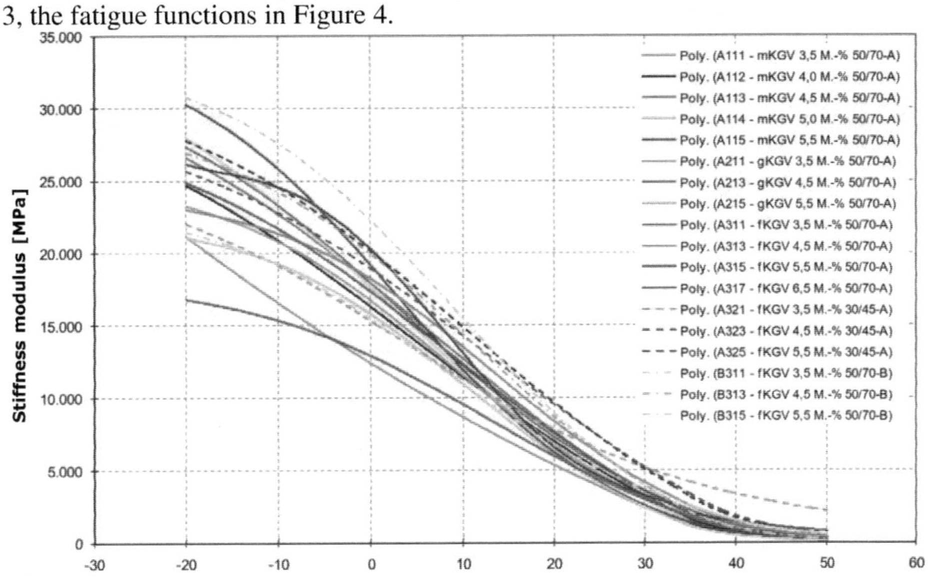

Figure 3. Stiffness-temperature function for all tested asphalt mixes

The mixture variants show significant differences in stiffness. The "softest" asphalt at 10 ° C has only 7,500 MPa of the stiffest asphalt twice the amount of 15,000 MPa.

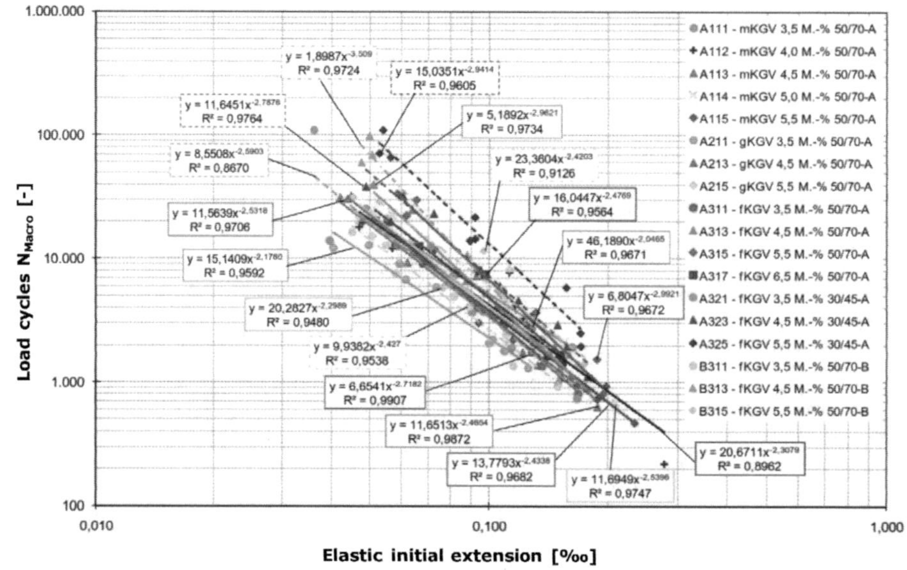

Figure 4. Fatigue function for all tested asphalt mixes

There are significant differences in the fatigue functions. The asphalt with the lowest fatigue resistance could bear only 2,000 load cycles at 0.1 ‰ elastic initial extension until failure of the test specimen, whereas the asphalt with the highest fatigue resistance could withstand about 12,000 load cycles.

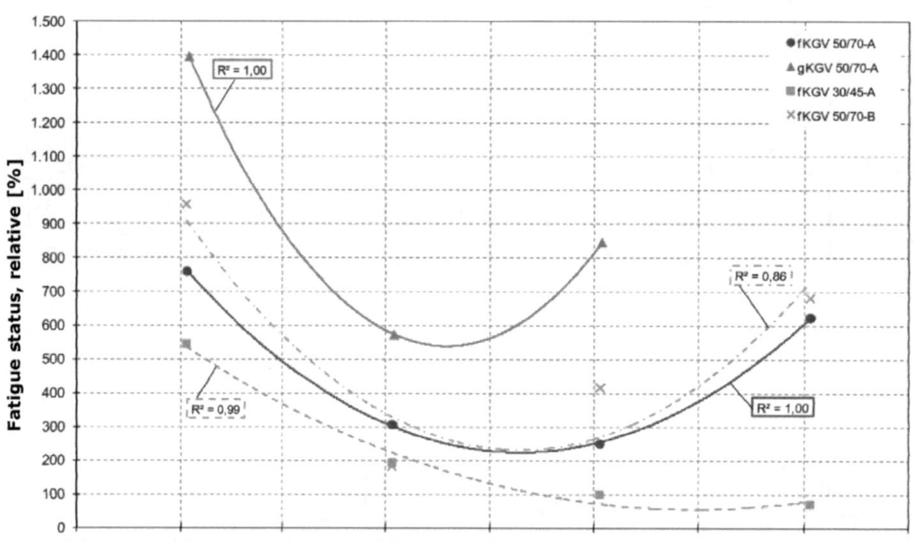

Figure 5. Layer thickness adaptation to achieve the same calculated service life depending on the binder content and the grain size distribution

Figure 5 shows the layer thickness adjustment, which is necessary to obtain the same calculated service life. The calculations were made for the pavement variant according to Figure 1. The traffic load was assumed in such a way that no layer thickness adjustment is necessary for variant fKGV 30/45-A with a binder content of 5%. Two main conclusions can be drawn from these results. The binder content, particle size distribution and binder viscosity have a significant influence on the fatigue resistance of the asphalts and thus of the pavement even with the slightest variations. It is obviously possible to determine the optimal binder contents for given particle size distributions and binder viscosities by the methodology used here. The absolute differences in the required layer thickness adjustments show very clearly that based on the conventional empirical method of dimensioning and setting the mix compositions only on the basis of void considerations, a desired service life cannot be achieved accurately. Detailed considerations for optimizing the asphalt are absolutely necessary, at least for the area with high and very high loads and frequencies.

The construction of asphalt pavement in Germany generally consists of a surface course, binder course and base course. The interlayer bond allows transfer of shear and normal stresses between the asphalt layers. This interlayer bond is achieved through interlocking of the aggregates at the interface, the friction between the surfaces of the two asphalt layers and adhesion between the applied tack coat and asphalt binder of the two layers.

Sometimes, the shear stresses at the layer boundaries can be extremely high. This must be absorbed in such a way that the horizontal sliding of the layer boundary surfaces against each other is largely excluded. Missing or reduced interlayer bond results in a complete different stress state within the whole pavement structure. This may result in a significant reduction in service life of the pavement due to premature material fatigue.

A stiff and durable interlayer bond ensures all layer of the asphalt pavement can act as a single unit when the pavement is loaded resulting in minimal or no displacement between the layers and their interface. Research concerning this issue has been carried out at the Institute of Urban Construction and Road Construction at Technische Universität Dresden. The aim was to experimentally determine the effectiveness of the interlayer bond using relationship functions between shear stiffness, temperature, stress frequency and normal stress. By using the Finite Element (FE) program SAFEM [2], implemented in the German method for the analytical design of asphalt pavements according to [4], further conclusions about the effect of the interlayer bond on the service life of the pavement (fatigue criteria) can be drawn.

A new improved apparatus for cyclic testing of the interlayer bond has been used to simulate as accurately as possible the real load conditions that exist during the service life of the pavement structure. The effects of temperature, shear stress (cyclic), load frequency and the normal stress were considered. The shear stiffness was then experimentally determined for different nature and degree of contamination of the surface of the layers, bitumen emulsion types and amounts. The tests initially revealed a clear dependence of the shear stiffness on the temperature, the frequency of the shear stress and the normal stress. Less clear, but notable was the dependence of the shear stiffness on the surface properties of the layers and on the degree of contamination. Initially, it was determined that bitumen emulsion produces a very low shear stiffness (in no case above 100 MPa/mm, even at low temperatures). The shear stiffness at the interlayer bond was 10 to 100 times lower than the asphalt layers at comparable temperatures and load frequencies. At higher temperatures and low or non-existent normal stress, the shear stiffness tends to zero. Based on the experimental results a sigmoidal master functions for describing the shear stiffness for different states could be derived. The function with the lowest shear stiffness was filtered out as a "bad case" and similarly the function with the highest shear stiffness was called "good case". Furthermore, shear fatigue tests were carried out at Braunschweig Pavement Engineering Centre (ISBS). The tests showed, after the fatigue-related breakage of the interlayer bond, only friction/interlocking is effective as a function of the normal stress. From these experiments, a function was set up as a basis for the calculations for the worst case. All functions were then implemented in the FE program SAFEM. The fatigue status of the asphalt pavement investigated with interlayer bonds (varying from full bound to no bound) could then be determined using the analytical pavement design procedure according to the RDO Asphalt (Fig. 1) in order to investigate the effects of the interlayer bond on the fatigue behavior of asphalt structures. For this purpose, five different bond configurations were assumed between the surface course and binder course, as well as between the binder course and the base course layer:

- Full interlayer bond (theoretical bond which has never been reached in the tests, upper and lower asphalt layer are rigidly connected; does not apply to in-service asphalt pavements)
- "Good Case": best measured interlayer bond produced on test specimens in the laboratory for each layer combination.
- "Bad Case": worst measured interlayer bond produced on test specimens in the laboratory for each layer combination.
- Completely fatigued interlayer bond: normal stress-dependent and temperature-independent bond determined in long-term laboratory tests.
- Completely missing interlayer bond (theoretical bond which has never been reached in the tests, upper and lower asphalt.

The calculations were carried out for a construction according to Fig. 1 and for a very high traffic load and frequency (Fig. 6). Fatigue status 100% means the end of the calculated service life. The traffic load was chosen in a way that 30 years of service life can be achieved for the "good case". It is interesting that full interlayer bond would ensure a computational service life far beyond 30 years under otherwise identical conditions, whereas all other realistic cases are far below 30 years.

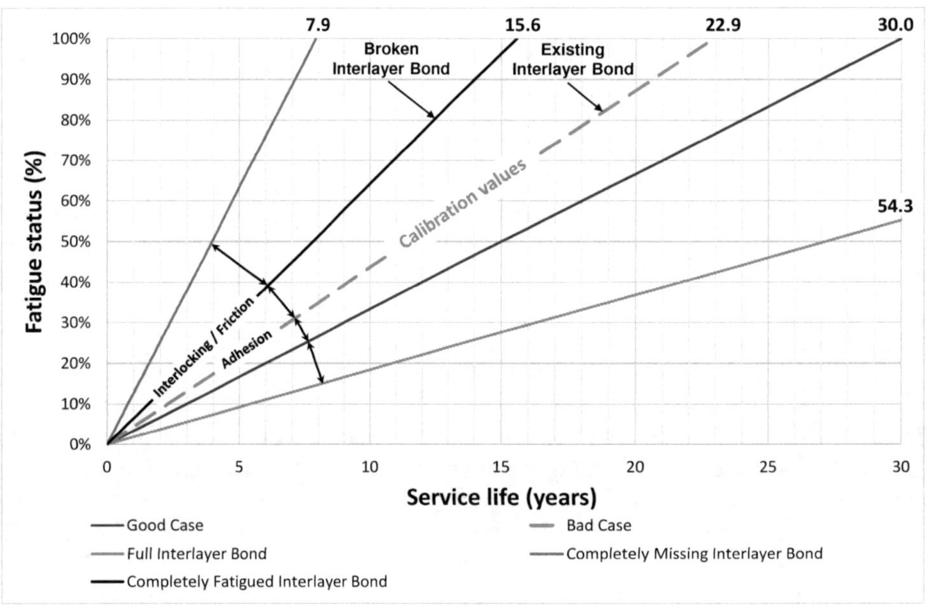

Figure 6. Layer thickness adaptation to achieve the same calculated service life depending on the binder content and the grain size distribution

These results show clearly that the interlayer bond, as one of the most important factors influencing the service life of asphalt pavements, can be estimated realistically only through detailed consideration of the structure and on the basis of test results.

4. Outlook

As already mentioned, the illustrated examples are based on simplified assumptions. There is an immediate need for improved methods of characterizing material behavior, shear stiffness etc., as a basis for improved modelling of the pavement. In addition, there is a need to couple this sub-module for the pavement with partial models for the tire and for the vehicle. As a result, interactions between the pavement, tires and vehicle can be presented in addition to improving the understanding of the relationships in the sub-models.

References

[1] Forschungsgesellschaft für Straßen- und Verkehrswesen 2012 RStO 12: Richtlinien für die Standardisierung des Oberbaus von Verkehrsflächen, *Köln*

[2] Wellner F and Hristov B 2015 Zyklische Schersteifigkeits- und Scherermüdungsprüfung zur Bewertung und Optimierung des Schichtenverbundes in Straßenbefestigungen aus Asphalt, *Forschungsvorhaben Nr. 17634 BG, Abschlussbericht, Dresden*

[3] Forschungsgesellschaft für Straßen- und Verkehrswesen 2013 TL Asphalt-StB, Technische Lieferbedingungen für Asphaltmischgut für den Bau von Verkehrsflächenbefestigungen, *Köln*

[4] Forschungsgesellschaft für Straßen- und Verkehrswesen 2009 RDO Asphalt 09, Richtlinien für die rechnerische Dimensionierung des Oberbaus von Verkehrsflächen mit Asphaltdeckschicht, *Köln*

[5] Forschungsgesellschaft für Straßen- und Verkehrswesen 2009 AL Sp-Asphalt 09, Arbeitsanleitung zur Bestimmung des Steifigkeits- und Ermüdungsverhaltens von Asphalten mit dem Spaltzug-Schwellversuch als Eingangsgröße in die Dimensionierung, *Köln*

[6] Forschungsgesellschaft für Straßen- und Verkehrswesen 2013 ZTV Asphalt-StB, Zusätzliche Technische Vertragsbedingungen und Richtlinien für den Bau von Verkehrsflächenbefestigungen aus Asphalt, *Köln*

[7] Wellner F and Weise C 2012 Schadensakkumulation bei unterschiedlicher Beanspruchung unter Berücksichtigung von Heilungs- und Affinitätseinflüssen an ausgewählten Asphalten, *Forschungsvorhaben Nr. WE 1642/5-1, Abschlussbericht, Dresden*

IOP Conf. Series: Materials Science and Engineering 1075 (2021) 012022 doi:10.1088/1757-899X/1075/1/012022

Feasibility study of low density polyethylene modified bituminous mix in DBM layer of flexible pavement

SK S Islam[1*], A Ghosh[1], GDRN Ransinchung[1] and T K Roy[2]

[1]Civil Engineering Department, Indian Institute of Technology Roorkee, Roorkee 247667, India.
[2]Civil Engineering Department Institute of Engineering Science and Technology, Howrah 711103, India.
Corresponding author email: ssohelislam@ce.iitr.ac.in

Abstract. An efficient and adequate system of transportation is inevitable for the development of any country. A high traffic intensity in terms of overloaded commercial vehicle, introduction of new axle configuration, significant variations of daily and seasonal temperature of the pavement are significant factors contributing to early development of distress symptoms like ravelling, rutting, cracking, bleeding, shoving and pot holing of bituminous surfacing, which demand a proper strengthening of pavements. The early development of distress in the pavement with the conventional mixes revealed the need of design specification based on performance tests. To overcome such kind of distress, properties of bitumen and bituminous mixes can be improved to meet the requirement of pavement with the incorporation of certain modifiers. The present study evaluates the Low-Density polyethylene (waste milk pouches) modified bituminous mix properties corresponding to the optimum binder content by conducting different performance test. Experimental studies demonstrated that waste milk pouches (6% by weight of the binder) modified bituminous mix gives the best result in terms of stability and durability. Experimental results indicated that using the waste milk pouches in bitumen is not only a cost-effective mix but also a very good solution in terms of improving bitumen pavement mix properties. The present paper aims to highlight the change in properties by addition of modifiers in bitumen and develop a procedure to estimate the reliability of a particular mix design. By considering these aspects, statistically it has been proved that the bituminous mix obtained experimentally, is also reliable to be used in real life scenario.

1. Introduction

The Rapid industrialization aided by changing production-consumption pattern has emerged as the primary reason for sustainable impedance. Depletion of natural resources and generation of enormous waste from both industry and agricultural practices with subsequent inadequate disposal capacity are the major areas of concern throughout the world. Extensive research is being undertaken considering the re-use of these by-products as raw materials in different construction industry. Additional aspects converging to the recycle of these waste products is its significant utility in enhancing the performance of pavements as "Bitumen Modifiers".

Mass production of plastic nowadays in different forms to accommodate societal demands has resulted in environmental pollution [1,2]. Plastic bags are made up of polyethylene (PE) mainly of chemical formula $(C2H4)n$, Re-use of these non-biodegradable products contributes to a sustainable solution for its efficient and effective disposal. Polyethylene has gained prominence as one of the most effective polymer additives increasing the service life of pavements [3]. A detailed review of polymer consumption in day-to-day scenario focused on the abundance of milk pouches thereby incepting its incorporation in the present study as one of the modifiers in conjunction with RHA. This paper evaluates the reliability of the mix obtained from optimal percent of bitumen (when conventional binder is replaced by RHA) and varying percentages of plastic thereby promoting the usability of plastics in pavement construction to prevent premature failures.

Since the year 2000, the concept of using wastes as bitumen additives has gained repute [4]. Mohammad et al. (2007) reported the addition of PE as the best alternative to conventional mix by achieving the incremental stability of the mix [5]. The advantage of stabilizing the stone mastic asphalt (SMA) mix in flexible pavements with shredded waste plastic was developed by Bindu et al. (2010) [6]. An analysis by Chen et al. (2002) portrayed that incorporating polymers into asphalt binders attenuated the major causes resulting in asphalt pavement failures [7]. Rice husk ash has been identified as another significant waste that can be used as filler material in hot mix asphalt instead of conventional filler yielding better Marshall Stability value as compared to conventional mix Sargin et al. (2013) [8]. Ahmad et al. (2017), Mishra et al. (2017) delineated a comprehensive study on how the addition of processed plastic in conventional bituminous mix contributed to increasing Marshall Characteristics [9, 10].

The objective of this present study is to establish the applicability and efficacy of the waste material containing LDPE in the binder course as a bitumen modifier. Considering their deterministic role in the performance enhancement of polymer-modified bituminous blends, the study was focused on the various performance analysis of the mix.

2. Materials and methodology

2.1. Materials

2.1.1. Plastic waste.

Plastic is integrated with our everyday life. Constant use of plastic not conforming to the specified standards has a detrimental effect on the environment. Hence, any use of plastic as a raw material in the construction process is gaining eminence to promote sustainability. In this study, waste plastic in the shredded form with size ranging between 2-8mm has been used for the present investigation. The source was waste milk pouches made of polyethylene that was abundantly available as household waste. The properties of which are listed below in table 1.

Table 1. Properties of shredded plastic waste.

Properties	Description
Melting Point	130^0C
Plastic Type	Shredded thin plastic milk pouches
Plastic material	Low Density Polyethylene
Density (gm/cm^3)	0.94

2.1.2. Aggregate.

Aggregates serve as the integral component of the pavement structure, bearing stresses due to wheel load as well as resist wear due to abrasive action of traffic. Proper choice of aggregates along with the correct proportioning affects the performance of pavements. The mineral aggregate used in this study were obtained from Pakur (Jharkhand, India) quarry. Representative samples have been examined in the laboratory and results are presented (table 2).

Table 2. Physical properties of aggregate.

Properties	Value	Specification
Impact Value	21	Max 24%
Crushing Value	17.5	Max 30%
Specific gravity (coarse aggregate)	2.57	2-3
Specific gravity (fine aggregate)	2.68	-
Flakiness Index	16.2	Max 35%
Elongation Index	12.5	Max 35%
Los Angles Abrasion Test	18	Max 30%
Water Absorption test	0.7	Max 2%
Stripping Value (%)	2	Max 5%

2.1.3. Rice husk ash.

Rice husk ash (RHA) is the by-product of the rice milling industry. Around 20 – 22% rice husk is produced in India as agricultural waste from paddy cultivation [11]. Roughly 25% of this accumulated husk becomes ash when burnt. It has an excellent pozzolanic property and is non-plastic in nature. In this study, RHA was collected from local rice mill of Burdwan district, West Bengal, India. The properties of which are detailed in table 3, along with their permissible limits as per IS 73 Specifications [12].

Table 3. Properties of rice husk ash.

Properties	Value
Specific Gravity	1.95
Plasticity Index	Non-Plastic
Silica Content (%)	90-95

2.1.4. Bitumen.

A hydrocarbon product solid at room temperature and highly viscous at a temperature above 100 degrees Celsius, bitumen acts as the major constituent in flexible pavement construction. For this study, 60/70 penetration grade bitumen has been selected. The physical properties of the same are given in table 3 along with their permissible limits as per IS specification.

Table 4. Physical properties of bitumen.

Properties	Value	Specification
Penetration (at 25°C) (1/10th of mm)	64	50-70
Ductility at 27°C, cm	81	>75
Specific gravity at 27°C	1.01	>0.99
Softening Point °C	50	46-54
Viscosity at 60°C, poise	2482	>2400
Viscosity at 135°C, cSt	371	>350
Flash Point Test °C	306	-
Fire Point Test °C	315	-

2.2. Methodology

The mix has been designed as per the Marshall Method. Aggregates conforming to Grading-II for Dense Bituminous Macadam layer have been used for this research work. In this study, total tests were performed in three phases. In the first phase, Optimum Bitumen Content (OBC) was calculated by varying the percentage of bitumen content using 2% Rice Husk ash (RHA) as a filler instead of conventional cement. In the second phase, the test specimens were prepared by varying the percentage of plastic content ranging from 2% to 8%. The plastic content increment was standardized as 2% by weight of OBC (obtained in the first phase) for determining the Marshall Parameters along with other relevant performance tests. From the results obtained on testing, the optimum plastic content was concluded. In the third phase, the bituminous mix was prepared with optimum waste plastic content (as obtained) by weight of OBC to determine the Reliability of the mix using statistical tools.

2.3. Experimental framework

2.31. Marshall stability test

The Marshall Test is used to design bituminous paving mixes by determining two important properties; stability and flexibility. Strength as obtained by Marshall Stability of the mix is defined as the maximum load sustained by a compacted mix at a standard test temperature of 60°C. The flexibility is measured in terms of flow value which may be regarded as its obverse property that determines the reversible response of the wearing course under the action of traffic loads. The optimum binder content (OBC) of the asphalt mixes were determined by evaluating the mixes with binder content from 4.75 wt.% to 5.75 wt.% at an increment of 0.25 wt.%, in accordance to the procedure of MS-2 guidelines [13].

2.3.2. Static Indirect Tensile Strength Test

This test is used to evaluate the resistance of bituminous mix against cracking and assess its sensitivity to moisture damage. To assess the susceptibility of the coating of aggregate and bitumen binder to moisture damage, tensile strength is determined according to ASTM D 4867 [14]. Tensile strength ratio (TSR) was thereby calculated as the ratio of average indirect tensile strength of conditioned specimens to the indirect tensile strength of un-conditioned specimens. The load at failure of respective specimens was recorded and the indirect tensile strength (ITS) was calculated from the following equation.
Indirect tensile strength (ITS) = $2P/\boldsymbol{\pi td}$
Where, P is load (kg); d is the diameter of the specimen (cm), and t is the thickness of the specimen (cm).

2.3.3. Resilient Modulus Test

The repeated loading indirect tensile strength test was performed as per ASTM D4123 on compacted plastic waste modified bituminous mix for finding resilient modulus values at different temperatures [15]. The test was conducted by applying a compressive load in the form of haversine wave at 25°C, 35°C and 45°C for different Bituminous mixes. The samples were conditioned for 5 hours in an environmental chamber at a specific temperature and then subjected to a repeated loading pulse width of 100 ms, and pulse repletion of 100 ms.

2.3.4. Rut Depth Test

Wheel Tracking Device (WTD) was used to assess the rutting resistance of the bituminous concrete mixes. It is a destructive test and has a distinct feature of direct contact between the loaded wheel and rectangular test samples. The test was conducted on a rectangular sample having dimension 300x300x50 mm. In this test 20,000 passes were applied on the test sample at 45°C and the resulting rut depth was duly measured.

3. Reliability analysis of the optimum mix

Arriving at an absolute certainty regarding the performance of a particular mix is unfeasible. Moreover, relying on large safety factors is not only impractical but also cost-intensive. Reliability analysis, therefore, ensures an acceptable level of risk in mix design, thereby giving us a probability regarding the satisfactory performance of the designed mix throughout its service period. It can be basically précised as the guide for consistent performance of the subject considered.

Usually, an acceptable asphalt mix may be expressed by limits on

- Volumetric parameters
- Marshall flow
- Marshall Stability.

Hence, the overall reliability of a particular mix must take into account the probability that all the limits are simultaneously satisfied. Although practically the parameters considered are statistically dependent on each other yet for simplification, it has been assumed to be statistically independent for the considered mix [16].

Hence the following expressions can be written as:

$$R\ (D) = \text{probability (volumetric limits satisfied by D)} \times \text{probability (flow limits satisfied by D)} \times \text{probability (stability limits satisfied by D)} \quad (1)$$

3.1 Probability analysis- governing limits

i. Volumetric limits are expressed in terms of limits on three parameters i.e. Void in Mineral Aggregate (VMA), % Air Voids (VA) and % Voids Filled with Bitumen (VFB). Let us consider a Cartesian space VMA x VA, theoretically the limits imparted represent a feasible area in this space. If we review the limits directed by MORT&H for Dense Bituminous Macadam, it provides the following criteria:
(a) Lower limit of VMA = 13% (b) limit of VA = 3% - 5% (c) limit of VFB = 65% -75%.
ii. Similarly for flow, the stipulated value of limits according to MORT&H specification is 2mm and 4 mm respectively for DBM [17].
iii. Finally for the stability which is specified only by the lower bound value is 9 KN for DBM as per MORT&H specification.

3.2 Determining the Probability Density Functions (PDF)

To develop the probability density function, corresponding to the obtained OBC 4.91% (by weight of aggregates), the significant variation of different Marshall parameters have been studied. It can be computed either by empirical or theoretical means. 30 Marshall Specimens were prepared at 4.91% OBC and tested for VMA, VA, flow, and stability. The PDF can be obtained by the following steps:

(a) Develop a frequency diagram for data
(b) fit a theoretical PDF to the data
(c) use Statistical Hypothesis test to determine Goodness of Fit.

4. Result and Discussion

4.1 Marshall stability

Figure 1 represents the relationship between stability and percentage of plastic used in bituminous mix. The stability value of the mix corresponding to 6% waste plastic modified mix provides the highest value which is 40 percent higher than the control mix (consisting of no plastic). The result is attributed to the increment of stiffness and flexibility of the modified mix by the incorporation of plastic. Beyond 6% plastic content, stability values slightly decrease thereafter with the increase of plastic additives. The results also portray that the value of Retained Stability increases with the increase of plastic content up to 6%, after which it decreases (figure 2).

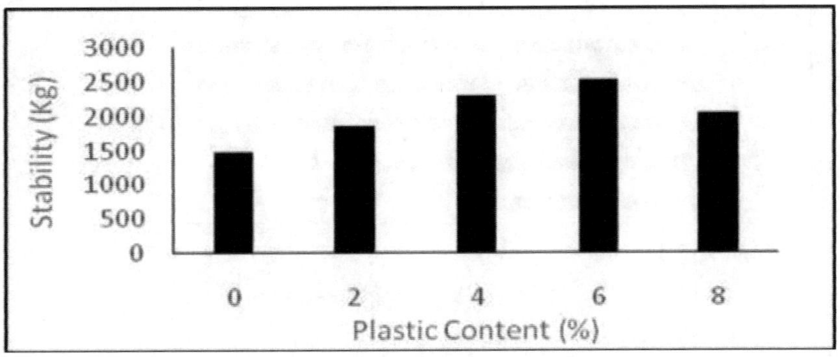

Figure 1. Marshall stability vs plastic content.

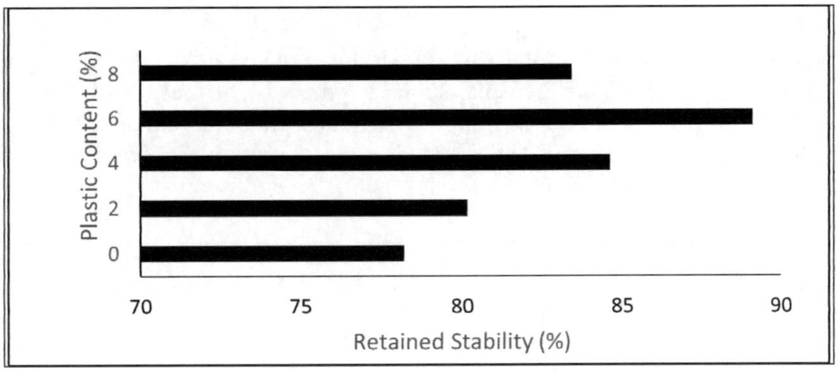

Figure 2. Plastic content vs retained stability.

4.2 Static Indirect Tensile strength
The Indirect Tensile strength test was performed for both unconditioned and conditioned samples for the varying percentage of plastic content. Results identify that for both unconditioned and conditioned specimens, the highest value of strength is obtained corresponding to 6% plastic content with a Tensile Strength Ratio (TSR) of 93%.

Table 5. Test results of Indirect Tensile Strength and TSR.

Plastic Content (%)	ITS (Unconditioned) (Kg/cm^{2})	ITS (conditioned) (Kg/cm^2)	TSR (%)
0	8.9	7.3	81.2
2	7.6	6.5	85.6
4	10.4	9.3	89.3
6	14.3	13.3	93.5
8	13.7	12.9	94.1

From figure 3, it is obvious that the use of a higher percentage of waste plastic in bituminous mix results in increased TSR which indicates that with an increase in plastic content, damage caused by ingress of water may be reduced to a large extent. Using 8% waste plastic as bitumen modifier provides the highest TSR value (94.1%).

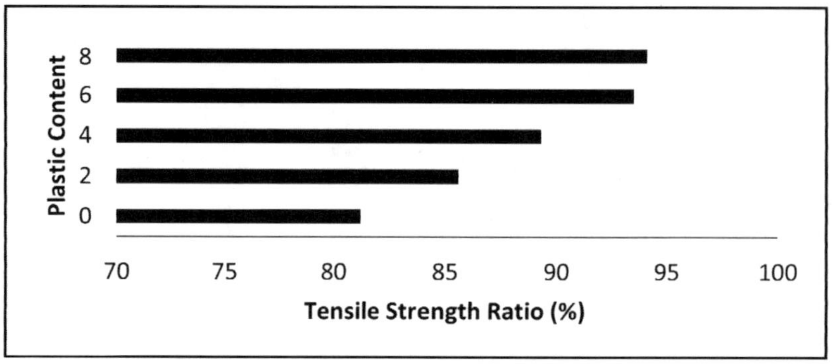

Figure 3. Tensile strength ratio vs plastic content.

4.3 Resilient Modulus Test

Figure 4 clearly shows the relationship of Resilient Modulus of the mix with varying plastic content with different test temperatures. Figure 3 displays the increment of stiffness modulus of asphalt mixes with the addition of plastic content and exhibited the highest modulus value for 6% plastic content followed by 8% for all test temperatures. The result signifies that the incorporation of plastic enhances the cohesion and stiffness nature of the mix.

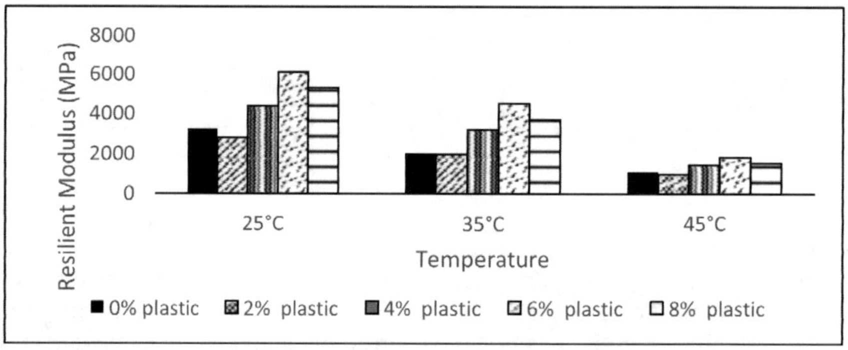

Figure 4. Resilient modulus vs temperature.

4.4 Rut depth study

Rut depth has a direct influence on traffic safety and gives a measure of general road conditions. Rutting- a distress associated with pavements is usually expressed in terms of rut depth. The rutting resistance of asphalt mixes was expressed in terms of their rut depth values as stated in figure 5. From the figure it can be understood that rut depth is maximum for the control binder. Upon the addition of the plastic content, the rut depth decreases significantly. It can be observed that at 20,000 cycles, rut depth was ≈ 7.2 mm, 6.1 mm, 5.6 mm, 3.7 mm and 4.1 mm for the mixes prepared with 0 wt. %, 2 wt.%, 4 wt. %, 6 wt. % and 8 wt. % plastic content, respectively. The decrease in rut depth in mixes prepared using waste plastic can be credited to the increase in stiffness, cohesion and elasticity of the mix.

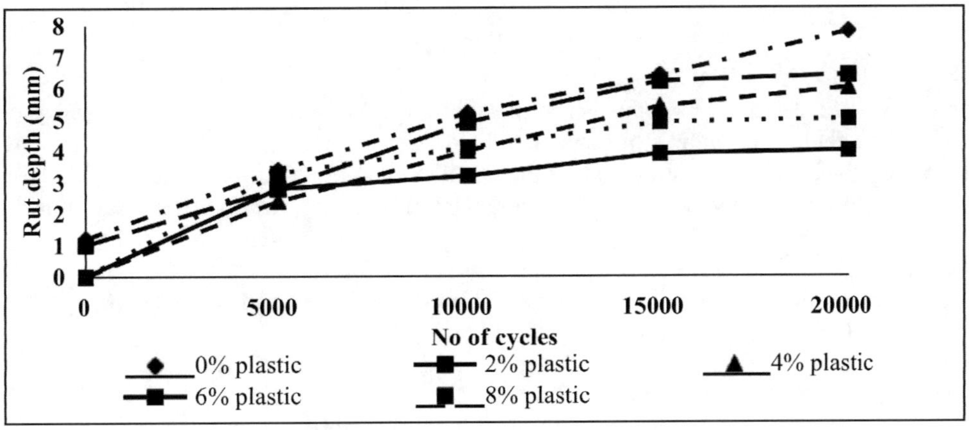

Figure 5. Rut depth vs number of cycles

4.5 Reliability Analysis

4.5.1 Frequency Diagram and Fitting a Standard Probability Density Function

After extracting the data required from observation, the frequency diagram was obtained by plotting the frequencies corresponding to the intervals. The obtained data has been divided into finite intervals and thereafter the frequency of a particular variable calculated considering its repetition in the particular interval. A standard distribution or PDF represents the frequency diagram so obtained. The method of moments has been used to determine the parameters of the chosen PDF. Computing the mean and standard deviation of the observed data conceived the fitted normal distribution.

Table 6. Results of fitted distribution for stability, flow, VMA, VA.

Variable	Observed Value		Fitted Distribution
	Mean	Standard Deviation	
Flow	3.34	0.8152	N (3.34,0.815)
Stability	22.6	6.772	N (22.5 , 6.772)
VMA	14.73	0.76003	N (14.73 ,0.760)
VA	4.39	0.8517	N (4.39 , 0.85)

4.5.2 Calculation of reliability

$$\iint \frac{1}{1.90} e^{-\frac{(m-14.73)^2}{1.76}} X \frac{1}{1.98} e^{-\frac{(a-4.39)^2}{1.26}} dmda = 0.75 \quad (2)$$

$$\int_{2}^{4} \frac{1}{1.835} e^{-\frac{(q-3.3)^2}{1.08}} dq = 0.79 \quad (3)$$

$$\int_{9.0}^{\alpha} \frac{1}{16.4} e^{-\frac{(s-22.6)^2}{86.07}} ds = 1.02 \quad (4)$$

From equation (1) and the probabilities obtained as above, Reliability = 0.75×0.79×1.02= 0.601.

So, from the current investigation, it is statistically proved that the proposed mix is quite (60%) reliable for field application.

4.6 Testing for Goodness of Fit

In this current work, a statistical hypothesis test like Chi Square has been used to assess the goodness of the obtained fit. It helps in finding out whether there is any consequential difference between the predicted and observed frequencies for each of the considered variables (flow, stability, VMA, and VA) thereby ensuring minimal effect on practical implementation with the current error in the approximation.

5. Conclusions

Taking into consideration the experimental investigations conducted and the detailed study involved, the following conclusions can be inferred.

1. Rice husk ash can be used as a mineral filler in binder course as cement replacement. By mixing rice husk ash in aggregate and LDPE containing waste milk pouches in bitumen as a modifier, Marshall Properties of dense bituminous concrete portrays a significant improvement.
2. By mixing 6% waste plastic as a modifier, retained stability and tensile strength ratio (TSR) value was highly improved as compared to bituminous mix consisting of only RHA. Results imply that there was an enhancement in the moisture sensitivity of the mix.
3. The values of Modulus of Resilient at 35°C and 45°C for 6% plastic waste modified mix are respectively 72 % and 53% higher than the only RHA added control mix. The reason may be the production of a denser and stiffer mix due to a decrease in air voids. Therefore, use of the rice husk ash-plastic waste composite was responsible to increase the indirect tensile strength of the bituminous mix.
4. It was observed that among the varied percentage of modifiers, 6 % waste composite reduced rutting in bituminous mix during wheel track testing by a considerable amount. So, it can be particularly concluded that 6% plastic mix gives the best-suited result in terms of minimizing permanent deformation.
5. Using 2% RHA and 6% plastic in the bituminous mix, the increased percentage of reliability value obtained clearly indicated a better and stable pavement in the construction industry. Thereby, the current investigation provides a viable solution in terms of solid waste management focusing on environmental sustainability and presenting a stable and cost-effective pavement.

Acknowledgement

The authors extend their gratitude to the Ministry of Human Resource and Development (MHRD, India), for financial assistance through student scholarship.

References

[1] Kumar S and Gaikwad S. 2004 Gupta K.R.(Ed): *Urban Development Debates in the New Millennium, Atlantic Publishers and Distributors*, New Delhi pp 100-111
[2] Manual on Municipal Solid Waste Management 2000 Government of India.
[3] Vasudevan R 2010 *Int. Journal Pavement Res. Technol* **3** 34-42.
[4] Verma S S 2008 *The Indian Concrete Journal* 43-44.
[5] Mohammad T A 2007 *American Journal of Applied Sciences* 390-396.
[6] Bindu C S and Beena D K S 2010 *International Journal of Engineering and Technology* **2** 379-387.
[7] Chen H 2010 *Fuel* **89** 1616–1622.
[8] Sargin S 2013 *Construction and Building Material* **48** 390-397.
[9] Ahmad M 2017 *ARPN Journal of Engineering and Applied Sciences* **15** 4351-4356.
[10] Mishra B 2017 *American Journal of Engineering Research* **6** 253-262

[11] Gidde M R and Jivani A P 2007 *Indian Literature Review*, Bharati Vidyapeeth Deemed University College of Engineering.
[12] IS: 73 2003 Paving Bitumen-Specification (Second Revision)
[13] Asphalt Institute 1997 Mix design methods for asphalt concrete and other hot-mix types: manual series No. 2 (*MS-2*) 6th ed.
[14] ASTM D6931-12 2012 Indirect Tensile Strength for Bituminous Mixtures *ASTM, West Conshohocken.*
[15] ASTM D4123-82 1995 Standard Test Method for Indirect Tension Test for Resilient Modulus of Bituminous Mixtures *ASTM, West Conshohocken.*
[16] Chakroborty P, Das A and Ghosh P 2010 *Journal of Transportation Engineering* **136** 31-37.
[17] MORTH 2013 Specifications for Road and Bridge Work *Ministry of Road Transport and Highway* New Delhi, India.

IOP Conf. Series: Materials Science and Engineering 1075 (2021) 012023 doi:10.1088/1757-899X/1075/1/012023

Non-Destructive Assessment of The Quality of Asphalt Laboratory Samples

M Zargar[1,*] and F Bullen[2]

[1,2]Department of Engineering and Build Environment, Faculty of Health, Engineering and Sciences, University of Southern Queensland, Toowoomba, Queensland, 4350, Australia

[1]Australian Road Research Board, 21 McLachlan St, Fortitude Valley, Queensland, 4006, Australia

[*] Corresponding author: Majidzargar2009@gmail.com

Abstract. In addition to instrumentation accuracy and test repeatability, the correlation of asphalt test data is very dependent on targeted "known" properties of asphalt samples or cores, such as air voids. In the case of air voids, sample preparation targets the desired sample attribute, however an assessment of the true air voids and their distribution within asphalt samples remains a challenge. Current assessment is usually via visual inspection and measuring the total air voids within the sample. However, such an approach fails to evaluate important factors such as air voids distribution and aggregate structure. In the case of testing regimes that are localised on test samples, the true air voids at the test location is essential for proper data correlation. In this study, an Ultrasonic Wave Transmission (UWT) technique was developed and applied as a non-destructive test to better assess the true localised air void properties of asphalt samples. The data correlation outcomes significantly changed the performance rankings of asphalt mixes and indicated that the UWT technique has excellent potential as a supplementary tool for evaluating asphalt laboratory samples.

1. Introduction

1.1. Non-destructive laboratory assessment of air voids of asphalt samples

An asphalt mixture (asphalt) is a complex distributed system of coarse aggregates, bitumen mastics (bitumen binder, fine aggregates and mineral powder) and air voids. The complex distribution of internal air voids and aggregate structures of asphalt samples as seen in figure 1 have been investigated in the laboratory using various non-destructive techniques.

Dubois et al. [1] successfully used a gamma-densitometer in 2010 to assess the influence of the compaction process on the air voids homogeneity of asphalt samples. Other non-destructive techniques such as X-ray Computed Tomography (CT) and image analysis methods [2-5] have been applied by researchers. The research outcomes have shown that air voids within compacted asphalt are not uniformly distributed [2, 4].

While X-ray methodologies have historically been used by researchers to investigate the complex distribution of air voids in asphalt mixtures, both cost and complexity of application do not support their widespread use. The research reported here describes a simple technique using ultrasonic assessment for laboratory use to predict the air voids content and their distribution in manufactured asphalt samples. This technique will be useful for better evaluating some tests such as fatigue and creep where the applied load (dynamic or static) is localised at a specific part of the asphalt sample.

1.2. Research plan

This research was planned in two phases. In phase one, which was presented in 2017 [6], the UWT technique was developed to measure the localised air voids in asphalt samples. In phase two, the Creep slope of the same asphalt samples was evaluated , and the results analysed using both the air voids of the whole sample (Using AS/NZS 289) and the air voids in the centre of the sample under the applied load (using the UWT technique developed in phase one).

The research reported here summaries the outcome of phase one and moves on to illustrate the benefits of measuring and application of localised air voids using UWT technique for analysing asphalt tests such as creep.

Figure 1. Asphalt structure [7]

2. Experimental Details

This research focuses on the importance of knowledge of localised air voids on the laboratory assessment of permanent deformation of manufactured asphalt samples using a confined creep approach.

2.1. Materials and sample preparation

Three commonly used hot mix asphalt (HMA) types used in Queensland Australia, were selected for use in the study. The aggregate gradation as seen in figure 2 was maintained constant. The aggregate and filler characteristics are shown in tables 1 and 2. Three different types of bitumen binders were used at 5%. They were Conventional (C320), Multigrade (M1000), and an SBS Polymer Modified (PMB-A5S) and their physiochemical properties are provided in table 3.

An asphalt shear box compactor using a range of compaction efforts was used to produce asphalt slabs with targeted air voids from 1% to 10%. Samples with a 150mm diameter were cored from the compacted asphalt slabs and then cut and trimmed to 50 mm height. Specimens were washed, dried and preconditioned at the test temperature of 25°C for 48 hours before being subjected to the air voids measurement test and the UWT test.

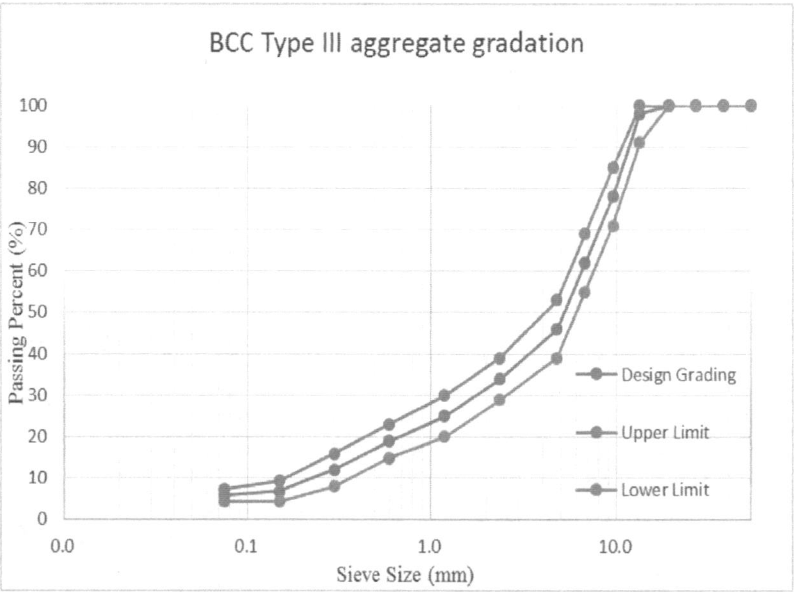

Figure 2. BCC Type III aggregate gradation

Table 1. Aggregate characteristics

Test Method	Description	Limits	7mm Test result	9mm Test result	14mm Test result
AS 1141.11	Grading	-	Conforming	Conforming	Conforming
AS 1141.15	Flakiness	<30%	11.6	12.4	7.8
Q 214 B	Water Absorption	Max 2%	0.68	0.39	0.5
Q 214 B	Particle density (Dry)	t/m3	2.666	2.668	2.672
Q 215	Crushed Particles	Min 80%	100	100	100
Q 217	Weak Particles	Max 1%	0.7	0	0
Q 205 B	10% Fines	Min 150 kN	296	272	239

Table 2. Filler characteristics

Test Method	Description	Limits	Baghouse Test result	Rockflour Test result	Combined BH/RF Test result
AS 1141.11	600 μm Grading (AS 2357 Limits)	100	100	100	100
AS 1141.11	300 μm Grading (AS 2357 Limits)	95-100	100	100	100
AS 1141.11	0.075 μm Grading (AS 2357 Limits)	75-100	92.8	97.1	97.1
AS 1141.17	Voids in Compacted Filler	Min 38%	48	49	46
AS 1141.7	Apparent Particle Density	TBR	2.706	2.756	2.729
	Products Conforms	(Yes/ No)	Yes	Yes	Yes

Table 3. Properties of bitumen binder class C320, M1000 and PMB-A5S

	Property	Test Method	Limits	Test result
C320	Viscosity at 60°C (Pa.s)	AS 2341.2	260-380	328
	Penetration at 25°C, 100g, 5s (pu)	AS 2341.12	Min 40	49
	Softening point (°C)	AS 2341.18	Report	51.2
	Viscosity at 135°C (Pa.s)	AS 2341.2	0.4-0.65	0.53
Multigrade	Viscosity at 60°C (Pa.s)	AS 2341.2	reported value	910
	Viscosity at 135°C (Pa.s)	AS 2341.4	1.5 max	0.78
	Softening Point (°C)	AS 2341.18	Report	58.5
	Flash Point, °C	AS 2341.14	250 min	348
PMB-A5S	Consistency at 60°C (Pa)	AG: PT/T121	5000 min	11254
	Stiffness at 25°C (kPa)	AG: PT/T121	30 max	19
	Viscosity at 165°C (Pa.s)	AG: PT/T111	0.9 max	0.64
	Softening Point (°C)	AG: PT/T131	82-105	103

2.2 Experimental procedure

2.2.1 UWT technique to measure air voids distribution

The UWT technique used to measure the air voids in the centre of the sample is briefly described below and the reader is referred to the paper by Zargar [8] for more details.

1. A Pundit 7 apparatus was applied according to BS EN 12504-4 standard to generate an ultrasonic wave and to determine the ultrasonic pulse velocity of the asphalt sample. Two 54-kHz piezoelectric crystal transducers (transmitter and receiver) were placed in parallel at each side of the specimen to measure the ultrasonic wave transit time at 17 different locations at 25°C as shown in figure 3. The transit time for the compression wave (P-wave) to pass the length of the asphalt sample was recorded to calculate the ultrasonic pulse velocity using the below formula:

$$V = l/t \tag{1}$$

   Where, V = ultrasonic pulse velocity (km/s)
   l = length of specimen (mm)
   t = transit time (ms)

2. The air void content of whole samples was measured using the specimen's bulk specific gravity (Saturated Method) and the asphalt's theoretical maximum specific gravity (Rice Method) according to Australian Standard (AS/NZS 2891).
3. The linear correlation between average ultrasonic velocities at 17 points and air voids of the whole sample were used to generate the regression equation.
4. Air voids at the centre of samples were calculated using 'ultrasonic velocity in the centre' applied in the regression equation (obtained in stage 3). It is this value that has been used in the current research.

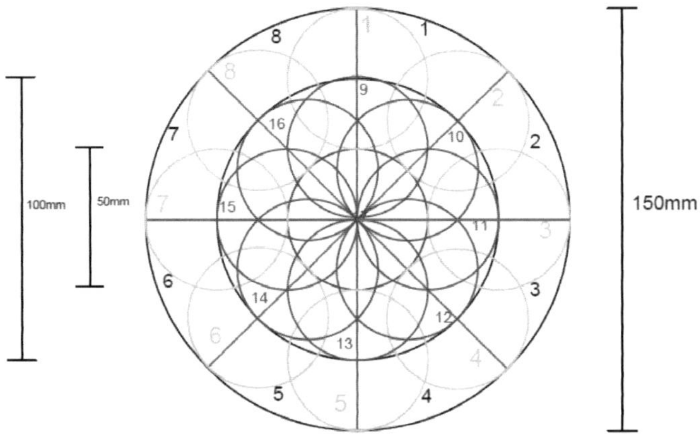

Figure 3. Sample divisions (17 points) for Ultrasonic measurements

2.2.2 Confined Dynamic Creep Test

The Confined Dynamic Creep Test (CDCT) is a development of the current Australian creep test (AS 2891.12.1995) based on research conducted by Ahmadinia [9]. The CDCT test used to measure Creep slope of sample is briefly described below and the reader is referred to the Thesis entitled "Redesigning the Austroads creep procedure for the evaluation of permanent deformation of asphalt pavements in Australia" by Ahmadinia [9] for more details.

1. The CDCT test (figure 4) used a Universal Testing System (UTS) manufactured by Industrial Process Controls (IPC) Global Limited Company to measure the creep deformation for confined samples. In this test stress responsive confinement for samples is provided by an asphalt annulus, PVC ring (2.5mm) and resin infill.
2. The test conditions used were:
    - Total stage number: 1 stage
    - Compressive stress: 750kPa
    - Seating stress: 20 kPa
    - Loading period: 100 milliseconds (ms)
    - Pulse repetition period: 1000 milliseconds (ms)
    - Test temperature: 50°C
    - Termination pulse count: 40,000 cycles
    - Platen size: 50mm diameter
    - Sample size: 150mm diameter * 50mm thickness
3. Creep slope measured from the steady state of creep (figure 5) for all samples

Figure 4. Confined dynamic creep test

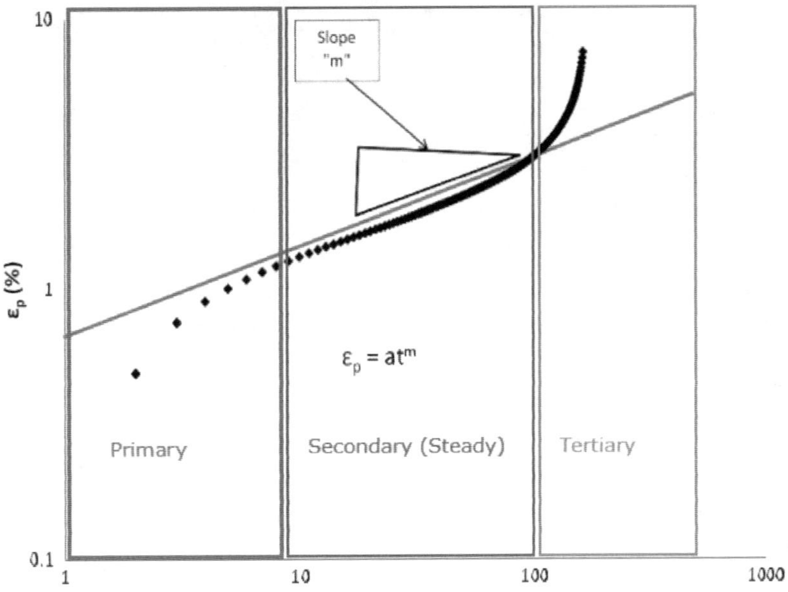

Figure 5. Schematic diagram of creep stages and slope

3. Experimental Results and Discussion

3.1. Findings from phase one of the project

In phase one of this study reported in 2017 [6], asphalt samples with 14mm nominal aggregate size were manufactured with three types of bitumen: Class 320 (C320), Multigrade (M1000), and an SBS Polymer Modified Bitumen (PMB-A5S), and with different air voids were analysed using the UWT technique. The results for ultrasonic velocity (the overall average of the 17 locations) and air voids content (1% to 10%) for all asphalt mixes (C320, M1000 and PMB-A5S), calculated at the reference temperature of 25°C, are summarised in figure 6.

It was concluded from the results that the higher the air void content in the sample, the lower the expected ultrasonic velocity. The trend lines indicated that there is a close correlation between air voids and average ultrasonic velocity. The data demonstrated the potential use of the non-destructive UWT tests to estimate localised air voids contents in laboratory compacted asphalt using the relationship between air voids and ultrasonic velocity.

The air voids at the centre of all three types of laboratory compacted asphalts were estimated as described earlier in section 2.2 and compared with the air voids of the whole sample. The difference between the air voids of the whole sample and localised air voids in the centre can be seen in figure 7. The data indicates that the air voids at the centre of the sample vary from those of the whole sample for all three types of asphalt.

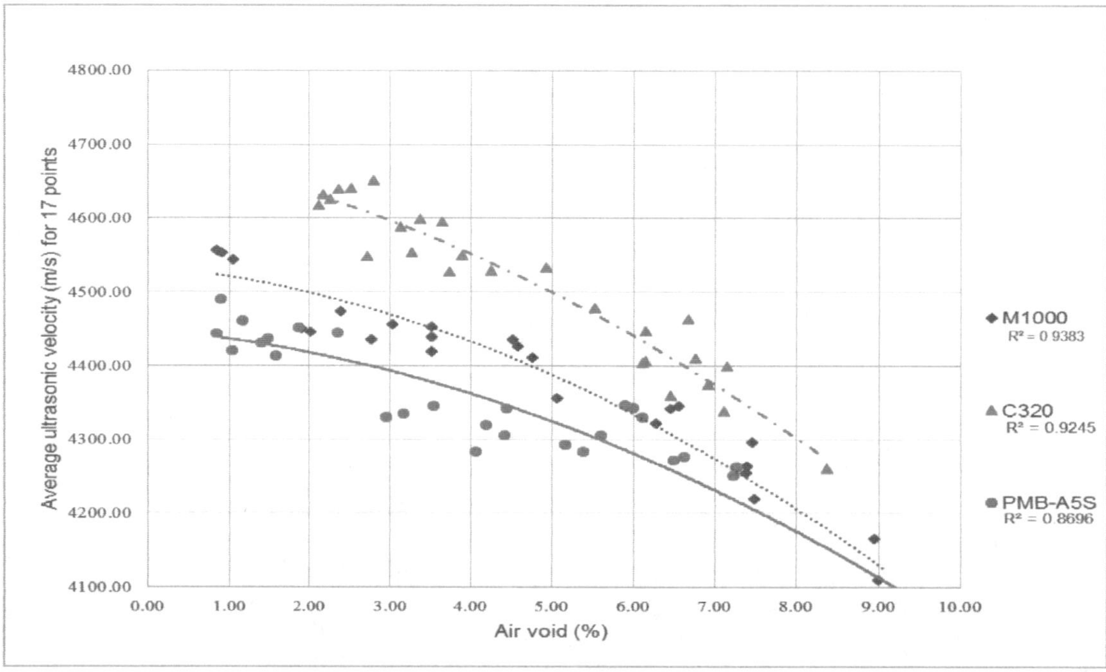

Figure 6. Ultrasonic velocity versus air voids (entire sample) content for mixtures with M1000, C320 and PMB in average 17 points [6]

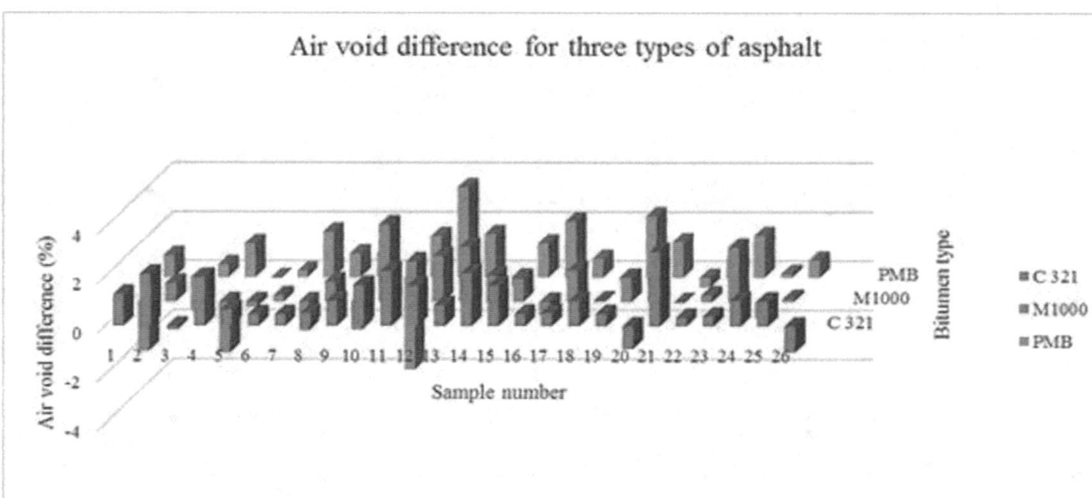

Figure 7. Air Voids difference (air void of whole sample minus localised air voids in centre)

3.2 *Findings from phase two of the Project*

In phase two, the Creep slope of asphalt samples was measured (figure 5), and the results correlated with the air voids of the whole sample (Using AS/NZS 289) and the air voids in the centre of the sample directly under the 50mm diameter load platen (figure 4), determined using the UWT technique.

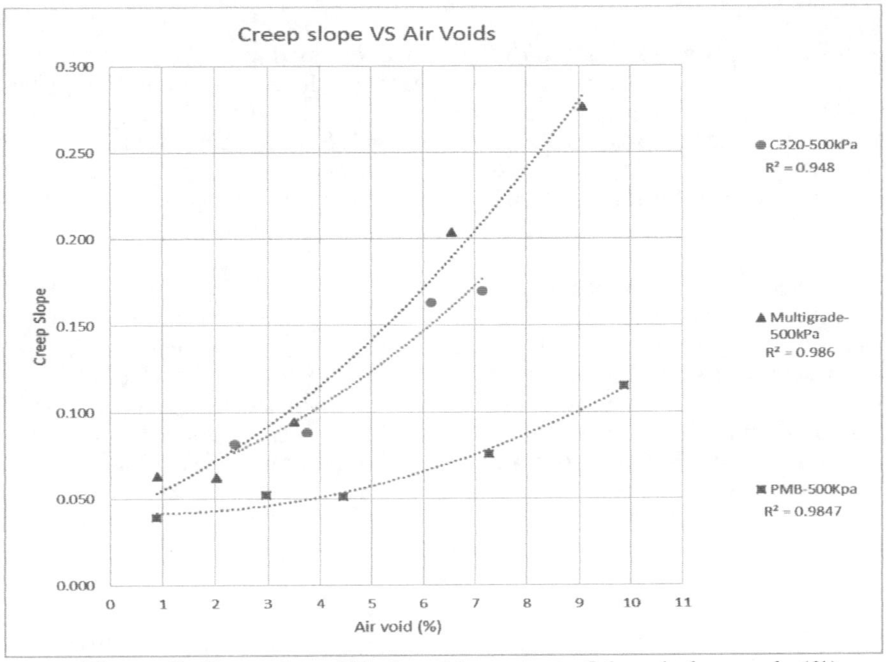

Figure 8. Creep slope VS air voids content of the whole sample (%)

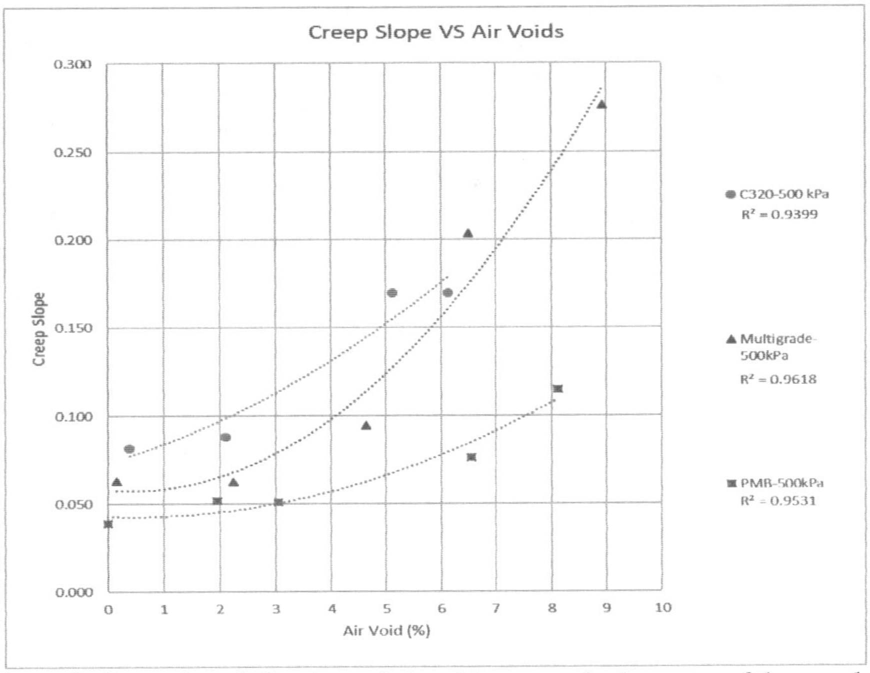

Figure 9. Creep slope VS estimated air voids content in the centre of the sample (%)

Correlation between creep slope and overall air voids content

In figure 8, the creep slope is plotted versus the air voids content of the entire sample for the three types of asphalt. The data shows a good correlation between increasing air voids contents, and creep slope for each mix type. The data could be interpreted as ranking the PMB mix as the most rut (creep) resistant, followed by C320 and then closely by Multigrade M1000.

Correlation between creep slope and air voids in centre of the sample

The creep slope versus air voids at the centre of the sample for the three types of asphalt is plotted in figure 9. Again, close correlation is observed between increasing air voids, and increasing creep slope for each type of mixture. The data could be interpreted as ranking the mix types, and PMB is again ranked as the most rut resistant. However, the data now suggests that the Multigrade M1000 is the next best performer with C320 being ranked last. Such a change in mix ranking is significant and is a cause for further investigation.

4. Conclusion

The research has shown that the UWT technique can be applied as a fast and effective non-destructive tool to estimate the air voids content and distribution in asphalt mixes. It was found that the air voids are non-uniformly distributed in laboratory compacted asphalt samples. The UWT technique can be applied to assist researchers in the analysis and application of asphalt test data involving air voids as an experimental variable.

The usefulness of the ultrasonic technique to measure air voids and their distribution has been illustrated using the newly developed Confined Dynamic Creep Tests (CDCT) that were performed in conjunction with this research. A change in the ranking of mix types occurred when asphalt sample creep was based on localised air voids under the applied loads, instead of whole of sample air voids. It is believed that the use of UWT techniques allows for more accurate ranking of mixes by enabling the use of local air voids content at the point of stress application as a test variable.

References

[1] Dubois V, De La Roche C and Burban O 2010 Influence of the compaction process on the air void homogeneity of asphalt mixtures samples *Construction and Building Materials,* **24,** 885-897.

[2] Castillo D and Caro S 2014 Probabilistic modeling of air void variability of asphalt mixtures in flexible pavements *Construction and Building Materials*, **61**, 138-146.

[3] Masad E, Muhunthan B, Shashidhar N and Harman T 1999 Internal structure characterization of asphalt concrete using image analysis *Journal of computing in civil engineering*, **13**, 88-95.

[4] Masad E, Jandhyala V, Dasgupta N, Somadevan N and Shashidhar N 2002 Characterization of air void distribution in asphalt mixes using X-ray computed tomography *Journal of materials in civil engineering*, **14,** 122-129.

[5] Partl M N, Bahia H U, Canesterari F, De La Roche C, Di Benedetto H, Piber H and Sybilski D 2012 Advances in interlaboratory testing and evaluation of bituminous materials: state-of-the-art report of the RILEM technical committee 206-ATB *Springer Science & Business Media*, **9**.

[6] Zargar M, Banerjee S, Bullen F, Ayers R 2017 An Investigation into the Use of Ultrasonic Wave Transmission Techniques to Evaluate Air Voids in Asphalt *Global Civil Engineering Conference, Springer* 1427-1439.

[7] Tashman L, Masad E, D'Angelo J, Bukowski J and Harman T 2010 X-ray Tomography to Characterize Air Void Distribution in Superpave Gyratory Compacted Specimens *International Journal of Pavement Engineering*, 1029-8436.

[8] Zargar M 2018 Development and application of a new multistage confined dynamic creep test for asphalt *PhD Thesis of University of Southern Queensland, Australia.*

[9] Ahmadinia E 2017 Redesigning the Austroads creep procedure for the evaluation of permanent deformation of asphalt pavements in Australia *PhD Thesis of University of Southern Queensland, Australia.*

Acknowledgements

This study was sponsored by the Brisbane City Council and the University of Southern Queensland. The authors wish to thank all laboratory technicians at the Eagle Farm asphalt plant and USQ, and the Director of Research at the Brisbane City Council.

Utilization of jarosite in precast concrete paver blocks

S Nandi[1,*] and G D R N Ransinchung[2]

[1] Department of Civil Engineering, Indian Institute of Technology Roorkee, Uttarakhand 247667 India
[2] Faculty of Civil Engineering, Department of Civil Engineering, Indian Institute of Technology Roorkee, Uttarakhand 247667 India
[*] Corresponding author: snandi@ce.iitr.ac.in

Abstract. Sustainability has become the major concern for the construction industry in the recent era. Concrete stands as a versatile construction material, but cement being its main constituent poses a threat to the environment consuming a huge amount of energy and emanating CO_2 and other greenhouse gases in its production process. On the other hand, the extraction of natural resources for construction works has been disrupting the environmental and ecological balance due to a huge accumulation of waste or by-products. So the major concern of the hour is the formulation of a mitigation strategy to fight this uncontrolled quantum of construction and its resulting consequences. The main objective of this present study is the utilization of jarosite, a hazardous solid waste, as a replacement of cement in concrete paver blocks. Six concrete mixes were prepared to keep the water-to-cementitious material ratio fixed at 0.43. The replacement of cement with jarosite was carried out at five replacement levels (0 – 25%) in an increment of 5%. Mechanical properties (compressive and flexural strength) and durability properties (water absorption and abrasion resistance) of the jarosite incorporated paver blocks were determined and compared to the control specimens. Experimental results indicate that mechanical and durability properties improve with the incorporation of jarosite, which could be attributed to a much developed packing of the jarosite concrete mix.

1. Introduction

The recent era obligates the construction industry to practice sustainability as a concurrent challenge in its production activities by exploring various eco-friendly raw materials or using solid wastes in concrete [1,2]. In this context, recycling indubitably serves as the most viable alternative to ameliorate the impact that the environment can suffer from the raw material consumption and uncontrolled generation of waste [3,4]. Concrete stands as a versatile construction material. Its characteristic to assess different waste materials by incorporating them in its production process has provided it with immense importance in the construction sector [3,5]. Being provided with a second life in concrete necessitates performance evaluation of these alternative materials since their properties may differ from the material they are used as a substitute for [6]. This will help in addressing the sustainability and durability issues both at the same time.

Jarosite is a solid residue universally produced in huge quantities during the zinc extraction process. Due to the existence of heavy metals like cadmium, lead, zinc, copper, chromium, and other oxides of metals and non-metals, it has been categorized as hazardous waste [7–9]. Moreover, the concentration of the toxic element in jarosite stands out much higher than the recommended permissible values of the United States Environmental Protection Agency (USEPA). About 0.25 million tonnes of jarosite waste is released from the industrial sector per annum in India [9]. Limited literature is available regarding the jarosite recycling and its utilization. The use of processed jarosite waste with small additions of lime and Portland cement has found applications in civil engineering. They may be used as construction materials in road bases and subbases, dams and airfields, as a replacement of conventional materials and even finds its place in the production of tiles, bricks, and other similar products [10]. Development of models simulating the coupling effect of diffusion and precipitation at the interface of chemically contrasting wastes advocates the potential use of the jarosite waste to form a self-sealing layer at combined jarosite/fly ash disposal sites [11]. Past studies have affirmed the possible use of sand and coal combustion residues (CCRs) as admixtures immobilizing the transport of heavy metals and toxic elements present in jarosite wastes [5,9,12,13].

The influence of jarosite replacing cement in precast concrete paver blocks is not studied yet. Paver blocks have recently emerged out to be an attractive economic and engineering alternative compared to asphalt and rigid pavements. Besides this, its improved mechanical and durability properties, easy maintenance, and aesthetically appealing surfaces have made it fit for application in parking lots, pedestrian crosswalks, traffic intersections, container depots, and low volume roads [3]. Therefore, our present research objective was to assess the feasibility of using jarosite waste as a partial replacement of cement in manufacturing precast concrete paver blocks.

1.1. Research significance
The production process of Portland cement involves a massive amount of energy consumption and the generation of unwanted CO_2 emissions. The cement plants account for 5 – 7% of global CO_2 emissions, while one ton of cement production generates 900 kg of CO_2 to the atmosphere [14]. This uncontrolled CO_2 emission, which falls among greenhouse gases, can be catastrophic, leading to global warming [15,16]. As a mitigation strategy, much attention has been given to the development of a new generation of blended cement with the same engineering properties and resulting in energy saving [17]. Utilization of jarosite as a replacement of cement can abate CO_2 emissions and, at the same time, counter the disposal and contamination problems because of its hazardous nature. Further, the incorporation of jarosite in the manufacture of paver blocks will be first of its kind.

2. Experimental investigation

2.1. Materials
Ordinary Portland cement (OPC) of 43 grade, complying IS:8112 [18] requirements, procured from the local market, was used throughout the investigation. The cement used had a specific surface area of 3329 cm^2/gm determined as per IS:4031-Part-2 [19]. The jarosite used was availed from the Debari Zinc Smelter plant of Hindustan Zinc Limited located at Udaipur, Rajasthan, India. The jarosite samples collected were present as lumps, and it needed an amount of processing. The samples were first oven-dried at 50±5°C for 24 hours, milled into a fine powder and stored in air-tight containers for further use. The specific surface area of processed jarosite determined by the Brunauer-Emmett-Teller (BET) method was observed to be 102296 cm^2/gm, and it had a pH value of 4.19, showing its acidic nature. Specific gravity values of cement and jarosite reported in Table 1 have been investigated as per IS:4031-Part-11 [20].

The chemical composition of cement and jarosite samples were determined by X-ray fluorescence (XRF) technique. The elements present in oxide form in jarosite sample were as follows: Fe_2O_3 (37.23%), SO_3 (27.36%), SiO_2 (8.16%), Na_2O (3.76%), ZnO (2.90%), PbO (2.84%), Al_2O_3 (2.40%), CaO (1.04%), K_2O (0.55%), CuO (0.14%), TiO_2 (0.10%), MgO (0.09%), P_2O_5 (0.08%), MnO (0.07%), SrO (0.05%), CdO (0.04%), Cr_2O_3 (0.03%) and BaO (0.03%). The cement sample on the other hand chemically composed of the oxides of the following elements: CaO (63.56%), SiO_2 (19.65%), Al_2O_3 (4.58%), Fe_2O_3 (4.20%), MgO (3.21%), SO_3 (2.21%), K_2O (0.79%), Na_2O (0.65%), P_2O_5 (0.16%) and

MnO (0.07%). As observed from the XRF study, the oxides of sulphur (SO_3) and iron (Fe_2O_3) were present in substantial quantities in jarosite.

Table 1. Properties of raw materials used.

Property	Cement	Coarse aggregate (16 mm)	Coarse aggregate (20 mm)	Fine aggregate	Jarosite	SP[a]
Specifications	OPC Grade 43	Crushed sandstone	Crushed sandstone	River sand (Zone III)	–	–
Specific gravity	3.15	2.60	2.63	2.56	2.90	1.18
Water absorption (%)	–	0.63	0.55	1.42	–	–

[a] Superplasticizer

Two sizes of coarse aggregates, 16 mm graded, and 20 mm single size was used in this study. These two aggregates were blended (95% of 16 mm and 5% of 20 mm) to obtain graded aggregates of nominal size 20 mm conforming to IS:383 [21]. The fine aggregate used was procured from the Ganga riverbed located near Uttarakhand, India and affirmed to grading zone III as per IS:383 [21] specifications. The particle size distribution of natural coarse and fine aggregates can be visualized through figure 1. Specific gravity and water absorption of the different aggregates determined as per IS:2386 [22] have been presented in table 1. Potable water complying with IS:456 [23] requirements was used to prepare the considered mixes and curing the test specimens. A varying dose of sulfonated naphthalene-based super plasticizing admixture was used for all the concrete mixes. The dosage of the admixture needs to be determined through trial mixes.

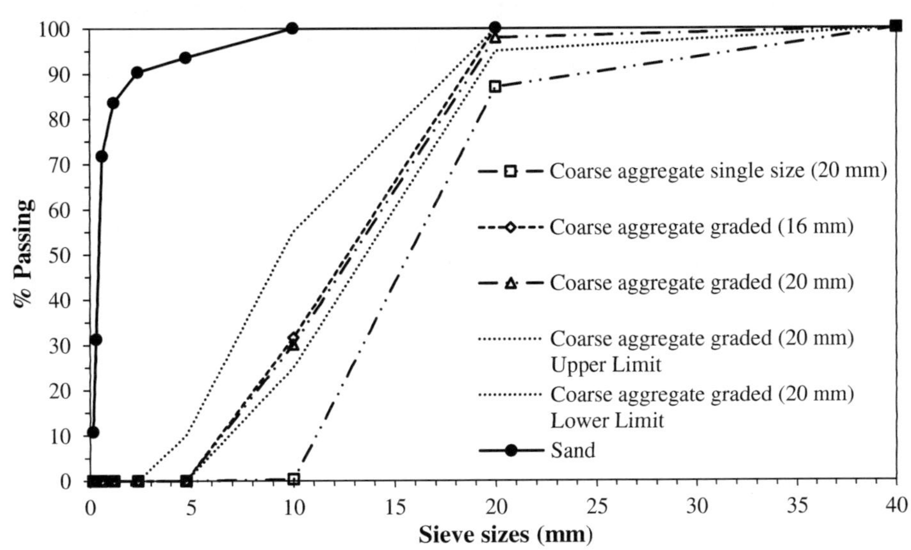

Figure 1. Grain size distribution of coarse and fine aggregates used in the study.

2.2. *Mix proportioning*

The mix design has been done following the provisions in IRC:SP:63 [24] and IRC:44 [25] adopting the water-to-cementitious material ratio (w/cm) as 0.43 throughout the study. Replacement of cement with jarosite (0 – 25%) in an increment of 5% has been carried out. A total of 6 mixes, including the control mix, was prepared for which the dosage of superplasticizer was adjusted accordingly to achieve the desired workability in terms of slump value. The concrete mix proportions have been summarized in table 2. A mix designation (MD) in the form “PBXJ” has been used to represent the various mixes. In "PBXJ", "PB" stands for paver block, "J" stands for jarosite, and "X" represents the different levels of replacement of cement with jarosite (0%, 5%, 10%, 15%, 20%, and 25%).

Table 2. Proportions of fresh concrete mixes.

MD	Water (Kg)	Cement (Kg)	Jarosite (Kg)	Fine aggregate (Kg)	Coarse aggregate 16 mm (Kg)	Coarse aggregate 20 mm (Kg)	SP (%)	Slump (mm)
PB0J	158.1	367.7	0.00	640.40	1167.91	62.18	0.7	17
PB5J	158.1	349.3	18.38	639.68	1166.60	62.11	0.8	20
PB10J	158.1	330.9	36.77	638.68	1164.78	62.01	1.0	16
PB15J	158.1	312.5	55.15	637.68	1162.96	61.91	1.2	19
PB20J	158.1	294.1	73.53	636.69	1161.14	61.82	1.4	17
PB25J	158.1	275.8	91.92	635.69	1159.32	61.72	1.6	19

2.3. Fabrication of paver blocks

The paver blocks were fabricated in steel molds having internal dimensions of 300 mm in length, 150 mm in width, and 150 mm in height using a wet-mixed method. A mixing procedure similar to that adopted for making concrete with lightweight aggregates was adopted. At first, the aggregates and half of the mix water were mixed in the drum mixer for 30 s. After this, the cement and jarosite were added to the wetted aggregates and mixed for another 30 s resulting in surface-coated aggregates. Finally, the superplasticizer mixed half of the remaining water was added and mixed continuously for 2 min to prepare the fresh concrete mix. The fresh concrete was placed into the molds in three layers of equal thickness, with each layer being tamped 25 times. Then the assembly was placed over a vibrating table and vibrated for 5 seconds. While the specimens were vibrated, a trowel was used to remove the excess material and provide a good surface finish. The paver block specimens were demolded 24 h after the casting and cured in water until the testing date.

2.4. Experimental methods

The initial slump of fresh concrete mixes using a slump cone test was carried out as per ASTM C143 [26] specification. The density of the freshly mixed concrete was calculated according to the guidelines presented in ASTM C138 [27]. Compressive and flexural strength tests of the 300 x 150 x 150 mm paver block specimens at 7 and 28 days of age were conducted according to the procedure laid down in IS:15658 [28]. For the determination of water absorption of the paver blocks IS:15658 [28] specification was followed. The abrasion resistance was determined on concrete specimens of size 300 x 300 mm in plan and depth 100 mm according to the guidelines prescribed in ASTM C779 [29], where Procedure A involving the revolving-disk type abrasion machine was followed. The water absorption and abrasion resistance test was performed on concrete specimens moist cured for 7 and 28 days. Finally, the microstructure study of powdered jarosite sample and jarosite inclusive concrete specimens were carried out using scanning electron microscopy (SEM).

3. Results and discussion

3.1. Effect of jarosite on fresh properties

The effect of the incorporation of jarosite on the initial slump value of the fresh concrete mixes is shown in table 2. It may be noted that the slump value of the mixes could be kept within a range of 16 – 20 mm by varying the dosage of superplasticizer from 0.7 – 1.6%. The decreased workability of mixes with an increase in the replacement level of jarosite could be attributed to the much higher specific surface area of jarosite particles in comparison to cement. As depicted in table 3, the reductions in fresh density values were noted to be 0.8%, 1.6%, 2.2%, 2.6% and 3.3% in comparison to control mix (PB0J) for the mixes containing 5%, 10%, 15%, 20% and 25% of jarosite respectively.

The lower specific gravity of jarosite as compared to cement possibly accounts for this reduction in the density of the fresh concrete mixes.

Table 3. Fresh density of considered concrete mixes.

MD	PB0J	PB5J	PB10J	PB15J	PB20J	PB25J
Fresh Density (kg/m^3)	2479	2459	2439	2426	2415	2396

3.2. Effect of jarosite on mechanical properties

3.2.1. Compressive strength. Figure 2 illustrates the effect of jarosite content on the compressive strength of the paver blocks. As depicted in this figure, with increasing of jarosite content up to 15%, the paver compressive strength increases at both 7 and 28 days curing age. At the initial days of moist curing (7 days), a higher increment in compressive strength values was observed for different mixes than the latter days (28 days). Compared with the control mix, the addition of jarosite at 5%, 10% and 15% resulted in a compressive strength gain of 3.6%, 13.9% and 24.4% at 7 days, while the increment values were noted to be 3.4%, 11.7% and 19.1% at 28 days; respectively. However, the addition of jarosite of more than 15% resulted in a reduction in concrete compressive strength. For instance, using of jarosite at 20% and 25% reduced the concrete strength by 4.9% and 4.4% (with respect to control) at 28 days. The increase in compressive strength could be attributed to the development in the microstructure of concrete mix with subsequent incorporation of jarosite particles. The fine jarosite particles favor the occurrence of pozzolanic activity and fill the micropores within the concrete matrix. On the other hand, the potential reduction in the cementing materials and dilution effect occurring due to the increasing replacement of cement with jarosite results in a decrease in compressive strength.

Comparing to specifications of paver blocks, according to IS:15658 [28], all the hardened jarosite concrete mixes fulfill the strength criterion for applicability in medium-traffic areas like low volume roads, city streets, small and medium market roads. It may be noted that the concrete mix with 15% jarosite replacement level even fulfilled the strength criterion of heavy traffic applications.

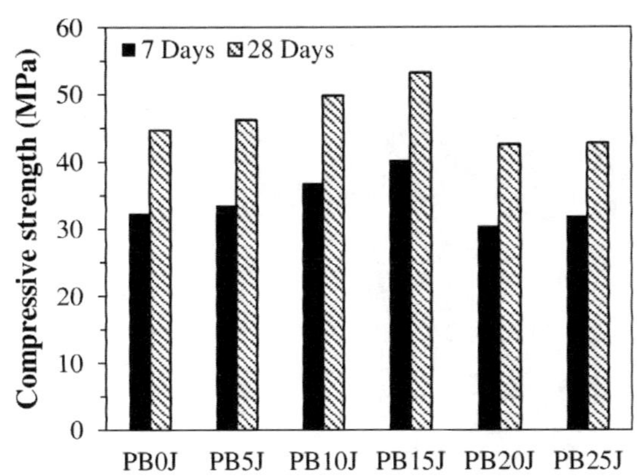

Figure 2. Effect of jarosite replacement on compressive strength.

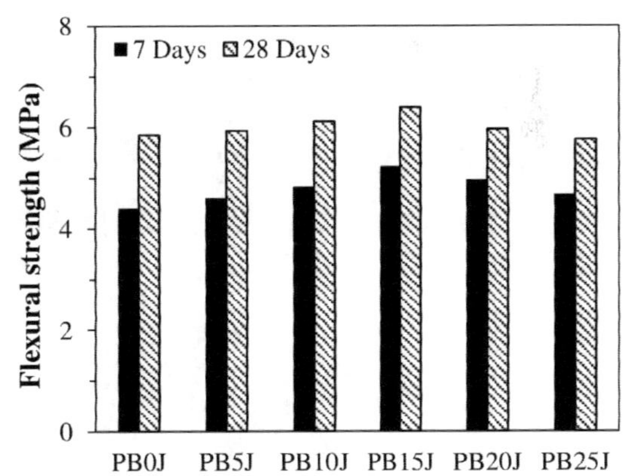

Figure 3. Effect of jarosite replacement on flexural strength.

3.2.2. Flexural strength. The flexural strength result of the jarosite concrete paver blocks follow a similar trend as the compressive strength and is presented in figure 3. The flexural strength increases by the addition of jarosite up to 15% and then decreases with further addition. As compared with the control mix, the addition of jarosite at 5%, 10% and 15% resulted in a strength gain of 1.4%, 4.7% and 9.4%, respectively, at 28 days. However, unlike compressive strength results, though there is a reduction in flexural strength at 20% jarosite replacement level, the strength is 2% higher than control concrete. Moreover, the 28-day flexural strength value of 20% and 25% jarosite concrete mix are comparable to that of the control concrete. These results may be explained with the arising heterogeneity in concrete mixes at higher jarosite incorporation levels [5].

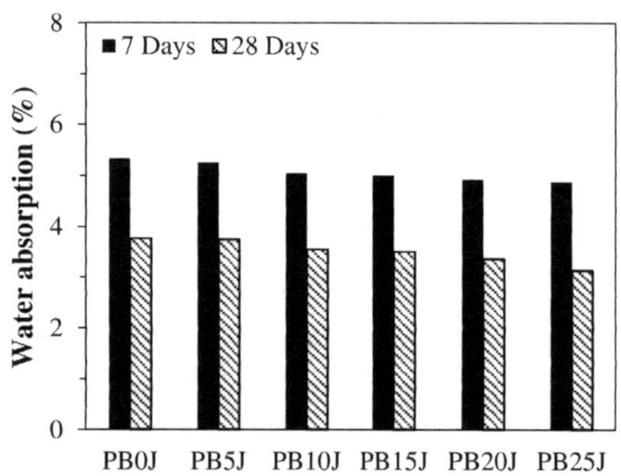

Figure 4. Effect of jarosite replacement on water absorption.

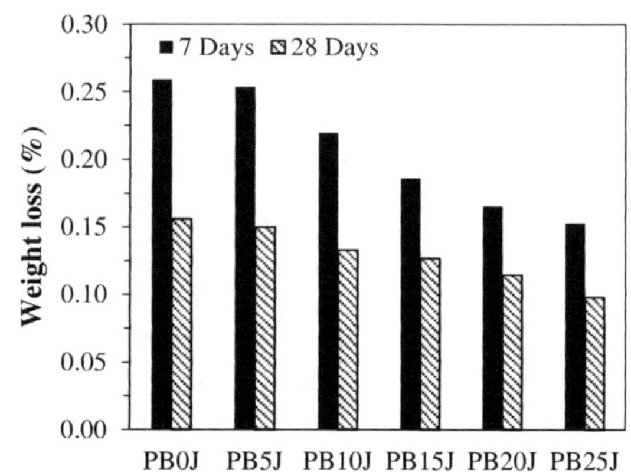

Figure 5. Effect of jarosite replacement on abrasion resistance.

3.3. Effect of jarosite on durability properties

3.3.1. Water absorption. The test results for water absorption of jarosite inclusive paver block samples is presented in figure 4. The figure depicts a decrease in water absorption percentage with an increase in jarosite content and is in line with previous findings [5]. Comparing with control mix, addition of jarosite at 5%, 10%, 15%, 20% and 25% resulted in a reduction in water absorption by 1.5%, 5.3%, 6.2%, 7.6% and 8.5% respectively after 7 days of moist curing age. At 28 days of moist curing, a higher reduction in water absorption value was found. This decrease in water absorption value may be possibly due to the higher surface area of jarosite particles filling the pores and results in a densely packed concrete matrix. Moreover, the water absorption values fall within the permissible limits stipulated for paver blocks as per IS:15658 [28] specifications.

Figure 6. SEM microstructure of jarosite particles.

3.3.2. Abrasion resistance. The effect of jarosite inclusion on the abrasion resistance of paver blocks is illustrated in figure 5. The abrasion test results are presented in terms of weight loss of specimens on exposure to abrasive grit. As compared to control specimen, the reduction in weight loss of 7 days cured specimens were found to be 2%, 15.1%, 28.1%, 36.2% and 40.9% for jarosite replacement levels of 5%, 10%, 15%, 20% and 25% respectively. Similarly, the abrasion resistance increased for the 28 days cured samples with subsequent jarosite inclusion levels compared to the control specimen. The abrasion resistance of the hardened paver block mixes can be correlated to the water absorption results. The pore filling effect due to the higher specific surface area (102296 cm^2/gm) of jarosite particles helps improve the abrasion resistance of the mixes.

3.4. Microanalysis using SEM

The microstructure characteristics of jarosite particles, as illustrated in figure 6, show them to be irregularly shaped, non-uniform, and with multiple humps. This is in line with the past literature studies of jarosite particles [9,12,13,30]. SEM images of the jarosite concrete mix specimens designated PB0J and PB15J have been presented in figures 7 and 8, respectively. PB0J mix containing no jarosite (control) was spotted to have a greater void (V) and portlandite (CH) compared to the PB15J mix containing 15% jarosite replacement level. PB15J mix was observed to have a denser microstructure with lesser voids (V), dense clouds of calcium silicate hydrate (CSH) gel, and fibrous hydration products, which confirms that increasing the percentage of jarosite has a pore filling effect and offers a denser concrete matrix.

Figure 7. SEM micrograph of PB0J concrete specimen.

Figure 8. SEM micrograph of PB15J concrete specimen.

4. Conclusions

Based on the present laboratory investigation, it can be concluded that the manufacture of paver blocks using jarosite stands a feasible option. Further, the following inferences can be drawn from the present study.

- The inclusion of jarosite decreased the workability of concrete mixes. Further, a reduction in fresh density value due to the replacement of cement with jarosite was found.
- Enhancement in compressive and flexural strength (mechanical properties) was observed with a subsequent increase in jarosite replacement level up to 15%. However, though there is a reduction in mechanical properties, cement can be replaced with jarosite up to 25% in paver blocks as the compressive and flexural strength values fall within a similar range to that of the control mix.
- As far as the effect of jarosite on durability properties of concrete are concerned, an improvement was observed with subsequent jarosite replacement levels.
- The study of the microstructure of hardened concrete confirms the formation of a denser matrix with jarosite incorporation.

Acknowledgment
The authors would like to acknowledge Hindustan Zinc Limited, Udaipur, Rajasthan, India to provide jarosite to carry out the present laboratory investigation.

References

[1] Thomas B S and Gupta R C 2016 Properties of high strength concrete containing scrap tire rubber *J. Cleaner Prod.* **113** 86–92
[2] Pelisser F, Zavarise N, Longo T A and Bernardin A M 2011 Concrete made with recycled tire rubber: effect of alkaline activation and silica fume addition *J. Cleaner Prod.* **19(6–7)** 757–63
[3] Uygunoğlu T, Topcu I B, Gencel O and Brostow W 2012 The effect of fly ash content and types of aggregates on the properties of pre-fabricated concrete interlocking blocks (PCIBs) *Constr. Build. Mater.* **30** 180–7
[4] Fattuhi N I and Clark L A 1996 Cement-based materials containing shredded scrap truck tyre rubber *Constr. Build. Mater.* **10(4)** 229–36
[5] Mehra P, Gupta R C and Thomas B S 2016 Properties of concrete containing jarosite as a partial substitute for fine aggregate *J. Cleaner Prod.* **120** 241–8
[6] Tripathi B and Chaudhary S 2016 Performance based evaluation of ISF slag as a substitute of natural sand in concrete *J. Cleaner Prod.* **112** 672–83
[7] Theodore L and Reynold J 1987 *Introduction to Hazardous Waste Management* (New York: John Wiley and Sons Inc.)
[8] Leclerc N, Meux E and Lecuire J M 2003 Hydrometallurgical extraction of zinc from zinc ferrites *Hydrometallurgy* **70(1–3)** 175–83
[9] Asokan P, Saxena M and Asolekar S R 2006 Hazardous jarosite use in developing non-hazardous product for engineering application. *J. Hazard. Mater.* **137(3)** 1589–99
[10] Mymrin V A, Ponte H A and Impinnisi P R 2005 Potential application of acid jarosite wastes as the main component of construction materials *Constr. Build. Mater.* **19(2)** 141–6
[11] Ding M, Schuiling R D and Van der Sloot H A 2002 Self-sealing isolation and immobilization: a geochemical approach to solve the environmental problem of waste acidic jarosite *Appl. Geochem.* **17(2)** 93–103
[12] Pappu A, Saxena M and Asolekar S R 2006 Jarosite characteristics and its utilisation potentials *Sci. Total Environ.* **359(1–3)** 232–43
[13] Mehra P, Gupta R C and Thomas B S 2016 Assessment of durability characteristics of cement concrete containing jarosite *J. Cleaner Prod.* **119** 59–65
[14] Benhelal E, Zahedi G, Shamsaei E and Bahadori A 2013 Global strategies and potentials to curb CO2 emissions in cement industry *J. Cleaner Prod.* **51** 142–61
[15] Zhang X F, Zhang S Y, Hu Z Y, Yu G, Pei C H and Sa R N 2012 Identification of connection units with high GHG emissions for low-carbon product structure design *J. Cleaner Prod.* **27** 118–25
[16] Mahlia T M 2002 Emissions from electricity generation in Malaysia *Renewable Energy* **27(2)** 293–300
[17] Mehta P K 1980 Investigations on energy-saving cements *World Cement Technology* **11** 166–77
[18] IS 8112 1989 Specification for 43 Grade Ordinary Portland cement *Bureau of Indian Standards*
[19] IS 4031 1999 Methods of physical tests for hydraulic cement Part 2 Determination of fineness by specific surface by Blaine air permeability method *Bureau of Indian Standards*
[20] IS 4031 1988 Methods of physical tests for hydraulic cement Part 11 Determination of density *Bureau of Indian Standards*
[21] IS 383 1970 Specifications for coarse and fine aggregates from natural sources for concrete *Bureau of Indian Standards*
[22] IS 2386 1997 Part I to VI Methods of test for aggregates for concrete *Bureau of Indian Standards*

[23] IS 456 2000 Plain and reinforced concrete - code of practice *Bureau of Indian Standards*
[24] IRC SP 63 2004 Guidelines for the use of interlocking concrete block pavement *Indian Roads Congress*
[25] IRC 44 2008 Tentative guidelines for cement concrete mix design for pavements *Indian Roads Congress*
[26] ASTM C143 2015 Standard test method for slump of hydraulic – cement concrete *Annual Book of ASTM Standards*
[27] ASTM C138 2017 Standard test method for density (unit weight) yield, and air content (gravimetric) of concrete *Annual Book of ASTM Standards*
[28] IS 15658 2006 Precast concrete blocks for paving – specifications *Bureau of Indian Standards*
[29] ASTM C779 2012 Standard test method for abrasion resistance of horizontal concrete surfaces *Annual Book of ASTM Standards*
[30] Senapati P K and Mishra B K 2014 Rheological characterization of concentrated jarosite waste suspensions using couette & tube rheometry techniques *Powder Technol.* **263** 58–65

IOP Conf. Series: Materials Science and Engineering 1075 (2021) 012025 doi:10.1088/1757-899X/1075/1/012025

Design Mixture of RAP-HMA Pavements: A Review

A Sharma[1], G D Ransinchung[1,*], P Kumar[1] and D S N V A Kumar[2]

[1]Indian Institute of Technology Roorkee, Roorkee-247667, Uttrakhand, India
[2]Bitumen R&D of Shell Technology Centre, Bengaluru-562149, India
[*] Corresponding author: g.n@ce.iitr.ac.in

Abstract. Roads are the national assets of a country. Once they are non-functional, they can be recycled to serve the next design life. Flexible pavements are the predominant type of pavement all over the world. The distressed flexible pavement material can be termed as Reclaimed Asphalt Pavement (RAP) material. The most common percent of RAP in the recycling process is 20% only. For higher incorporation of RAP material, there is uncertainty in the performance of RAP- HMA mixture. Design Mixture of RAP incorporated HMA technology depends on criteria's like the viscosity of aged RAP extracted binder, dosages of recycling agents, overheating of RAP material and method of intrusion of RAP material in plant production units etc. The fields cum laboratory studies have found rutting, fatigue and moisture susceptibility as major concerns in this technology. The uses of rejuvenators have been found to improve the fatigue resistance of asphalt mixture but its over-dosages may impact rutting resistance. Hence, a balanced mixture design method which involves performance evaluation of recycled mixture is required. This paper discusses the mixture design of RAP-HMA pavement.

1. Performance of RAP-HMA and its concern

Hot Mixture Asphalt (HMA) technology is widely adopted across the world for pavement construction. In the U.S.A., 90% of pavements are constructed using HMA technology. Around 90 million tons of Reclaimed Asphalt Pavement (RAP) material is generated every year in the U.S.A. and 90% of it, is recycled. According to the FHWA survey in 17 states of the U.S.A., the performance of RAP incorporated HMA is equivalent to conventional virgin HMA. Most of the highway agencies in the USA allows RAP incorporation up to 30% in top lifts. But the most common percentage of RAP is only 10- 20% in top lifts, this is because of the dilemma of performance prediction in terms of cracking and rutting of RAP-HMA[1].

High RAP proportion increases stiffness and makes the mixture less workable, by which pavement may crack and deform prematurely to its service life. Due to the unknown performance of RAP-HMA, state transport agencies have reluctant behavior towards the adaptation of RAP-HMA technology. One more reason for the reluctant behavior of state transport agencies is due to an unknown degree of blending between RAP and virgin material (aggregates and binder). Figure 1 demonstrates the most common distresses in flexible pavement and their probable reasons.

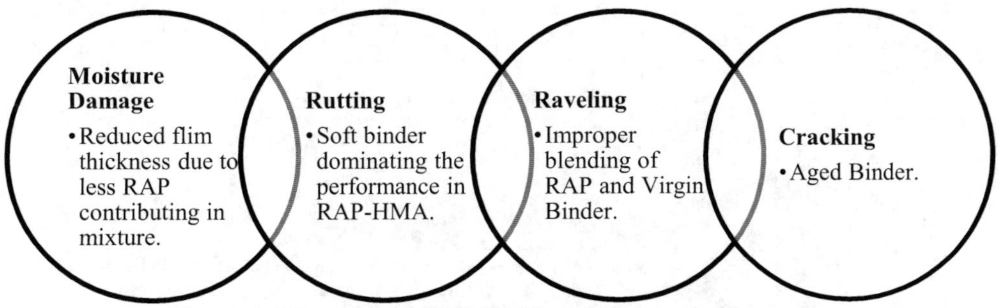

Figure 1. Distresses in RAP-HMA and their probable reasons

The RAP-HMA mixtures are susceptible to moisture-induced damage due to exposed aggregate surfaces of recycled mixtures. Improper coating of the recycled mixture is the result of a lesser degree of blending between two differently aged binders. Blending chart based on Performance Grade (PG) or viscosity can be used for determining appropriate virgin binder or recycling agent for particular RAP percent [2]. Rutting in recycled mixtures may occur due to over-dosages of the softer grade of binder or the recycling agents [3]. The concept of Balanced Mix Design does justification with the determination of Optimum Bitumen Content (OBC). If the OBC of the recycled mixture was evaluated on the rutting and cracking criteria, rutting can be prevented. The raveling of RAP-HMA pavements happens due to improper coating between RAP and virgin aggregate [4]. The reason behind it is the incompatibility of two binders, some chemical tests may perform to check the compatibility of the aged and virgin binder. Although, mechanical test as striping value may be performed to judge the compatibility between RAP and virgin aggregates. For it, certain RAP material of specific size (e.g. <19 mm) can be mixed with a certain weight of virgin aggregate of size (19 to 26.5 mm), percent of active aged binder in RAP, can be visually observed after mixing of these two materials. This approach will give some basic idea of the aggregate coating by RAP material [5]. Thermal cracking of recycled mixture is again based on the properties of the final blend of RAP-HMA. The blend must have required elasticity to resist cracking under traffic and adverse climatic conditions. In this paper, performance parameters, as shown above, have discussed and comments have been made based on the mechanical performance of RAP-HMA. The increased cost of bitumen and ban on mining of natural aggregates push to find alternate construction material such as RAP. RAP replacement level up to 100% can curtail cost per ton of asphalt by 50 to 70% as shown in Figure 2 below [6].

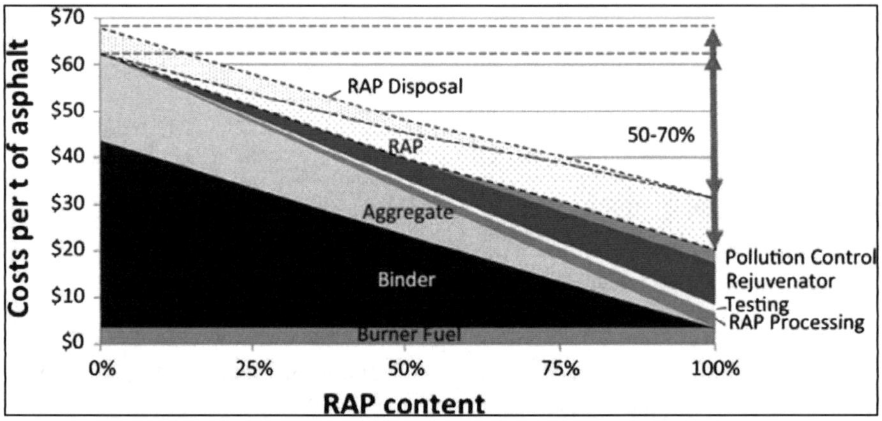

Figure 2. Cost of construction v/s percent RAP content [7]

2. Mixture design of RAP-HMA

Conventional Mixture Design consists of a selection of ingredients (bitumen, aggregate and filler), and determination of OBC based on volumetric parameter determination (VMA, VFB etc.). A recycled bituminous mixture having more than 20% RAP replacement, requires certain performance tests in addition to the conventional mix design method to find balanced OBC. The limit of percent RAP in HMA depends on certain factors as illustrated below:

- Gradation of RAP material.
- Stiffness of extracted RAP binder.
- Mixture Design of Reclaimed Asphalt Pavement Material.

  Following design method is required for RAP-HMA:

  Step 1: Determine the gradation of RAP material.

  Step 2: Based on pavement layer gradation (e.g. Dense Gradation) and percent RAP incorporation level (e.g. 20%), the requirement of virgin aggregate is decided.

  Step 3: Bitumen content of RAP material is determined using the centrifuge method of bitumen extraction according to ASTM D2172, 2017 [8]. Absolute viscosity of RAP extracted bitumen is determined at 60°C using Canon Manning Viscometer according to ASTM D2171, 2018 [9].

  Step 4: After calculating the combined aggregate gradation, percent bitumen demand of combined aggregates is calculated.

Here, we have to check the slab in which our RAP replacement level lies i.e. Tier 1 (< 15% RAP), Tier 2 (15- 25% RAP) and Tier 3 (>25% RAP).

Table 1. Mix design guidelines based on RAP replacement level [10]

Tier	Amount of RAP	Recommendation
1	<15%	No changes in the mix design process.
2	15% to 25%	One grade lower than conventional Performance Grade (PG).
3	>25%	Refer blending chart provided in ASTM D4887, 2016

The tentative guidelines of Hot Mixture Recycling based on percent RAP incorporation as shown in table 1 also supported by the study [11], according to which recovered RAP binder can be blended with a virgin binder and its properties are evaluated to satisfy the binder requirement of new mix recycled mix. Later these guidelines were incorporated in the AASTHO T 323 [12].

According to a study [13], there is no need to change mix design in case RAP percent is less than 15% whereas, when it comes in the range of 15% to 25%, one grade drop is necessary for virgin asphalt binder and when more than 25% RAP is incorporated, it is imperative to evaluate PG of RAP recovered binder, virgin binder and recycling agents. In another study it is found that up to 25% RAP can be incorporated without the drop of virgin binder grade and for 25% to 40%, one grade drop is required [14]. The blending chart can be used to determine the maximum amount of RAP incorporation. The most important aspects for achieving a successful mix design with RAP material is to first achieve required gradation i.e. blending of virgin and RAP material gradation and second is the achieving required binder grade of the recycled binder. To truly determine the participation of aged binder into the new recycled asphalt mix, binder replacement terminology is more justified. As the stiffness of RAP incorporated mix is dependent on RAP age and percent bitumen content, it is imperative to calculate the contribution of aged binder in a recycled mixture [15].

$$Percent\ Binder\ Replacement = \left(\frac{A*B}{C}\right) \quad (1)$$

where A = Percent bitumen content in RAP, B = Percent RAP incorporation, C = Total binder content of the mixture.

Step 5: The target grade of recycled binder should be checked and required adjustment in dosages of recycling agent and RAP replacement should be made recovery of recycled binder should be performed in accordance to ASTM 5404, 2012 [16].

Step 6: To determine the OBC of the recycled mixture, Marshall Method or Superpave Design method may be adopted and Job Mix Formulae is prepared.

Conventional design procedure to determine OBC of RAP-HMA is the volumetric analysis. The Voids in Mineral Aggregate (VMA) of natural aggregate are determined to calculate available void space to be filled with bitumen and air. But in RAP material, aggregates are coated with an aged binder, calculation of VMA is not simple. There are three different approaches available to determine the specific gravity of RAP as shown in Figure 3 below.

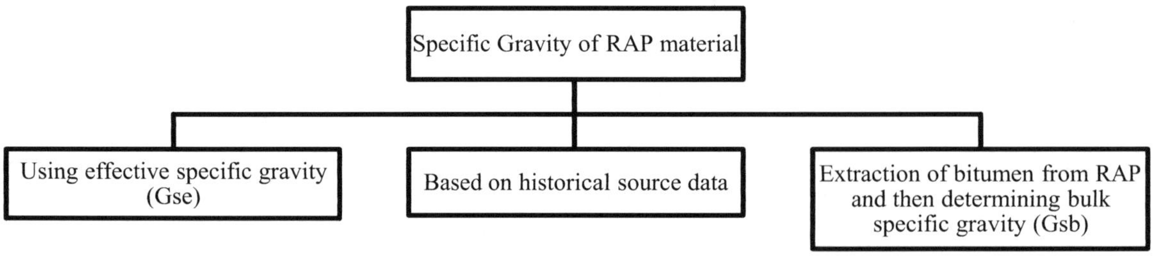

Figure 3. Approaches to determine the specific gravity of RAP-HMA

Calculation of "Gse" through Maximum Theoretical Specific Gravity "Gmm" of RAP is recommended. Further, by assuming the absorbed asphalt, Gsb of RAP can be calculated which can be used in the mix design process. The use of "Gse" in place of "Gsb" is preferable as the first is more repeatable than later. However, the use of "Gse" results in low VMA value which leads to underestimating the OBC and affect the durability of pavement. Hence, the more justified method is to use "Gsb" by assuming absorbed asphalt binder by aggregates (Pba) by a blend of RAP and virgin aggregates. For more than 20% RAP, use of Gsb is recommended by assuming a "Pba" of 1.2% [17].

The Mechanical performance of RAP-HMA is essential to determine OBC in addition to volumetric properties of RAP-HMA. Figure 4 & 5 indicates a broad view of RAP sampling and OBC determination through conventional as well as Balance Mixture Design method. The flow chart is prepared to demonstrate the difference between conventional as well as Balance Mixture Design Method. Certain mechanical performance tests are necessary to judge the unknown performance of aged bitumen coated aggregates.

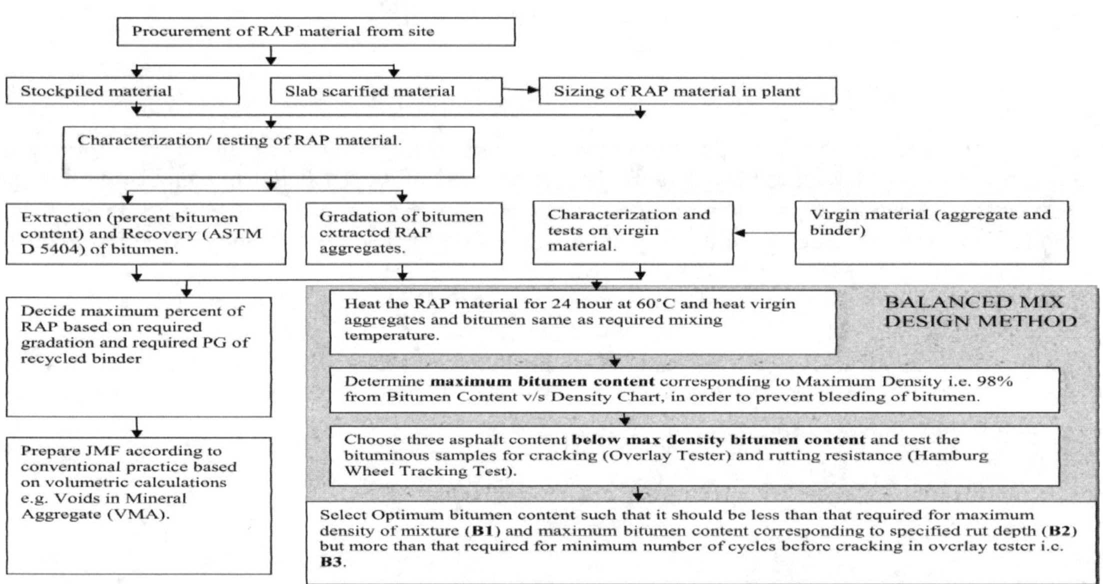

Figure 4. Hot mixture asphalt design process (conventional and balanced)

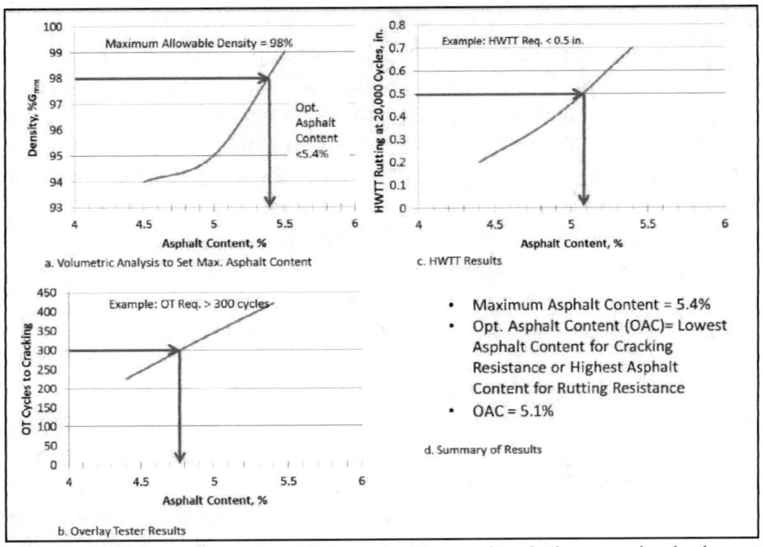

Figure 5. Optimum bitumen content determination using balance mix design criteria [1]

According to the study [1], there are four assumptions applied to Balance Mix Design Method, which are:

- Total bitumen content is the sum of % RAP bitumen content and externally added virgin bitumen content in the new recycled mixture.
- The minimum number of cycles to pass Overlay Test Criteria is 300 cycles.
- Minimum allowable rut depth in Hamburg Wheel Tracking Test (HWTT) is 12.5 mm.
- OBC has chosen so that it should be less than maximum bitumen content corresponding to maximum allowable rut depth (12.5 mm) and bitumen content corresponding to 98% of maximum density whereas it should be more than that minimum bitumen content, required for achieving 300 cycles in Overlay Tester.

On increasing RAP binder, high-temperature stiffness increases for mixtures containing 40% RAP [14]. Whereas in a study conducted by [18], in total 28 field mixtures containing up to 25% RAP were evaluated and it was found that PG of the recycled blend was same as of virgin binder or increased by one grade both for high and low temperatures. In another study in which RAP mixtures up to 50% prepared and recycled blends were tested using Amplitude Sweep Test (AST) in DSR, it was found that the best approach to resist fatigue cracks in recycled mixtures is to use a softer grade virgin base binder [19]. There is need to examine percent RAP incorporation level for achieving a particular PG grade e.g. an aged binder from 15% RAP incorporation level meet the base requirement of PG 58-28 but on increasing RAP up to 50%, even with a softer grade of binder PG 58-35, it could not achieve PG 58-28 criteria [20].

2.1 Field cum laboratory performance of RAP-HMA

The performance of Recycled Hot Mixture Asphalt (RAP-HMA) is at least equal or better than the conventional HMA. The properties of RAP incorporated HMA are influenced by aged binder and percent of RAP material in the asphalt mixture. Recycled Asphalt Mixtures have a lesser rate of ageing compare to virgin asphalt mixture. According to the Georgia Department of Transportation, AC-20 grade of the virgin binder is used for recycling purpose where AC-30 is used for conventional Hot Mixture Asphalt (HMA). After recycling, the target viscosity of recycled blend should be in between 6000 to 16000 Poise after

Rolling Thin Film Oven (RTFO) ageing on the specimen. The upper limit of RAP incorporation is 40% and 25% for continuous and batch type mixture plant. The author has shown a case study under which they had considered 5 pavement locations where conventional and Recycled pavements have constructed using similar materials, construction practice and equipment. After an average life of 2 years, It was found that none of the sites has significant distress in terms of rutting, raveling and alligator cracking. It proves that RAP incorporated bituminous mixtures have performed equally to the conventional bituminous mixtures. As air voids are the important criteria of bituminous mixture design, excess air voids lead to rapid ageing of pavements whereas very low air voids result in distresses like bleeding of pavement in the summertime. In this study, it was found that the air voids of recycled bituminous mixtures are less than control bituminous mixtures which is an encouraging finding to use RAP-HMA. In another study on Ranchi Ring Road, Jharkhand, India, 80% RAP was used in Hot in-place Recycling where the fresh binder grade was VG-10 and rejuvenator was Savsol (5% by weight of binder). The target binder grade was VG-40, it was observed that pavement performed satisfactory.

The main problem associated with RAP incorporated asphalt mixture is the embrittlement or durability of mixture. The second concern with RAP incorporated asphalt mixture is the degree of blending. In a study, [21] eighteen samples of RAP incorporated asphalt mixtures were procured from plants where the maximum percent of RAP was 40%. The performance test results indicate that stiffness and so cracking increases as RAP percent increases whereas rutting and moisture resistance improved. Reheating of RAP-HMA in laboratory increases the stiffness of mix. Dynamic Modulus of asphalt mixtures was examined and it was found that there is no significant difference between the stiffness of RAP-HMA and virgin HMA mixtures [22].

Figure 6. Production factors influencing the degree of blending between RAP and virgin aggregates

Production factors affect the performance of RAP-HMA. As shown in figure 6, it can be understood that there is a major role of Plant Type i.e. Batch Plant or Continuous Flow type plant. Second factor i.e. production temperature which decides the workability of asphalt mixture and controls the stiffness of RAP-HMA. Mixing time should be so that all the RAP and virgin aggregates coat properly with a virgin binder and the aged RAP binder can be activated within this duration. Mixing time depends on the percent RAP incorporation.

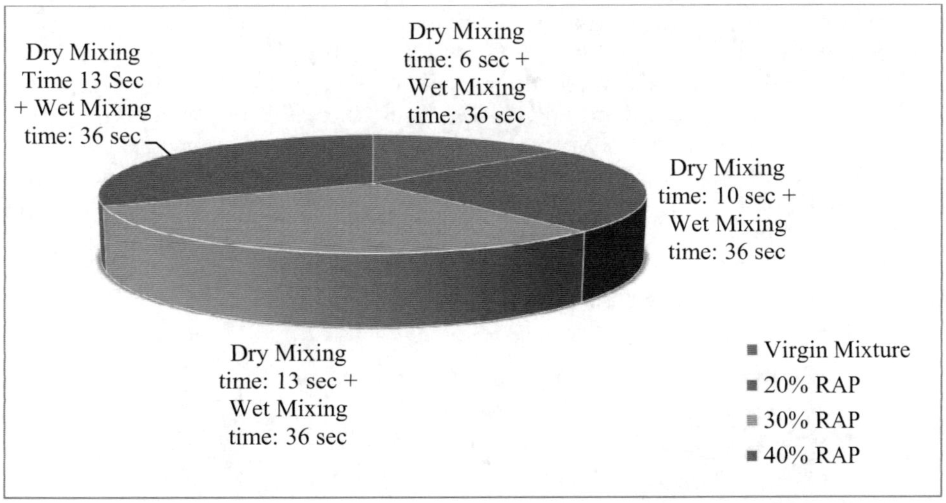

Figure 7. Dry mix time and wet mix time variation with percent RAP incorporation [21]

It is clear from Figure 7 that as RAP percent increases the time required to activate the aged binder is more. The blend of aged binder and virgin binder should be so that its properties should match with required binder properties of new asphalt mixture as per climatic and traffic loading conditions.

The discharge temperature is related to the allowable hauling distance of recycled mixture, the care should be taken that the mix remains workable as well as there should not be over the ageing of the binder. Storage of RAP material is generally done as stockpiling practice. Conical stockpile is preferred over rectangular stockpile as less surface area is exposed to the environment in case of conical stockpile and change in moisture is less. RAP material should be stored according to its source e.g. if it came from parking lot rehabilitation, most of the material may have finer gradation with rich bitumen content as compared to if it would have come from a National Highway having heavy traffic volume. RAP Source is again a very important factor as the life of pavement or ageing decide the RAP properties and the required dosages of rejuvenators. It is observed that RAP variability is lesser than Reclaimed Asphalt Shingles (RAS). RAS has more binder content than RAP and have fibrous material due to which the calculation of bitumen content by oven ignition method become erroneous. Moreover, ageing of RAS is more in comparison to paving bitumen [1].

3. Performance of recycled blends

RAP extracted and recovered binder are characterized according to their stiffness using Dynamic Shear Rheometer (DSR). The performance of recycled blends are analyzed on the assumption that there is perfect blending between an aged binder and virgin binder which is not true in all the cases. There is a degree of blending between these binders which depends on mixing temperature and many other factors such as viscosity of aged and virgin binder, percent RAP replacement level, type of mixing, type of plant, mixing time etc. Aged binder and virgin binder may show the phase separation within the recycled blend. Degree of blending between two blends will depend on temperature, the viscosity of binders and their chemical composition.

To take full advantage of RAP material in the new bituminous mixture, Recycling Agents (RAs) can be added. According to study [23], polar aromatic compounds are useful as recycling agents in restoring the rejuvenation of aged RAP binder. Figure 8 (a and b) sketch the study [24] in which total 9 rejuvenators/

Recycling Agents (plant oils, waste-derived oils, engineered products and refinery base oil) were taken and blends were prepared using virgin binder PG 64-22, results of kinematic viscosity indicates that Paraffin Base Oil, Aromatic Extract and Waste Engine Oil (WEO) are effective in decreasing the viscosity of RAP extracted binder up to virgin binder PG 64-22 level whereas WEO bottoms, organic blend and distilled tall oils were observed to be ineffective in reducing viscosity. Refined Tallow, Organic blend, Distilled tall oil significantly decreased penetration at 25°C in comparison to Aromatic Extract, Paraffin Base Oil, Waste Engine Oil etc. All rejuvenators are indexed pass or fail based on their performance concerning "Dosages to Reach Virgin Penetration (%)", "Penetration Index (PI)", "Creep Compliance (1/GPa)", "Tensile Strength (kPa)" and "Fracture Energy (kPa)", in overall results, Organic Blend, Refined Tallow and Aromatic Extract found to pass.

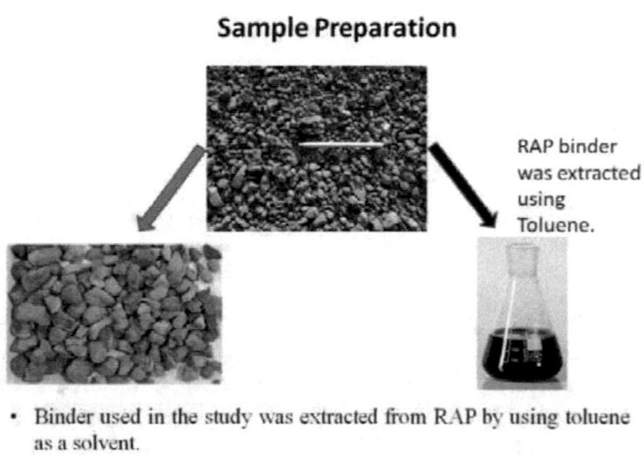

(a) Extraction of aged binder from RAP.

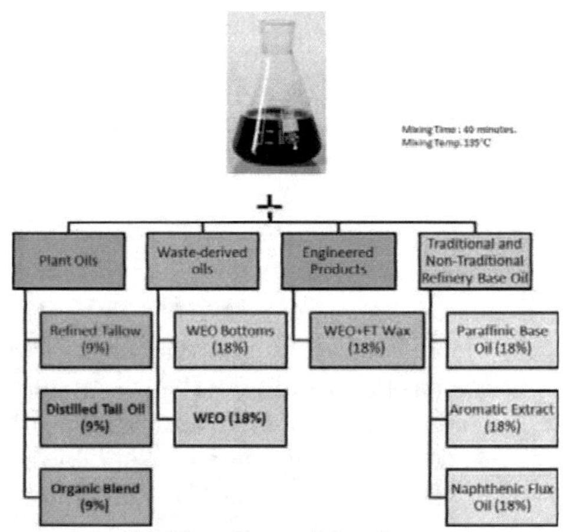

(b) Blending with rejuvenators.

Figure 8 (a and b). Extraction of aged binder and blending with rejuvenators

According to AASTHO MS-20, Recycling Agents are the organic compounds that heal the physical and chemical properties of aged asphalt. Although, various studies have shown that there are various commercial and non-commercial RAs are available. But the properties of RAs are reliable to a specific region only. For example, we cannot recommend a particular waste vegetable oil for all regions to recycle the aged asphalt material.

3.1 Critical temperature of RAP-HMA

RAP material is compatible with Warm Mix Asphalt (WMA) technology as aged binder becomes sufficient fluid at WMA mixing and compaction temperatures [25]. Similar findings were observed in some more studies [26], it was suggested that critical high-temperature grade and percent RAP has a linear relationship as shown in Fig.9 below.

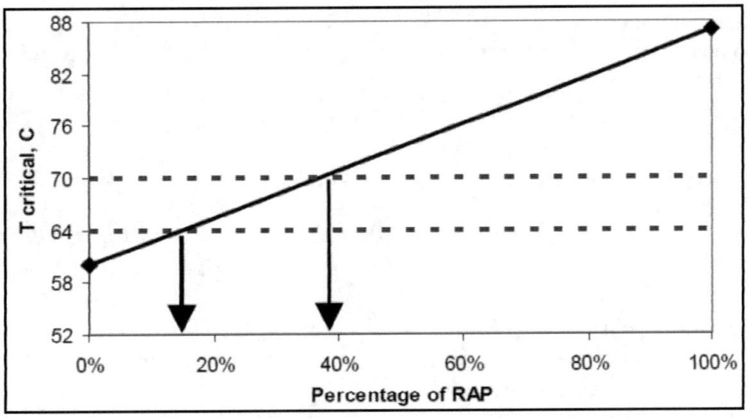

Figure 9. A linear relationship between critical temperature and percentage of RAP incorporation

Figure 9 is based on the method where virgin binder grade is known and the task is to determine the maximum and minimum amount of RAP incorporation. It is observed that RAP has upper PG as 82 and 100. A flow chart indicating the method of determining RAP amount is explained below.

Step 1: Set the target blend viscosity value or PG grading e.g. PG 64-22.

Step 2: Determine critical temperature and physical property of virgin binder grade and RAP recovered binder.

Step 3: Critical temperature and physical property of RAP recovered binder will be calculated as follows: (a) Original PG Grading. (b) RTFO aged PG Grading.

Step 4: Find percentage of RAP used using three critical temperatures,

$$\% \, RAP = \left(\frac{T\ Blend - T\ Virgin}{T\ RAP - T\ Virgin}\right) \tag{2}$$

above relation is true for a linear relation between %RAP and T critical of the binder.

Step 5: Determine High-Low Temperature RAP percentage range.

Step 6: Select allowable percent of RAP that satisfies all temperature limits.

After selection of a particular RAP percent, performance evaluation tests of asphalt mixture should be carried out to determine the performance of RAP-HMA and based on the test results, modification in the mixture design e.g. OBC can be made to improve the overall performance.

4. Conclusion

Flexible pavements are the predominant type of pavements all over the world. However, the rate of RAP recycling is limited to 30% only in Hot-in plant recycling. Higher percentage increase the environmental pollution and impact the durability of pavement. Use of rejuvenators found to increase the moisture resistance, cracking resistance but may impact the rutting resistance of recycled mixture. The balanced mix design process is discussed. It is based on determining OBC concerning density, Rutting and fatigue criteria. Further, the critical aspects of the mix design process were discussed. It is important to note that the determination of specific gravity of RAP material is not same as of in case of conventional hot mix design. Performance of recycled mixtures was also discussed. Rejuvenators like Tall oil, organic blend and aromatic extract found to improve the creep compliance, tensile strength, and fracture energy. Further, research is recommended on these rejuvenators.

Acknowledgements
This study was supported by a fellowship from the Ministry of Human Resource and Development (MHRD), India.

References

[1] Newcomb DE, Epps JA and Zhou F 2016 *Use of RAP & RAS in High Binder Replacement Asphalt Mixtures : A Synthesis*
[2] ASTM D4887 2016 Standard Practice for Preparation of Viscosity Blends for Hot Recycled Asphalt Materials BT - Standard Practice for Preparation of Viscosity Blends for Hot Recycled Asphalt Materials 5
[3] Mogawer WS, Booshehrian A, Vahidi S and Austerman AJ 2013 Evaluating the effect of rejuvenators on the degree of blending and performance of high RAP, RAS, and RAP/RAS mixtures *Road Mater. Pavement Des.*
[4] Bowers BF 2013 Investigation of Asphalt Pavement Mixture Blending Utilizing Analytical Chemistry Techniques 195
[5] Radhakrishnan V, Ramya Sri M and Sudhakar Reddy K 2018 Evaluation of asphalt binder rutting parameters *Constr. Build. Mater.* **173** 298–307
[6] Zaumanis M, Arraigada M, Wyss SA, Zeyer K, Cavalli MC and Poulikakos LD 2019 Performance-based design of 100% recycled hot-mix asphalt and validation using traffic load simulator *J. Clean. Prod.*
[7] Zaumanis M, Mallick RB and Frank R 2016 100% Hot Mix Asphalt Recycling: Challenges and Benefits *Transp. Res. Procedia* **14** 3493–502
[8] ASTM D 2017 Standard Test Methods for Quantitative Extraction of Asphalt Binder from Asphalt Mixtures 10
[9] ASTM D2171 2018 Standard Test Method for Viscosity of Asphalts by Vacuum Capillary Viscometer. *ASTM Int.* 10
[10] Kandhal PS, Foo KY and Mallick RB 1998 *A Critical Review Of VMA Requirements In Superpave. NCAT Report No. 98-1.*
[11] Mcdaniel R and Anderso RM 2001 Recommended Use of Reclaimed Asphalt Pavement in the Superpave Mix Design Method *Technician's Manual.*
[12] AASTHO M 323 2004 Standard Specification for Superpave Volumetric Mix Design.
[13] McDaniel RS, Soleymani H and Shah A 2002 Use of Reclaimed Asphalt Pavement (RAP) Under Superpave Specifications *A Regional Pooled Fund Project.* (West Lafayette, IN)
[14] McDaniel R, Kowalski K and Shah A 2012 *Evaluation of Reclaimed Asphalt Pavement for Surface Mixtures*
[15] Asphalt Pavement Association of Michigan 2014 *Using RAP to Stretch Your Pavement Dollars*
[16] ASTM D5404 2012 Standard Practice for Recovery of Asphalt from Solution Using the Rotary Evaporator. *ASTM Int.*
[17] Asphalt Institute 2014 MS-2 Asphalt Mix Design Methods ed M Buncher, M Anderson and A H Abedali 1–197
[18] Daniel J, Pochily J and Boisvert D 2010 Can More Reclaimed Asphalt Pavement Be Added? *Transp. Res. Rec. J. Transp. Res. Board* 2180 19–29
[19] Willis J R, Turner P, Padula F de G, Tran N and Julian G 2013 Effects of Changing Virgin Binder Grade and Content on High Reclaimed Asphalt Pavement Mixture Properties *Transp. Res. Rec. J. Transp. Res. Board* 2371 66–73
[20] Hajj E, Salazar L and Sebaaly P 2012 Methodologies for Estimating Effective Performance Grade of Asphalt Binders in Mixtures with High Recycled Asphalt Pavement Content *Transp. Res. Rec.*

J. Transp. Res. Board 2294 53–63

[21] Mogawer W, Bennert T, Daniel JS, Bonaquist R, Austerman A and Booshehrian A 2012 Performance characteristics of plant produced high RAP mixtures *Road Mater. Pavement Des.* 13 183–208

[22] Shah A, Mcdaniel RS, Huber GA and Gallivan VL 2007 Investigation of Properties of Plant-Produced Reclaimed Asphalt Pavement Mixtures 103–11

[23] Newcomb DE, Nusser BJ, Kiggundu BM and Zallen DM 1984 Laboratory Study of the Effects of Recycling Modifiers on Aged Asphalt Cement *Transp. Res. Rec.* 92

[24] Zaumanis M, Mallick R and Frank R 2013 Evaluation of Rejuvenator's Effectiveness with Conventional Mix Testing for 100% Reclaimed Asphalt Pavement Mixtures *Transp. Res. Rec. J. Transp. Res. Board* 2370 17–25

[25] Advanced Asphalt Technologies 2011 Special Mixture Design Considerations and Methods for Warm-Mix Asphalt: A Supplement to NCHRP Report 673 *A Manual for Design of Hot-Mix Asphalt with Commentary*

[26] Hajj EY, Sebaaly PE and Shrestha R 2009 Laboratory Evaluation of Mixes Containing Recycled Asphalt Pavement (RAP) *Road Mater. Pavement Des.* 10 495–517

Reviewing methods for determination of Dielectric Constant required to Calibrate GPR Study for Asphalt Layers

A Walia[1], R Rastogi[1], P Kumar[1] and S S Jain[1]
[1] Department of Civil Engineering, Indian Institute of Technology Roorkee, India
[2] National Institute of Technology, Delhi, India
[*] Corresponding author: ashishwalia12@gmail.com

Abstract. Ground Penetrating Radar (GPR) is a non-invasive technique developed substantially in the recent years. Initially, GPR technology was applied to natural ground to explore the sub-surface conditions and identify the anomalies in it. Now it has become a well-accepted technique to investigate any kind of materials such as wood, ice, concrete, asphalt etc. The calibration of GPR equipment is most important prerequisite criteria before data collection for accurate investigation of pavement subsurface features. During calibration, the velocity of radar waves will be calibrated in contrast to subjected material as whole examination is governed on the basis of it. The basic dielectric properties required as a prerequisite while using GPR system are dielectric permittivity, electrical conductivity along with magnetic permeability. The dielectric constant affects the velocity of transmitting wave as inverse proportion i.e. velocity decreases with an increase in dielectric constant and vice-versa. Considering the principle, the detailed study of pavements requires understanding and estimation of 'bulk' dielectric constant. The heterogeneous conditions of pavement actually obscure the straightforwardness of analysis and therefore estimation of a 'bulk' dielectric constant value becomes essential. There are several methods to compute the dielectric constant of asphaltic layers, which are developed by different researchers, and these are discussed briefly in this paper. The paper tries to put forward the merits and demerits of these methods and to identify the most appropriate one suiting to Indian conditions.

1. Introduction

Ground Penetrating Radar (GPR) is a non-invasive technique developed substantially in the past few years. Being non-intrusive equipment, it facilitates subsurface testing non-destructively along with providing more appropriate and reliable information. The GPR system generally consists of transmission and receiver antenna with controlling unit to generate the subsurface profile of any surface needs to be assessed. At the inception of GPR technology, it was primarily applied to natural grounds though now it is a well-accepted technique to investigate any kind of materials such as wood, ice, concrete, asphalt etc.

The application of GPR in road infrastructure having an eminent role nowadays predominantly because of its ability to conduct testing in non-destructive manner. Furthermore, the data collected by this NDT equipment can be collected in an abundant amount which will be never feasible with traditional destructive testing due to several limitation including cost and time consumption. Therefore, the GPR technology is an inevitable activity which would not only collect data at faster rate but also avoids the post data collection consequences which will definitely diminishes the structural integrity and reduces an overall pavement performance. In case of flexible pavement, nowadays GPR is generally used to collect layer thickness, presence of moisture or air voids, subsurface cracking, subsidence prone area, quality assurance of pavement constructions or any other kind of anomalies needs to be detected. It can also be used to detect the slab thickness, depth to dowel bars and their orientation, moisture content etc. in rigid pavement structures.

GPR application is entirely based on the medium characteristics or particularly the dielectric properties of the medium through which radar waves will transmit. The basic dielectric properties prerequisite to be considered while using GPR system are dielectric permittivity, conductivity along with magnetic permeability [1]. As most of the pavement materials are non-magnetic, the latter property can be neglected. Therefore, the electric property of the material is the only decisive parameter, which will affect the path of transmitted waves. Whenever a change in dielectric constant appears into the path of radar wave, a part of it will be reflected back which is of our interest. This reflected energy will be captured by receiver antenna and helps to generate the subsurface profile depending on its amplitude. Nevertheless, GPR is predominantly used in low electric loss material due to its susceptibility of signal attenuation when encounters with highly electrical conductive material like clays, water etc.

As abovementioned, the whole GPR practice governed by the change in dielectric constant or dielectric contrast. Though sufficient amount of radar signal reached at subgrade, the low dielectric contrast (due to contamination of base layer from migration of subgrade fine particles) between subsequent layers resulted in difficulty to distinguish layer interfaces [2]. The dielectric constant of different pavement materials reported to be in close range as mentioned in table 1 [3].

Table 1: Dielectric Constant of various materials incorporated in flexible pavements.

Material	Dielectric constant	Velocity (mm/ns)
Air	1	300
Asphalt mixture	4-10	90-160
Concrete	5-9	100-130
Aggregate	6-18	70-120
Water	81	33

From table 1 it can be concluded that the dielectric constant is directly affecting velocity of transmitting wave since velocity decreases with increase in dielectric constant. Since any pavement is a layered structure lying on naturally available or compacted subgrade and consists of combination of various materials mentioned above which makes it a heterogeneous medium. These heterogeneous conditions further obscure the analysis straightforwardness and therefore a 'bulk' dielectric constant value essential to be computed which will be used to investigate pavement characteristics accurately. The bulk dielectric constant of any pavement material relies on its material constituents (asphalt, aggregate etc.) and material conditions (like moisture content, air voids etc.). There are several methods developed by many researchers to compute the dielectric constant of pavement materials and will be discussed briefly in subsequent sections.

2. Need of GPR calibration

The calibration of GPR equipment is prerequisite before data collection for accurate investigation of road subsurface features. During calibration, the velocity of radar waves must be calibrated in contrast to subjected material as whole examination is governed on the basis of it. As explained in previous section, the effective dielectric property of heterogeneous materials (like layered pavement structure) needs to be computed as the different proportions of varying material characteristics influences the geometric features of any considered material. The presence of air and water significantly affects the overall dielectric constant of the asphalt mix as the presence of low-dielectric air reduces whereas high-dielectric water increases it [4]. Therefore, velocity calibration becomes the primary source of errors if not performed appropriately and leads to unrealistic results that can have severe consequences while applied in pavement assessment and management.

The dielectric constant is a function of several parameters including testing conditions as well as material characteristics and therefore fluctuates from one location to other with variation of these dependent parameters. Most of the current GPR analysis study follows either a fixed dielectric constant for the entire pavement section or estimates it from limited number of cores extracted from the pavement. Though few proposed methods are capable to determine the varying profile of dielectric constant, their field implementation is unfavourable because either involves combination of sophisticated equipment or required cumbersome analysis approach. As an outcome, the existing GPR data analysis technique interprets results that are either accurate or reliable on large scale. The prediction of accurate dielectric constant is one of the most challenging tasks in the data interpretation of GPR analysis. In old roads, the quotient of erroneous interpretation increases exponentially because of the variability present in both i.e. structural integration (excessive voids, layer de-bonding etc.) and material condition (aging, moisture presence, anomalies etc.). Therefore, these variabilities need to be taken into account while performing pavement assessment otherwise may have serious implications on planning of maintenance activities and prediction of residual life [1].

Once the dielectric constant of material is computed the propagation velocity of the radar wave through that material can be according to equation (1) given by [5].

$$v=c/\sqrt{\varepsilon} \quad (1)$$

Where v = velocity of EM wave in considered material
c = velocity of EM wave in air
ε = Dielectric constant of that material

The traditional method of velocity calibration in pavement structures is based on destructive approach where a core sample was collected from selected pavement site. However apart from its disadvantage of permanent damage, the obtained result is only point specific and cannot be considered as the representative of whole section. Here GPR technology outperformed the traditional approach by capturing the continuous profile of pavement structure and able to determine dielectric constant as a function of space for given attributes of GPR system.

3. GPR calibration procedures

The application of GPR is entirely based on the dielectric characteristics of pavement materials encountered with radar waves as the propagation velocity alters with change in dielectric properties. The asphalt mixture is a combination of primarily asphalt, aggregates, moisture and air etc. and hence the dielectric constant of the mix should be a value considering all the constitutive components to get the most appropriate solution. The dielectric constant of considered material provides the basic information required for road assessment including layer thickness, excess moisture and void detection, subsurface distress or other anomalies.

3.1 Dependency of calibration method on GPR system

GPR are generally available in two categories i.e. Air-coupled and Ground-coupled antenna system having their respective pros and cons, hence judicial decisions need to be taken while selecting the type of GPR system assigned to any project. Air-coupled systems have two horn antennas needs to be fixed at certain height from pavement surface and can collect data at traffic speed without hindering traffic movements. Whereas ground-coupled antenna system are in contact with ground and therefore bound to move at relatively lower speed. In the early 1990's, the utilization of GPR is limited to pavement thickness measurements particularly in Northern America. Initially air-coupled antennas were preferred for pavement assessment however due to its obvious limitation, researchers also appreciated ground-coupled GPR systems were in order to explore more depth of penetration along with more reliable and accurate results. Many researches have given their critical views about comparison between both kinds of systems according to features considered. The elevated antennas of air-coupled systems benefitted by reducing surface coupling effect that enables to clearly distinguish air-surface interface from other interfaces [6]. The benefit of higher penetration depth with ground-coupled antenna at a given frequency compared to air-coupled antenna however also acknowledged its speed limitation [7]. On the basis of type of GPR equipment employed, the procedure for dielectric characterization will also vary.

The most basic approach to determine the dielectric constant of asphalt pavements is to deduce its dielectric characteristics through core calibration. Although it seems to be an easy method, the collected information will be point specific apart from its common drawbacks like any conventional destructive testing method. In order to avoid damages incurred during destructive testing, [8] and [9] had established an approach predominantly know as surface reflection method to determine the dielectric constant of asphalt layers using only GPR data collected through air-coupled system by means of equation (2) to equation (4) given as

$$\varepsilon_A = \left[\frac{1+\left(\frac{A_{surface}}{A_{plate}}\right)}{1-\left(\frac{A_{surface}}{A_{plate}}\right)}\right] \quad (2)$$

$$\varepsilon_B = \left[\frac{F-R}{F+R}\right]^2 \quad (3)$$

$$F = \frac{4\sqrt{\varepsilon_A}}{1-\varepsilon_A} \quad (4)$$

Where ƐA is dielectric constant of asphalt layer, Asurface is maximum amplitude of the reflected signal from asphalt layer surface, Aplate is the maximum amplitude of the reflected signal from a perfect reflector (metal plate) placed underneath the antennae and R is the ratio of the maximum reflected amplitude from the base layer to the maximum amplitude of pulse reflected from pavement surface. These studies are only valid in air-coupled antenna system as it is difficult to distinguish between surface reflected signals and direct signals (transmitter to receiver) while using ground-coupled system. The abovementioned relation will be more effective when a homogeneous layer is present (same ƐA throughout its thickness) and the surface is free from any kind of distress. However, in practical the homogeneous condition is impossible in asphalt layer therefore the application of the described approach leads to erroneous results [10]. The unrealistic assumption of pavement layers along with presence of localized distresses and other anomalies makes analysis more inaccurate. Further in case of older pavements consisting of several overlays laid for strengthening purpose enhances the complication due to presence of materials with varying properties. Another model was provided to estimate dielectric constant based on amplitudes of reflected waves from pavement top surface and its interfaces for air-coupled antennas. But for ground-coupled GPR systems, the reflected amplitude based estimation method is not feasible due to lack of accuracy in measurement. In the latter case, the only option remains are core calibration to obtain dielectric constant by providing measured thickness and time parameters in the equation (5) given by [11].

$$\Delta d = \frac{c\Delta t}{2\sqrt{\varepsilon_t}} \quad (5)$$

Where Δd is two way travel time, c is speed of light in free space and Ɛt is dielectric constant of subjected material.

4. Methods for determining dielectric constant

4.1 Current Methodologies

There are many methods developed by several researchers to estimate dielectric constant for pavement structure by considering various computation techniques either in time domain or frequency domain analysis. Certain apparatus was also designed exclusively for measurement of dielectric properties of heterogeneous materials. Few studies have demonstrated the dependency of testing frequency on estimation.

A frequency domain model was developed in which reflected signal from layered medium are combined [12]. The layer thickness and corresponding dielectric properties were the model parameters involved. After initial assumption of layer thickness and dielectric properties, the theoretical time domain reflected signal was computed and then compared with measured reflected signals to compute mean square error. Subsequently an iterative change in the values of thickness and dielectric constant initiated until minimum square error was achieved. However, the calculation of inverse Fourier transform for each set of parameter and convergence of search algorithm to local minimum of mean square error leads to make it computationally intensive and inaccurate respectively. To account for the presence of thin bituminous overlays, a pavement model considering multiple layers was developed to estimate dielectric properties [13]. The data analysis computed the dielectric constant by iteratively detecting all the reflected pulses from the time domain signal. Nevertheless, the major drawback of this technique, due to assumption of zero conductivity, was underestimation of dielectric constant of subsurface layers.

An electromagnetic inversion technique in time domain was developed to determine dielectric constant for multilayered dielectric medium like pavement structure [14]. In this technique, the whole pavement was divided into N number of thin homogeneous layers having thickness lesser than resolution of GPR. Then it tries to accumulate the received signals on the basis of theoretical time domain model. Subsequently the Newton-Gauss iterative method was applied to estimate the best-fit dielectric constant profile that minimizes the mean square error between and measured and synthesized signals. The computationally intensive inversion method adopted in the study produced good results provided there is adequate input of material loss values. An automatic system was suggested to detect layers in a multilayered pavement system using likelihood test [15]. A probability density function (PDF) was determined for determining a reflected pulse using Markov model. The PDF for a given scan was computed from PDF of previous scan while the PDF for current scan was then computed based on Bayes theorem. The described approach was prolonged for multiple layers by iterating through layers and removing echo of individual layers detected form the total reflected signals. The iterative process was repeated until either the extraction of all known layers or the residual reflected signal drops below a fixed threshold value.

The design and calibration of an apparatus was presented for measurement of complex permittivity for heterogeneous materials [16]. It consists of a sample holder connected with coaxial cable to automatic network analyzer and data processing software. The proposed sample holder was capable to accommodate either field obtained asphalt core samples or laboratory prepared specimens lengthens up to 400 mm while having maximum aggregate size of 40 mm. The dielectric constants were measured for laboratory prepared samples over the frequency range of 0.1 MHz to 1.5 GHz. The measurements of dielectric constant also suggested that the influence of variation of moisture content was more compared to mix properties on measured values of dielectric constant. It was also claimed in another study that there is increase in dielectric constant with increase in moisture content [17]. A methodology was proposed to define dielectric constant as a function of frequency range considered in GPR system [2]. The dielectric properties are frequency dependent as proved by the study and showed the variation of dielectric constant with variation in frequency. The study of Virginia Smart Road Facility was incorporated in the study to validate the results obtained. [18] have also focused on the frequency dependency by showing a correlation between frequency bandwidth and sampling interval in time domain as given by equation (6).

$$\mathrm{dt} = \frac{1}{\mathrm{f}} \tag{6}$$

Where dt is time domain sampling interval and f is upper limit of frequency bandwidth. From the equation, it can be concluded that to detect thin layers in pavement structure a broad frequency bandwidth of GPR signals is required. The SVR (Support Vector regression) method was appreciated to estimate relative permittivity and time delay in order to determine pavement thickness [19]. The study results showed that SVR predictor gives a good performance with less time consumption. The proposed method was appreciated to accurately estimate pavement thicknesses along with capability to distinguish overlapping reflected signals.

4.2 CMP Methodologies

The Common Mid-point (CMP) method is the most conventional method used for calibration or dielectric constant estimation in GPR study. The separation distance between the transmission and receiver antenna increases in every step while keeping common mid-point in order to obtain several combinations of transmitted waves. The CMP dataset have couple of set of direct waves transmitted through air and ground along with radar waves reflected from pavement interface subjective to change in dielectric constant. However, the CMP method requires two separate units for transmission and receiver antenna, which is not quite common in GPR systems. There are several customized CMP methods proposed by several researchers incorporating various combinations of sophisticated equipment and discussed subsequently.

A modified CMP methodology was adapted, generally used for seismic wave's analysis, to establish the dielectric constant of asphalt layers by integrating a 900 MHz ground-coupled system with a 1 GHz air-coupled system [7]. The methodology estimates the reflection times from the bottom of asphalt layer by both ground-coupled and air-coupled systems to determine dielectric constant. The core thickness data was also collected for validation purpose by comparing with computed layer thickness based on determined dielectric constant according to developed methodology. The thickness prediction error was around 6.8 % whereas ranges from 1% to 15 % and mainly attributed due to inaccurate detection of surface reflection from ground-coupled system as well as small separation distance between bi-static antennas of air-coupled system. An improved modified CMP method by substituting the ground-coupled system with the air-coupled system was established and named it as extended CMP (XCMP) method [6]. The method has undertaken a pair of 2 GHz air-coupled system which not only enhanced the survey speed but also upgraded the result's reliability as shown by the validation study performed on tailor-made pavement sections consisting of steel plates as perfect reflector. Their algorithm resulted in a set of equations to determine dielectric constant and conclusively thicknesses by collecting time difference between surface reflections and bottom reflections i.e. $\Delta t1$ and $\Delta t2$ for both the GPR systems individually. The study had also shown the inability to determine thin layer thickness due to the small difference between $\Delta t1$ and $\Delta t2$. In further advancement, the XCMP method was integrated with a 3-D stepped frequency GPR system to develop a 2-D thickness profile at higher speeds accounting dielectric constant calibration without coring [18]. The proposed method has incorporated a multi array radar system comprised of 11 bow-tie antennas having an effective scan width of 1.57 m and enabled to transmit a broad frequency spectrum ranges between 200 MHz to 3 GHz. The proposed method produces an error of 3% to 8% while validating with core thickness data however absolute error was less than construction tolerance limits of 5 mm.

Since all the above mentioned methodologies considering CMP technique comprises of different amalgamations of GPR systems, to estimate dielectric constant with considerable accuracy in spite of employing complex system, which further makes analysis procedure more cumbersome.

5. Discussion and Conclusion

Few methods discussed above have a common problem that, only a few values are obtained along the interested length of pavement, where dielectric constants are used to analyze GPR data for pavement assessment. Often, a single GPR data profile is gathered for a length of pavement, and this data is calibrated to just one dielectric constant value which considered as 'best fit' value for the entire data file and also referred as "representative value". Nevertheless, there will be variations in the dielectric properties along the length of the pavement, caused by deviations in material and environmental condition, therefore leads to uncertainties and errors in results reported for locations that are not beneath the calibration points.

Another set of improvised methods, capable to estimate dielectric constant profile as a function of space and ascertained to be very beneficial in terms of obtaining more reliable results as exploited the whole section for calibration purpose. These methods are either performed with the aid of combination of GPR equipment or involved cumbersome computation techniques leads to economically and computationally expensive solutions. Certain factors affecting the precision and consistency are related to technological and engineering limitation of GPR equipment, however, it is important to noteworthy that the inculcation of erroneous results can be evolved substantially from both the way in which the technology exploited as well as the way data analysis performed.

References

[1] Evans R, Frost M, Stonecliffe-Jones M and Dixon N 2007 *Transp. Res. Rec.* **2037** 128–35.
[2] Al-Qadi I L, Lahouar S and A. Loulizi 2001 *NDT E Int.* **34** 427–34
[3] Porubiaková A and Komačka J 2015 *Procedia Eng.* **111** 660–5
[4] Al-Qadi I L, Leng Z, Lahouar S and Baek J 2010 *Transp. Res. Rec.* **2152** 19–27
[5] Balanis C A 1989 *Advanced Engineering Electromagnetics* (New York: John Wiley & Sons)
[6] Leng Zand and Al-Qadi I L 2014 *NDT E Int.* **66** 90–8
[7] Lahouar S, Al-Qadi I L, Loulizi A, Clark T and Lee D 2002 *Transp. Res. Rec.* **1806** 81–7
[8] Florida Department of Transportation 1993 Development of a Procedure for the Automated Collection of Flexible Pavement Layer Thicknesses and Materials: Penetrating Radar Technology (Florida: FHWA)
[9] Wimsatt A J, Scullion T, Ragsdale J and Servos S 1998 Proc. Structural Materials Technology III: An NDT Conf. vol 3400 p 372
[10] Saarenketo T and Scullion T 2000 *J. Appl. Geophys.* **43** pp. 119–38
[11] Skolnik M I 1980 *Introduction to radar Systems* (New York: McGraw-Hill Book Company)
[12] Lau C L, Scullion T, and Chan P 1992 *Transp. Res. Rec.* **1355** 99–107
[13] Kurtz J L, Fisher J W, Skau G, Armaghani J and Moxley J G 1997 Radar Sens. Technol. II vol 3066 p 11
[14] Spagnolini U 1997 *IEEE Trans. Geosci. Remote Sens.* **35** 454–63
[15] Spagnolini U and Rampa V 1999 *IEEE Trans. Geosci. Remote Sens.* **37** 383–94
[16] Shang B J Q, Umana J A, Bartlett F M and Rossiter J R 1999 *J. of Transp. Engg.* **125** 347–56
[17] Davis J L, Rossiter J R, Mesher D E, and Dawley C B 1994 Proc. 5th Int. Conf. on Ground Penetrating Radar p 319
[18] Zhao S and Al-Qadi I L 2016 *NDT E Int.* **78** 29–36
[19] Bastard C L, Wang Y, Baltazart V and Dérobert X, "*IEEE Geoscience and Remote Sensing Letters* **11** 873–7

IOP Conf. Series: Materials Science and Engineering 1075 (2021) 012027 doi:10.1088/1757-899X/1075/1/012027

Establishing stress boundaries for various loading and pavement configuration

Yazeed Malhas[1], Ratnasamy Muniandy[1,*], Salihudin Hassim[1] and Fauzan Mohd Jakarni[1]

[1]Department of Civil Engineering, Faculty of Engineering, Universiti Putra Malaysia, 43400 UPM Serdang, Selangor, Malaysia.
[*] Corresponding author: ratnas@upm.edu.my

Abstract: This article is aimed to understand the relationship between stress and pavement components and finding the boundaries of the vertical and lateral stress distribution. It is carried out on four-layer pavement structure to understand the load distribution behavior through different type of materials in each layer with different resilience modulus values and different thickness. This analysis supports us to improve the pavement design accuracy and realistic, which will reduce the road maintenance cost and increase the pavement service life. Several simplifying assumptions regarding tire load applied location has been used. Therefore, 25,000 different vertical stress values in form of 125 different set of data (difference loading and pavement configuration) under five different loading conditions were analyzed using the KENPAVE software (using mechanistic empirical design method). Conclusions from the data and plot analysis show that pavement layer load distribution is affected unevenly as the effect of thickness is greater than the impact of the strength of the layer. Moreover, the results showed that there is a theoretical boundary for load distribution in the lateral direction at the bottom of the surface layer between (8 in - 14.5 in), bottom of base layer (10 in - 19.5 in) and bottom of sub-surface layer (12 in -22 in). The stress distribution might be used as an indicator for engineers to determine the pavement behavior under an applied load.

1. Introduction

Expansion of the heavy industry sectors have resulted in more changes in terms of the characteristics of the heavy vehicles types and configurations. The changes are to cope with the increase rate in economic and industrial activities, heavy goods and construction materials movements and the individuals travelling demands Truck traffic also is a major factor in pavement design because truck loads are the primary cause of pavement distress. Different axle configurations of track give a different level of pavement damage [1].

Highways are constructed in a series of layers of different materials to distribute high surface stresses to relatively low bearing subgrades. Generally, the highest quality materials are located near the surface, the load carrying capacity of such a highway is dependent upon the load-distributing characteristics of the layered system. To create a rational economical design of a layered pavement system it is necessary to understand the load-distributing characteristics of the system.

Most pavements consist of a surface course, a base course, a subbase course and a subgrade. The thickness of layers, types of materials and other variables offer an infinite number of possible combinations [2].

A Number of simplifying assumptions must be made in order to arrive at a theoretical mathematical description of the stresses induced in the components of a roads by tire loads on the surface. Under such idealized assumptions a highway pavement can be symbolized by a layered system.

Therefore, unparalleled expansion of major expressway systems makes imperative the development and use of adequate methods for evaluation, design and construction control of flexible pavement systems. The problem to be solved by the engineer in roads design or airport construction deals mainly with layered soil deposits.

In foundation engineering, serious difficulties are encountered, where a soft compressible clay layer is sandwiched at some depth between an upper layer of sand and underlying layer of sand or rock.

Whether the surface layer is stronger or weaker than the underlying layer, Road foundation design of most important structures in civil engineering deals with the evaluation of stresses and strain in multi-layered systems.

This investigation was undertaken to provide a general analysis of stresses and strain distribution through a multi-layered system for the engineer. The analysis is intended as a useful tool which can be directly applied to the analysis of actual conditions encountered in layered soil structures [3].

Therefore, by using various types of loads and layer configuration and using Kenpave program (which use Mechanistic-Empirical Methods) the results will give better prediction for the load behavior and effect (stress, strain and distribution boundaries) through the pavement layers. Also, the results will be more accurate when size of data increase.

Therefore, the investigation builds on 25000 value of stress and strain. Then the stress results plotted in tables and figure to be analyzed and extract the lateral distribution boundary for each set of data.

2. Objectives

This investigation was undertaken to provide a general analysis of stresses and strain of a multi-layered system, for civil engineers. The analysis is intended as a useful tool which can be directly applied to the analysis of actual conditions encountered in pavement layered structure, Numerical evaluation for these data of certain cases is given in a form of influence curves of practical problems. This investigation used Kenpave software as the analyzing program for the pavement therefore area of study will be affected by the limitation of the software.

3. Methodology

Kenpave software was developed by Yang H. Huang [4]. It is a Microsoft- windows based version that combines the old KENLAYER flexible pavement software and KENSLABS rigid pavement software[5]. This software allows the use of linear elastic, nonlinear and viscoelastic properties of the materials for the different layers. The software can handle up to 19 layers and performs damage analysis. The interface between the different layers can be specified as either un bounded or fully bonded.

3.1. Input Parameters in Kenpave Computer Program

3.1.1. coordination (vertical and horizontal) (in)

For this study lateral and vertical coordination has been entered for analyzing:
The pavement divided to 8 vertical locations:
1-At the top of each layer
2-At the bottom of each layer
3-At the middle of each layer
Moreover, at each vertical location there will be 25 different points in the lateral direction the distance between each point and the next is ¼ contact radius.

3.1.2. Thickness (in)

Table 1 show The thicknesses for each layer obtained from JKR [6] and AASHTO [7] limits for layer thickness for The Structural Design of Flexible Pavement. Subgrade is the natural layer in pavement structure and has infinity thickness. Therefore, the input data for subgrade thickness in KENLAYER should be "infinity".

Table 1. Thickness values used in the study [7]

thickness (in)		
Asphalt layer	Base layer	Sub-base layer
4	6	6
5	7	7
6	8	8
7	9	9
8	10	10

3.1.3. Poisson's Ratio

Poisson's ratio can vary from initially 0,5. Generally, stiffer materials will have lower Poisson's ratio than softer materials. The Poison's Ratio keyed into kenpave [4] is:
1-Asphalt layer 0.45.
2-Base layer 0.40.
3-Sub-base layer 0.40.
4-Subgrade 0.35.

3.1.4. Elastic Modulus (psi)

The modulus of elasticity is stress divided by strain for a slowly applied load. In kenpave computer program, each layer has unique number of modulus of elasticity. The elastic modulus for each layer is obtained as follow:
1-Asphalt concrete elastic modulus E is between (450000 – 250000) psi
2-Crushed aggregate base elastic modulus E is between (45400 – 32000) psi – CBR(100-60)
3-Granular sub-base elastic modulus E is between (32000 – 15000) psi – CBR(60-20)
4-Subgrade elastic modulus E is between ((0.07 – 0.03) x E for surface layer) psi

3.1.5. Traffic loading

In this study for simplification of the analysis, the dual wheels are converted into an equivalent 70-80-90-100-120 kN single axle load (ESAL).
The contact area is important to be determined so the axle load can be assumed to be uniformly distributed. In this study, single axle with dual tires need to be considered and each tire is assumed to have circular contact area. The tire spacing is assumed with a typical distance between dual tires of 30 cm (distance between center lines of tires) [8] . The contact pressure in this study is (80 psi) and contact radius is calculated as follows:

$$a = \sqrt{\frac{p}{q\pi}} \qquad (1)$$

Where are:
a- radius.
P- Total load on the tire.
q- tire pressure.

IOP Conf. Series: Materials Science and Engineering 1075 (2021) 012027 doi:10.1088/1757-899X/1075/1/012027

3.1.6. Output Parameters of Kenpave

The output from this program are stress, strain and displacement. For a single and multiple load groups, a maximum of nine and ten responses can be obtained respectively. Only the vertical stress and vertical strain values are used for this study.

3.1.7. Analyzing the results using excel

All data were tabulated and used to plot relationship graphs between vertical stress and vertical strain against the lateral distribution distance in order to observe the ultimate distance the stress or strain will reach at the interconnecting point between the two tangential drown between curves.

After all the lateral distance values has been measured, a new relationship will was created using Excel analyzing tools in order to understand how each variable will affect the stress distribution between pavement layers.

4. Test results and discussion

The results had tabulated into five deferent tables depending on deferent arrangement of layer, resilience modulus and layers thickness.

Each table is divided into four main columns, column one shows the load value in KN which is similar in all tables 70,80,90,100 and 120kn.

Second column shows the thickness in (inches) of layers and it is divided into 3 miner columns (surface layer, Base-layer, Sub-Base layer).

third column show the strength of material or resilience modules in Psi and contain four miner column for each layer (surface layer, base layer, subbase layer and subgrade).

Fourth column show the maximum lateral distribution distance from the center line in inches for each layer and contain four miner column for (top of surface layer, top of base layer, top of Sub-base layer and top of Subgrade). The tables are arranged as follow[1]:

1-Highest strength with lowest thickness,
2-Lowest strength with lowest thickness,
3-Lowest thickness with lowest surface layer strength and highest strength for bottom layers,
4-Lowest thickness with highest surface layer strength and lowest strength for bottom layers,
5-Lowest thickness with highest strength for A.L & SUB base and lowest strength of B.L.

[1] Due to the large size of data, only illustrative examples will be presented in this paper

IOP Conf. Series: Materials Science and Engineering 1075 (2021) 012027 doi:10.1088/1757-899X/1075/1/012027

Table 2. Values of lateral stress distribution distance results from (highest strength with lowest thickness)

Load(KN)	Thickness(in)			Strength (Psi)				Maximum lateral distance from the centre line (in)			
	Asphalt layer	Base layer	Sub-base layer	Asphalt layer	Base layer	Sub-base layer	Subg-rade layer	Asphalt layer	Base layer	Subbase layer	Subgrade layer
	4	6	6	450000	45400	32000	31500	3.6	9	13	16
	5	7	7	400000	42250	28200	24000	3.6	10	12.5	16
70	6	8	8	350000	38900	24000	17500	3.6	11	14.5	17
	7	9	9	300000	35500	19900	12000	3.6	11	13.5	16
	8	10	10	250000	32000	15000	7500	3.6	9.5	16	17.5
	4	6	6	450000	45400	32000	31500	4	10	13.5	16.5
	5	7	7	400000	42250	28200	24000	4	11.5	14	17
80	6	8	8	350000	38900	24000	17500	4	12	14.5	17.5
	7	9	9	300000	35500	19900	12000	4	11.5	14	17
	8	10	10	250000	32000	15000	7500	4	12.5	17.5	18.5
	4	6	6	450000	45400	32000	31500	4.4	10	14	16.5
	5	7	7	400000	42250	28200	24000	4.4	12	15	17.5
90	6	8	8	350000	38900	24000	17500	4.4	12	16.5	18.5
	7	9	9	300000	35500	19900	12000	4.4	12	14	17
	8	10	10	250000	32000	15000	7500	4.4	12.5	18	21
	4	6	6	450000	45400	32000	31500	4.8	10	14.5	17
	5	7	7	400000	42250	28200	24000	4.8	12	15	19
100	6	8	8	350000	38900	24000	17500	4.8	13.5	17	20
	7	9	9	300000	35500	19900	12000	4.8	12	16.5	20
	8	10	10	250000	32000	15000	7500	4.8	13	19.5	22
	4	6	6	450000	45400	32000	31500	5.2	11	15	18.5
	5	7	7	400000	42250	28200	24000	5.2	12	15.5	20
120	6	8	8	350000	38900	24000	17500	5.2	13.5	18	21
	7	9	9	300000	35500	19900	12000	5.2	14	17.5	21
	8	10	10	250000	32000	15000	7500	5.2	13.5	19.5	22

4.1. Relationships between pavement layer thickness and the stress lateral distribution distance in comparison to the relationship between resilience modules and the stress lateral distribution distance

According to the relationship graphs (Figures 1,2,3,4,5 and 6) between (resilience modulus, layer thickness) vs lateral stress distribution distance in 'inches' under 80&120 kN load for the pavement layers (Surface layer, Base layer, Sub-base layer), it can be concluding that the lateral distribution distance increase by increasing the layer thickness and the resilience modules moreover, the plots show that the distribution distance is significantly affected by the thickness under high load values.

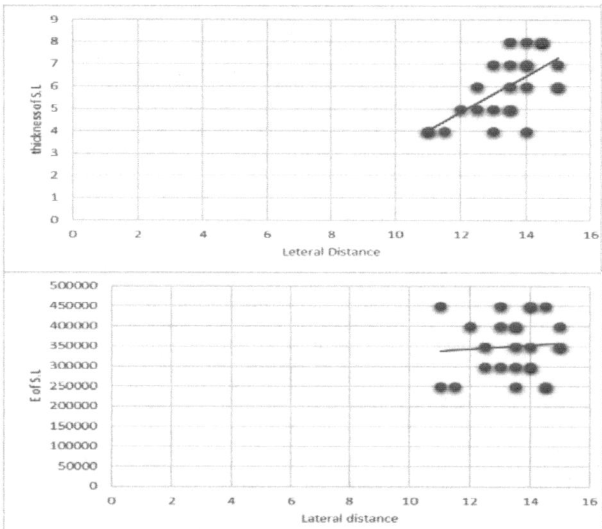

Figure 1. Relationship between (resilience modulus, layer thickness) vs lateral stress distribution distance under 80kn for surface layer

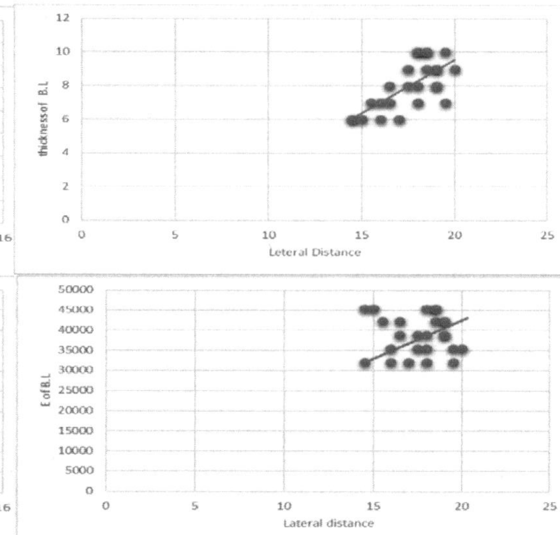

Figure 2. Relationship between (resilience modulus, layer thickness) vs lateral stress distribution distance under 80kn for Bas-layer

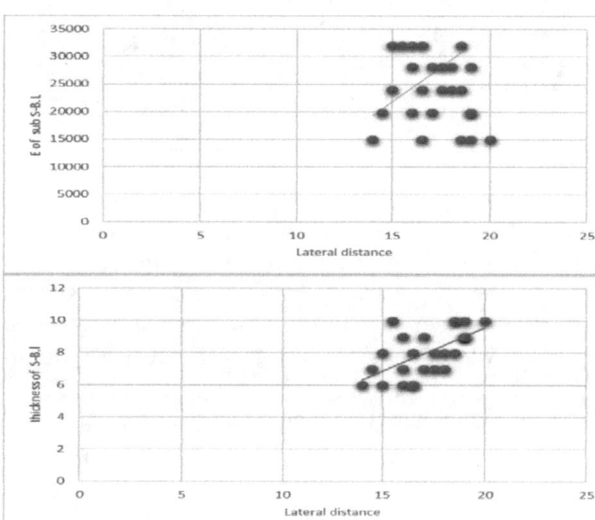

Figure 3. Relationship between (resilience modulus, layer thickness) vs lateral stress distribution distance under 80kn for Sub-Bas layer

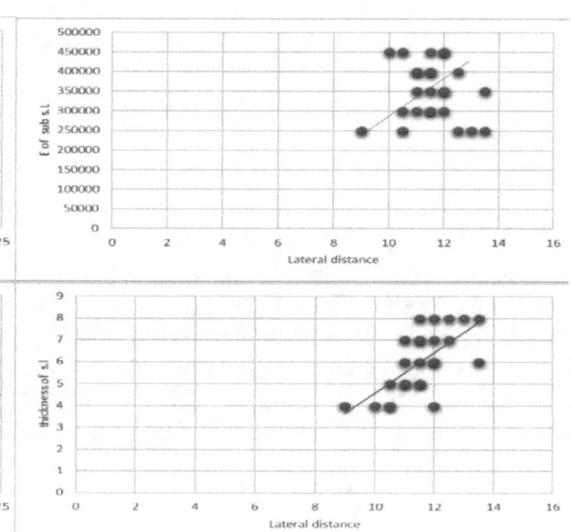

Figure 4. Relationship between (resilience modulus, layer thickness) vs lateral stress distribution distance under 120kn for surface layer

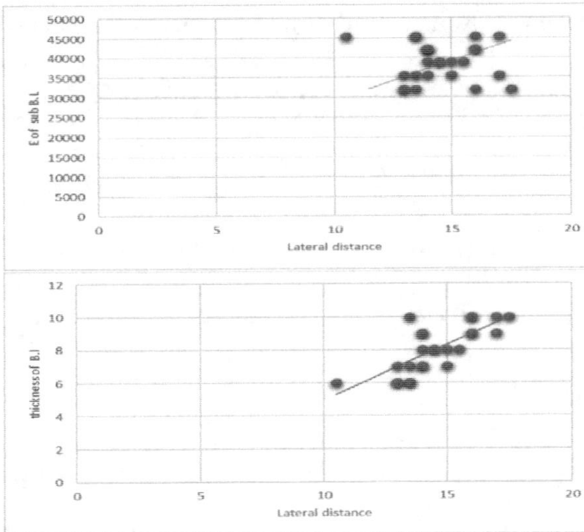

Figure 5. Relationship between (resilience modulus, layer thickness) vs lateral stress distribution distance under 120kn for Base layer

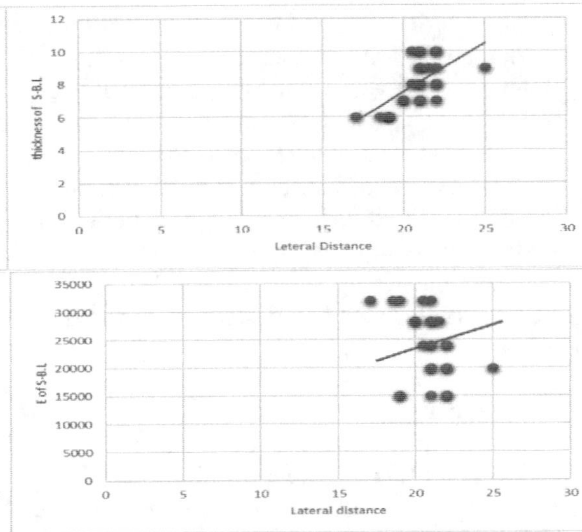

Figure 6. Relationship between (resilience modulus, layer thickness) vs lateral stress distribution distance under 120kn for Sub-Base layer

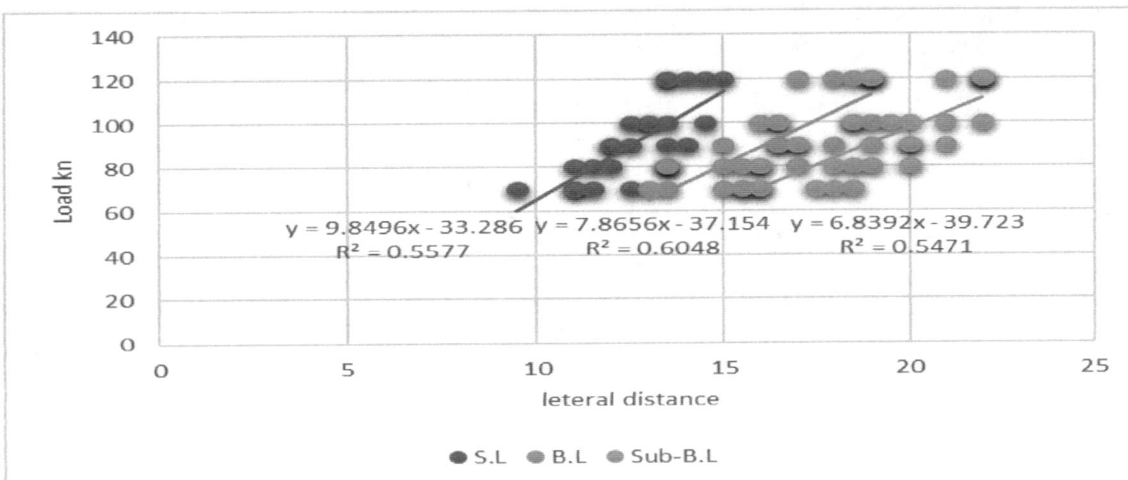

Figure 7. Rlationship between applied load vs lateral stress distribution distance for Surface layer, Base layer, Sub-Base layer

From Figure 7 which show the relationship between LOAD in KN at Y-axis and Lateral distribution distance X-axis for (surface layer and Base layer and Sub-base layer) it can be seen that the lateral distribution distance increase by increasing the applied load value on the pavement.

5. Summary and conclusion

The study was built on 125 sets of data for three main variables applied load, layer thickness and strength of material which presented as (resilience modulus) and CBR for the sub-grad. Therefore, after using Kenpave software (linear method of analysis) to get values for stress and strain, the result was presented in 25,000 values for different point location. the finding has been tabulated and plotted in-different type of graph in order to understand the relationship between each factor and how it effects on the lateral stress disruption distance therefore, as results:

1) Boundaries for vertical stress and vertical strain for each layer of the pavement are as follow:
   1-Top of surface layer is related to the contact radius and between (3.6-5.2) in
   2-Bottom of surface layer between (8-14.5) in
   3-Bottom of base layer between (10-19.5) in
   4-Bottom of Sub-base layer between (12-22) in

2) Vertical stress value is significantly affected by the layer thickness more than the strength of the layer Under the influence of large values of loads.

3) Vertical strain value in the lateral direction is significantly affected by the resilience modulus for each layer.

6. References

[1] I. Zaabar and K. Chatti, “Estimating Vehicle Operating Costs Caused by Pavement Surface Conditions,” *Transp. Res. Rec. J. Transp. Res. Board*, vol. 2455, no. 1, pp. 63–76, Jan. 2014.
[2] M. Rogers, *Highway engineering*. Blackwell Science, 2003.
[3] M. Kesaree Charyulu, “Theoretical stress distribution in an elastic multi-layered medium,” Iowa State University, 1964.
[4] Yang H. Huang, *Pavement Analysis and Design, 2nd Edition*, 2nd ed. 1993.
[5] E. Aburkaba, R. Muniandy, and N. Thamer, “Comparison of Flexible Pavement Performance Using Kenlayer and Chev PC Software Program,” *Aust. J. Basic Appl. Sci.*, vol. 7, no. 9, pp. 112–119, 2013.
[6] jkr, “PENARAFAN HIJAU JKR MANUAL New & Upgrading of Roads-KJ,” 2015.
[7] American Association of State Highway and Transportation Officials, “AASHTO Guide for Design of Pavement Structures,” 1993.
[8] European Tyre and Rim Technical Organisation - Pneumatic tyres, rims and valves Standards Manual,2020

IOP Conf. Series: Materials Science and Engineering 1075 (2021) 012028 doi:10.1088/1757-899X/1075/1/012028

Development of structure model of hollow slab for solar pavement based on light-guide concrete

X D Zha[1], H W Hu[1], Z W Wang[1], H P Chen[1] and C J Zhang[1]

[1]School of Traffic and Transportation Engineering, Changsha University of Science and Technology, Changsha 410114, Hunan, China

Corresponding author: 993561963@qq.com

Abstract. To develop a reasonable structure type of solar pavement, a kind of structure model of hollow slab was proposed for solar pavement based on light-guide concrete in this paper. It is composed of two parts: a rectangular hollow slab of light-guide concrete with an arch chamber as well as a solar panel which is installed in the arch chamber. Each light-guide body is compounded of a convex lens, a prism and a light-guide fiber. Through building a three-dimensional finite element model for the pavement structure of hollow slab, the mechanical responses of hollow slab were numerically simulated, thus the geometric sizes of arch chamber were optimized according to the requirements of bearing capacity. By establishing a simulation analysis model of light path for the light-guide body, the total solar energy of single day transmission for the light-guide body were numerically simulated, then the geometric sizes and optical parameters of light-guide body were optimized according to the condensing and light-guide performances. Accordingly, the steel shaping die was designed and manufactured, and the hollow slab model of light-guide concrete was prepared by taking self-compacting concrete as slab matrix. The mechanical properties and the power generation efficiencies were tested for the model. The results show that the developed hollow slab of solar pavement based on light-guide concrete has high strength, and its working voltage and maximum output power increase with the increase of the number of light-guide bodies. Therefore, the model can meet the dual function requirements of safe driving and clean power generation.

1. Introduction

Solar pavement is a kind of pavement structure which paves a surface layer of power generation with photovoltaic solar cells on new or existing pavement. It can achieve the purpose of clean power generation on the basis of satisfying the normal traffic function to provide the power support for smart transportation. This idea was first put forward by S. Brusaw and his wife in 2006, and named as "Solar Roadways" or "Intelligent Highway". In March 2014, they built the world's first prototype solar parking lot, and the structure of newly designed SR3 is a hexagonal solar road panel with "upper and lower float glass panels + middle solar cells" [1]. In 2014, A. B. Northmore et al. developed a kind of solar road panel in combination with Canadian traffic and environmental conditions, and the structure is an aluminum frame rectangular panel with "tempered glass panel + solar cells + fiberglass base panel" [2-3]. In November 2014, the world's first bike road with integrated solar panels was built in Netherlands, and named as "SolaRoad". The SolaRoad section is 70 m long and 1.7 m wide, which is a rectangular plate with the structure of "tempered glass panel + solar panels + concrete slab" [4]. In December 2016, the world's first solar paneled road with 1 km long was constructed in France, and named as "Wattway". The Wattway structure is a rectangular plate with "silicone resin coating + solar cells + polymer base panel" [5]. In 2017, A. S. Dezfooli et al. designed two prototypes entitled as "solar panel" (solar cell embedded in rubber and Plexiglas) and "solar pavement" (solar cell embedded between two porous rubber layers) [6]. In December 2017, China built the world's first expressway with solar pavement of 1080 m long on Jinan Ring Expressway of G2001 in Shandong Province, and named as "photovoltaic pavement". The structure is a rectangular plate of "tempered glass particle resin concrete + monocrystalline silicon photovoltaic panels + fiberglass polymer base panel" [7].

Overall, the researches on solar pavement at home and abroad are still in the exploratory stage at present. Most of existing researches adopted the rectangular or hexagonal solid panel structures, which are composed of three layers: a surface layer with transparence and skid resistance, a middle layer with photovoltaic cell and a bottom layer with water insulation and protection. However, because the solar pavement is obviously different from the traditional pavement in structure design, material property and construction technology as well as bearing capacity, pavement performance, power generation efficiency, durability and engineering cost etc., new structure types need to be further explored. Therefore, in 2010, we designed and prepared a kind of hollow slab structure which can protect solar cells [8-9], and further proposed a hollow slab structure of solar pavement based on light-guide concrete [10]. In this paper, the structure model was developed to provide a reference for the exploration of solar pavement.

2. Structure design and simulation analysis of hollow slab for solar pavement of light-guide concrete

2.1. Structure design

The hollow slab structure of solar pavement based on light-guide concrete is mainly composed of two parts: a rectangular hollow slab of light-guide concrete and a solar panel, as shown in figure 1. Wherein, the inner cavity of hollow slab adopts an arch chamber structure with parabolic vault, the solar panel is installed in it, and the slab matrix is made of self-compacting concrete. Each light-guide body is compounded of a condensing lens (compounding of a planoconvex lens and a truncated cone-shaped prism) and a cylindrical light-guide fiber, as shown in figure 2. The light-guide bodies are arranged in the hollow slab at equal intervals according to the chessboard style. The convex lens are installed on the surface of hollow slab as the light entrance end, and the light exit end of light-guide fiber is placed at the vault of arch chamber. The working principle of the structure is that: when the sunlight irradiates on the hollow slab surface, the solar rays are condensed, guided and projected onto the solar panel through the light-guide bodies, thereby the light energy is converted into the electric energy reserves by the photoelectric effect.

Where in figure 1, L, B and H are the length, width and height of hollow slab, respectively; b, l, t and h are the arch width, floor thickness, arch height and roof thickness of arch chamber, respectively, and that is $H = l + t + h$. Where in figure 2, d and r are the thickness and lower circular surface radius of planoconvex lens, respectively; a is the radius of light-guide fiber, the fiber can be appropriately bent, and its length can be adjusted according to its layout location in the hollow slab; c is the height of truncated cone-shaped prism, and the prism combines the planoconvex lens and the light-guide fiber into a whole, so that its upper and lower surfaces are separately connected with the lower circular surface of planoconvex lens and the top circular surface of light-guide fiber, i.e. its corresponding upper and lower surface radius are r and a, respectively; N is the refractive index of condensing lens; R and F are the upper spherical radius and focal length of planoconvex lens, respectively.

It can be seen that the structure has the following characteristics:

1) The concrete slab has sufficient bearing capacity, and the arch chamber can reduce the tensile stress of hollow slab concrete at the vault to meet the requirements of traffic load.
2) The spherical surface of condensing lens are exposed on the slab surface, and it is similar to the prompting blind road, which can improve the surface skid and wear resistance.
3) The solar panel is installed in the arch chamber, which can be obtained effective protection to avoid direct bearing and fracturing failure, and avoid overheating to reduce power generation efficiency.

4) The light-guide body has the function of condensing solar rays, which can improve the power generation efficiency of solar panel and reduce the number or size of panel.
5) The electronic components and connecting wires etc. can be installed in the arch chamber to meet the waterproof requirements.
6) The slabs are convenient in splicing and maintenance, appropriate in cost, and can be formed the solar power generation system.

Therefore, the structure can meet the dual function requirements of safe driving and clean power generation.

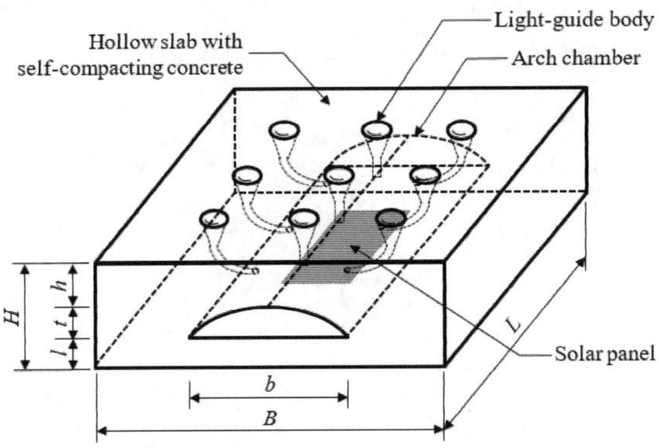

Figure 1. Hollow slab structure of solar pavement based on light-guide concrete

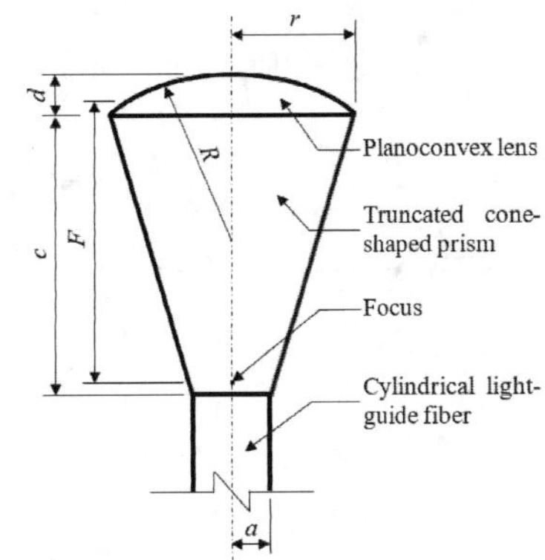

Figure 2. Elevation view of structure for light-guide body

2.2. Mechanical analysis of pavement structure of hollow slab

In order to determine the reasonable sizes of arch chamber, the three-dimensional finite element method was used for the numerical simulation in accordance with the hollow slab structure in figure 1. The whole pavement structure is regarded as a single hollow slab structure on linear elastic homogeneous foundation with the complete bonding between the layers and without counting the dead weight. The full constraints are applied to the bottom surface of foundation, and the horizontal displacement constraints are applied to the surrounding sides of foundation. The hollow slab is a plate with four free edges. The standard axle load of BZZ-100 is adopted and equivalent to the vertical uniformly distributed load of double-wheel rectangle of 18.6 cm × 19.2 cm with the contact pressure of 0.7 MPa, the double-wheel center distance of 31.4 cm and the total width of 50.0 cm.

According to our existing analysis results [8-10], both the length and width of hollow slab were taken as 50 cm, and both the length and width of foundation were taken as 3.5 m with the thickness of 5 m. Meanwhile, the worst loading position is that the double-wheel load is symmetrically arranged on the arch chamber of hollow slab. Considering the symmetries of structure and load, the quarter model was chosen to build the solid structure model, as shown in figure 3. Wherein, the solid element model was adopted as the SOLID65 element, and the interlayer contact was adopted as the surface-surface contact model. The mesh size of hollow slab was taken as 1 cm, and the mesh sizes of foundation were gradually increased from shallow to deep and from near to far.

On the basis of keeping the plane size of hollow slab unchanged at 50 cm × 50 cm, the value-taking schemes of geometric parameters of arch chamber were selected in table 1. The elastic moduli of concrete and foundation are E_c = 31 GPa and E_b = 200 MPa, and the Poisson's ratios are μ_c = 0.15 and μ_b = 0.30, respectively.

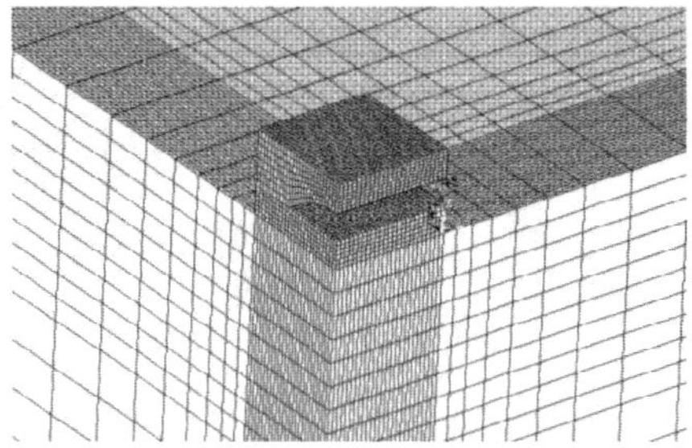

Figure 3. Three-dimensional finite element modelling for pavement structure of hollow slab

Table 1. Value-taking schemes of geometric parameters of arch chamber

Scheme \ Parameter	Arch width b (cm)	Floor thickness l (cm)	Arch height t (cm)	Roof thickness h (cm)
1	20	5.0	5.0	5.0
2	22	5.5	5.5	5.5
3	24	6.0	6.0	6.0
4	26	6.5	6.5	6.5
5	28	7.0	7.0*	7.0
6	30	7.5	7.5	7.5
7	32	8.0*	8.0	8.0*

The ANSYS software was used for the numerical simulation. The influence laws of various geometric parameter changes of arch chamber on the mechanical responses of pavement structure of hollow slab were analyzed in the order of $b \to l \to t \to h$, and the appropriate values of each parameters were determined separately. Among them, when a certain parameter change will be analyzed, its values are selected separately according to each schemes in table 1. The previously determined parameters are chosen according to the determined values, and the subsequently undetermined parameters are taken according to the schemes with "*" in table 1. It can be obtained by numerical simulation that the surface tensile stress at the floor of arch chamber corresponding to the center of arch width at the edge of hollow slab in the traffic direction is the largest (i.e. the maximum tensile stress σ_m), and the surface vertical displacement of slab corresponding to the center of arch roof at the edge of hollow slab is the largest (i.e. the maximum vertical displacement l_s). The stress and displacement of the two points are used as the analysis indexes of mechanical responses. Therefore, the relationship curves of σ_m and l_s with the changes of each geometric parameters of arch chamber are obtained in figure 4 and figure 5.

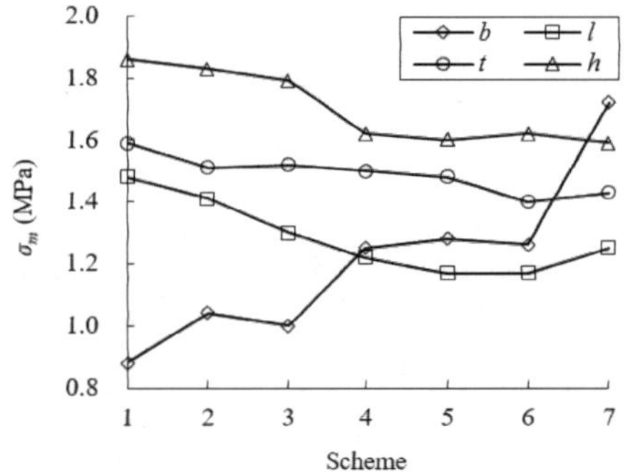

Figure 4. Change curves of σ_m under each schemes

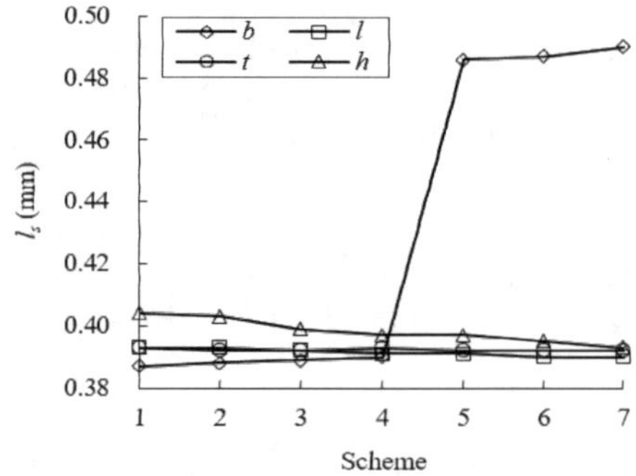

Figure 5. Change curves of l_s under each schemes

According to Specifications for Design of Highway Cement Concrete Pavement (JTG D40-2011), the value of allowable tensile stress is chosen as the control index to ensure the sufficient bearing capacity and durability for the concrete hollow slab. The grade of traffic load is set as extremely heavy, and the flexural tensile strength of self-compacting concrete is 5.0 MPa. By the checking formulas of structure ultimate state, the value of allowable fatigue tensile stress of hollow slab structure under the standard axle load can be calculated as 1.65 MPa, and the value of allowable maximum tensile stress is 4.18 MPa under the heaviest axle load (250 kN). Meanwhile, because of the smaller plane size of hollow slab, the temperature stress after checking is very small and negligible.

From Figures 4 and 5, it can be seen that σ_m and l_s basically show the significant increasing change trends with the increase of arch width b, and 26 cm of Scheme 4 is selected as the appropriate value of b by considering the reasonable force and increasing the size of arch chamber as much as possible. With the increases of floor thickness l and arch height t, σ_m tends to decrease firstly and then increase slightly, while l_s basically decreases gradually in a very small amplitude, thus 5 cm of Scheme 1 is selected as the appropriate values for l and t separately based on economic considerations. With the increase of roof thickness h, σ_m and l_s basically show decreasing trends, but when h is less than 6 cm, σ_m exceeds the value of allowable fatigue tensile stress and does not meet the bearing capacity requirements, thereby 7 cm of Scheme 5 is selected as the appropriate value for h to prepare easily the model. Therefore, considering the requirements of reliability and economy comprehensively, the suitable sizes of arch chamber for the hollow slab were determined as b = 26 cm, l = 5 cm, t = 5 cm and h = 7 cm, respectively, and the corresponding size of hollow slab is 50 cm × 50 cm × 17 cm. After checking, the maximum tensile stresses of the hollow slab under the standard axle load and the heaviest axle load are 1.60 MPa and 3.93 MPa, respectively, and all of them meet the requirements of allowable values. The results show that the structure sizes of hollow slab meet the design requirements.

2.3. Simulation of light path for light-guide body

In order to determine the reasonable structure sizes and materials for the light-guide body, the tracing method of light path was used for the simulation analysis. The solar rays are set to be parallel lines each other with wavelength of 500 nm and irradiance of 1000 W/m^2. The distance from the virtual emission surface to the center of light entrance end of convex lens surface is 10 cm, and the refractive index of air is taken as 1. The condensing lens is made of polymethyl methacrylate (PMMA) with refractive index N of 1.49. The light-guide body is a homogeneous body with total reflection around itself which does not absorb solar energy. The light entrance end of convex lens surface and the light exit end of light-guide fiber bottom are total transmission. There is full combination between the interfaces. Therefore, the TracePro software was used for the light path simulation, as shown in figure 6.

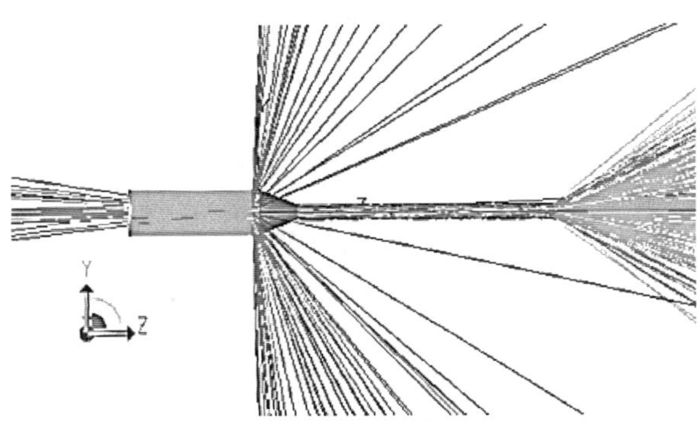

Figure 6. Light path simulation modelling for light-guide body

The design parameters and value-taking schemes for the light-guide body are selected in table 2, where f and n are the surface reflectivity of prism and the refractive index of light-guide fiber, respectively. In order to optimize the structure of light-guide body, the influence laws of various parameter changes on the light-guide performances of light-guide body were analyzed in accordance with the order of $a \rightarrow r \rightarrow d \rightarrow c \rightarrow n \rightarrow f$, and the appropriate values of each parameters were determined separately. Among them, when a certain parameter change will be analyzed, its values are chosen separately according to each schemes in table 2. The previously determined parameters are chosen according to the determined values, and the subsequently undetermined parameters are taken according to the schemes with "*" in table 2.

Table 2. Value-taking schemes of geometric and optical parameters for light-guide body

Parameter / Scheme	Light-guide fiber		Convex lens		Prism	
	Radius a (mm)	Refractive index n	Radius r (mm)	Thickness d (mm)	Height c (mm)	Surface reflectivity f
1	0.5	1.4	1.5a	0.1r	0.7F	0.4
2	1.0	1.5	2.0a	0.2r	0.8F	0.5*
3	1.5	1.6	2.5a	0.3r	0.9F	0.6
4	2.0	1.7*	3.0a*	0.4r	1.0F*	0.7
5	2.5	1.8	3.5a	0.5r*	1.1F	0.8
6	3.0	1.9	4.0a	0.6r	1.2F	0.9
7	3.5	2.0	4.5a	0.7r	1.3F	1.0

The incident angle of sunlight changes periodically every day and every year. The simulation results show that the received luminous flux of light-guide body is less when the incident angle is small, and it is difficult to condense and guide solar rays effectively. Therefore, the solar rays of different incident angles were set at intervals of 10 ° in the range of 30 ~ 90 °, and were irradiated onto the light entrance end of convex lens surface of light-guide body. After condensing and guiding solar rays, the transmitted luminous flux Φ_i at the light exit end of light-guide fiber bottom can be simulated at the corresponding incident angles. Then the total solar energy Q of single day transmission for the light-guide body can be computed and obtained, and it is used as the evaluation index of light-guide performance. Accordingly, the relationship curves of Q changed with each design parameters of light-guide body can be simulated and obtained in figure 7.

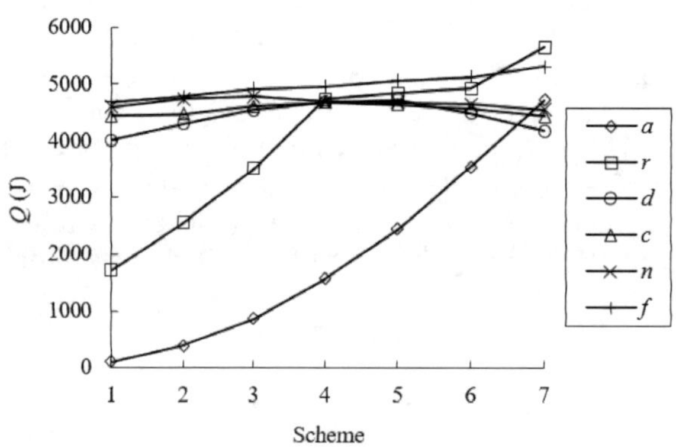

Figure 7. Change curves of Q under each scheme

From Figure 7, it can be seen that Q increases accelerate with the increase of light-guide fiber radius a, and the increase amplitude is most significant, thereby 3.5 mm of Scheme 7 is taken as the appropriate value for a. With the increase of convex lens radius r, Q increases in a stepped way, and 10.5 mm of Scheme 4 is chosen as the suitable value for r based on economic considerations. With the increase of convex lens thickness d, Q increases firstly and then decreases in a single-peak relationship change, and 4.2 mm of Scheme 4 is taken as the suitable value for d by considering the smoothness requirements that the maximum surface clearance of concrete pavement does not exceed 5 mm. The variation laws of Q with the increases of prism height c and refractive index n of light-guide fiber are both increased firstly and then decreased, so that 31.1 mm of Scheme 4 and 1.6 of Scheme 3 are taken as the appropriate values for c and n, respectively. The Q value increases gradually with the increase of prism surface reflectivity *f*. In order to increase the reflectivity, a reflective film can be pasted or coated on the prism surface, and the appropriate value of *f* is suggested as 1.0 of Scheme 7. Therefore, considering economy and convenient preparation comprehensively, the suitable design parameters of light-guide body structure were determined as a = 3.5 mm and n = 1.6 for the light-guide fiber, r = 10.5 mm and d = 4.2 mm for the convex lens, and c = 31.1 mm and f = 1.0 for the prism, thereby the corresponding total solar energy Q of single day transmission for the light-guide body reaches 5322.2 J.

3. Model preparation and performance test for hollow slab of solar pavement based on light-guide concrete

3.1. Preparation process

Because an arch chamber and multiple light-guide bodies are arranged in the hollow slab of solar pavement of light-guide concrete, it is difficult to vibrate and compact the normal concrete during preparation, and the self-compacting concrete was chosen for casting moulding. According to the above determined suitable materials and sizes for the hollow slab and the light-guide body, the corresponding steel die and light-guide body were manufactured, and the model of hollow slab for solar pavement of light-guide concrete was prepared. The specific process is as follows: assembling steel die → installing and fixing light-guide body → preparing and pouring self-compacting concrete → curing, demoulding and forming hollow slab of light-guide concrete → installing solar panel → installing and connecting solar power generation system. The prepared specimen is shown in figure 8 for the hollow slab of solar pavement of light-guide concrete.

Figure 8. Specimen preparation of hollow slab for solar pavement based on light-guide concrete

3.2. Mechanical properties testing

The tested compressive strength and flexural tensile strength of 28 day for the self-compacting concrete are 51.3 MPa and 5.5 MPa, respectively, which meet the strength standard of C40. Then the model specimens of concrete hollow slab were prepared and carried on the loading test in material testing system (MTS). Considering the dimension limitation of MTS worktable, the specimen size and the loading head area were reduced by 60% in proportion. According to the contact pressure of 0.7 MPa for the standard axle load, the maximum load was set to 8.7 kN, and the loading rate was 0.6 kN/s, as shown in figure 9. Therefore, the change curve of maximum vertical displacement with pressure was obtained in the center of specimen surface of hollow slab, as shown in figure 10.

Through the simulation of three-dimensional finite element, the maximum vertical displacement and maximum tensile stress in the hollow slab specimen are 7.14 μm and 1.08 MPa, respectively. From Figure 10, it can be seen that the test result of maximum vertical displacement under the standard axle load is 8.55 μm, which is 19.7% larger than the theoretical value. It is mainly caused by the error between the simplified theoretical model and the tested model, but the total deformation is very small. With the increase of pressure, the vertical displacement basically increases in a linear elasitic way, and the specimen can fully rebound and recovery to the original state after unloading.

Thus the rationality of simulation results was verified. The results show that the hollow slab of self-compacting concrete has good mechnical properties, and can meet the requirements of safe driving.

Figure 9. Loading test for hollow slab specimen of self-compacting concrete

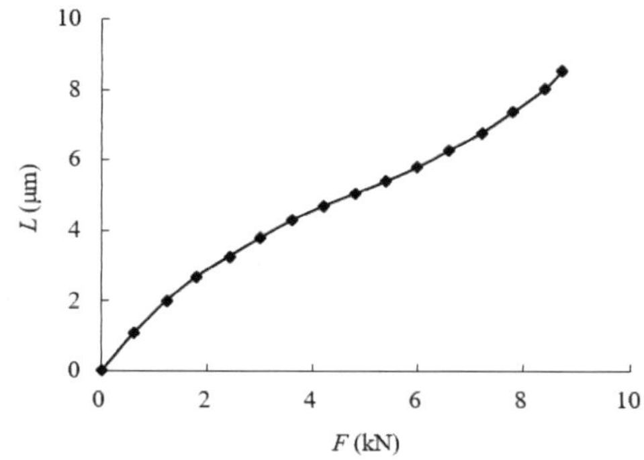

Figure 10. Change curve of maximum vertical displacement with pressure for specimen

3.3. Power generation efficiencies testing

A solar panel of 12 cm × 12 cm was installed into the arch chamber in the optimized hollow slab to prepare the model specimens of hollow slab for solar pavement of light-guide concrete. The five schemes for the number of light-guide bodies were selected separately, including 25 (Scheme A), 36 (Scheme B), 49 (Scheme C), 64 (Scheme D) and 81 (Scheme E). Through connecting the load of 0 ~ 48 Ω, the open-circuit voltage and the working voltage across load were tested under the direct sunlight of irradiance of 700 W/m². The change curves of output power P with working voltage U were obtained in figure 11, and the analysis results of optimal load and maximum output power were shown in table 3. Where the original solar panel refers to directly placed in the sun for power generation.

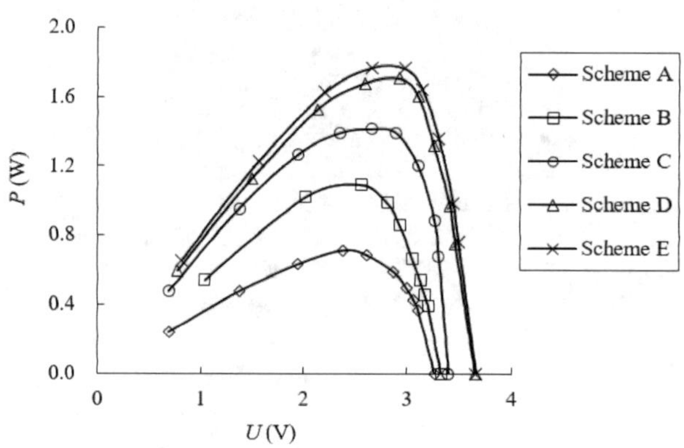

Figure 11. Change curves of output power with working voltage for hollow slab specimen

Table 3. Power generation efficiencies for hollow slab specimen

Scheme	A	B	C	D	E	Original solar panel
Number of light-guide bodies	25	36	49	64	81	/
Optimal load R (Ω)	8	6	5	5	4	5
Maximum output power P_m (W)	0.714	1.092	1.415	1.705	1.769	1.636

From Figure 11 and Table 3, it can be seen that the optimal load of specimen decreases with the increase of the number of light-guide bodies, while the open-circuit voltage, the output power and the maximum output power increase, but their increase amplitudes decrease gradually. When the number of light-guide bodies is 64, the maximum output power exceeds the generation power of original solar panel. It shows that the increase of light-guide bodies can improve the power generation efficiencies of solar panel and achieve the effective condensing effects. Therefore, it is suggested that the number of light-guide bodies should be set at 81 and not more than 100. Otherwise, due to the small distance between light-guide bodies, it is difficult for self-compacting concrete to meet the requirements of mouldability during the preparation.

4. Conclusions

According to the dual function requirements of safe driving and clean power generation, a kind of structure model of hollow slab were proposed for solar pavement based on light-guide concrete in this paper. Through mechanical analysis, light path simulation, model preparation and performance testing, it can be concluded that:

1. The maximum tensile stress and the maximum vertical displacement in the pavement structure of rectangular hollow slab with arch chamber cavity increase with the increase of arch width, but decrease with the increase of roof thickness, arch height and floor thickness. The influence of arch width is most significant, but the influences of other three parameters are relatively small. Therefore, based on the requirements of reliability and economy, the suitable structure sizes of hollow slab were determined.
2. The total solar energy of single day transmission of light-guide body increases with the increase of light-guide fiber radius, convex lens radius and prism surface reflectivity, but increases firstly and then decreases with the increase of convex lens thickness, prism height and light-guide fiber refractive index. Among them, the influences of light-guide fiber radius and convex lens radius are relatively significant. Therefore, by considering the condensing and light-guide performances, the suitable geometric sizes and optical parameters of light-guide body were determined.
3. The self-compacting concrete was used to prepare the model specimens of hollow slab for solar pavement of light-guide concrete. The model is high in strength, small in maximum vertical displacement and strong in elastic recovery ability, so that it has good mechanical properties, and can meet the requirements of safe driving. With the increase of the number of light-guide bodies, its working voltage and maximum output power increase continuously, thereby it has higher power generation efficiencies.

In summary, the research in this paper can provide a reference for the research and development of solar pavement, and its application effects need to be verfied in the future engineering.

Acknowledgments

This work was supported by National Natural Science Fundation of China (No. 51878077), Transportation standard (quota) program of the Ministry of transport of China (No. 2019-17-077), and Research and innovation program for graduate students in Hunan Province of China (No. CX20190645).

References

[1] Solar Roadways® 2019 Solar Roadways. http://www.solarroadways.com
[2] Northmore AB and Tighe SL 2014 Performance modelling of a solar road panel prototype using finite element analysis *Int. J. Pavement Eng.* **17**(5) 449-457
[3] Northmore AB 2014 Canadian solar road panel design: a structural and environmental analysis *MD Thesis of the University of Waterloo*
[4] SolaRoad® 2019 SolaRoad. http://www.solaroad.nl
[5] COLAS® 2019 Wattway. http://www.wattwaybycolas.com/en
[6] Dezfooli AS, Nejad FM, Zakeri H and Kazemifard S 2017 Solar pavement: a new emerging technology *Sol. Energy.* **149** 272-284

[7] Qilu Transportation Development Group® 2017 The first expressway photovoltaic pavement test section in the world is officially opened to traffic. http://www.qljfjt.com/detail_3835.html
[8] Zeng J 2013 Mechanical analysis of hollow slab pavement structure *MD Thesis of Changsha University of Science & Technology*
[9] Zha XD, Zhang CJ, Wu ZJ and Zhang QS 2016 Mechanical analysis and model preparation for hollow slab element of solar pavement *Acta Energiae Solaris Sinica.* **37**(1) 136-141
[10] Zhang CJ 2017 Research on preparation and performance of hollow slab model of solar pavement based on light guide concrete *MD Thesis of Changsha University of Science & Technology*

Construction technology and treatment effect of RoadMesh reinforcement in joint section between new and existing asphalt pavement

Q M Xiao[1], X D Zha[1], H W Hu[1], H P Chen[1], Z W Wang[1] and H Wu[2]

[1]School of Traffic and Transportation Engineering, Changsha University of Science and Technology, Changsha 410114, Hunan, China
[2]Xianning Traffic and Transportation Bureau, Xianning 437100, Hubei, China
Corresponding author: 1756023135@qq.com

Abstract. In order to verify the application effects of pavement reinforcement technology with RoadMesh steel wire mesh, and relied on the S415 Reconstruction and Extension Project from Xiapuzhen to Jietoutang in Tongshan County, Xianning City, Hubei Province, the application researches on construction technology and treatment effect of RoadMesh reinforcement were carried out for the joint section between new and existing asphalt pavement through scheme design, field construction, tests and analysis and economic comparison. The results show that the tensioning and fixing of RoadMesh are the key construction processes, and the arching and rolling of RoadMesh should be strictly prohibited during its construction. The asphalt pavement performances of RoadMesh reinforcement is better than that of unreinforcement, and the reinforcement effect between lower surface layer and base layer is better than that between base layer and subbase layer. The RoadMesh can effectively improve the integral bearing capacity of asphalt pavement structure and the resistances of reflection cracking and differential settlement in the joint section between new and existing pavement. Compared with the traditional reinforcement of geotechnical synthetic materials, the RoadMesh has higher performance price ratio and better long-term pavement performances. Therefore, the pavement reinforcement technology of RoadMesh can be widely popularized and applied.

1. Introduction

Due to the rapid development of road transport accompanied by the rapid socio-economic development, the large and loaded vehicles and the traffic volume have increased exponentially, which made it hard to meet the transportation demands for many early constructed highways. Under the combined effects of complex climatic and environmental factors, the distresses of pavement structure become more and more serious, so the reconstruction and extension projects are imperative. However, the problems of reflective cracking and uncoordinated deformation in the joint section between new and existing pavement have been one of the key technical problems in road engineering. According to many researches and applications, the reinforcement treatment technology is one of the most commonly used technologies to solve such problems [1-2]. Among them, the pavement reinforcement technology with RoadMesh steel wire mesh is a new type of pavement reinforcement technology that has been developed and applied to the highway reconstruction and extension projects in China over the past decade [3-4].

The RoadMesh steel wire mesh has a larger opening structure with unique hexagon braided steel mesh and penetrated transverse reinforced steel bar. The successful applications at home and abroad show that the unique three-dimensional occluding and wedging structure could be formed between the

RoadMesh and the mixtures of pavement structure layer. When the RoadMesh is laid between the structure layers of asphalt pavement, it can effectively enhance the capacities for cracking resistance, rutting resistance, fatiguing resistance and interlayer binding etc. in asphalt pavement structure. Its reinforcement performance is better than that of traditional geotechnical materials such as geotextile, geotechnical grid and glass fiber grid etc. Then the asphalt pavement performances can be significantly improved to extend the service life of asphalt pavement [5-7]. Therefore, China has gradually paid attention to the reinforcement technology since 2009, and achieved excellent application effect. Especially in recent years, the RoadMesh reinforcement technology has been applied more and more widely in the projects of directly repaving asphalt surface layer on the joint section between new and existing pavement and the existing cement concrete pavement [8-10].

Although the researches and applications of pavement reinforcement technology with the RoadMesh have been carried out at home and abroad, and some meaningful research results have been obtained, but it is still not deep enough and comprehensive enough. In order to further promote the popularization and application of this technology, and by relying on the S415 Reconstruction and Extension Project from Xiapuzhen to Jietoutang in Tongshan County, Xianning City, Hubei Province, the test road setion of pavement reinforcement with the RoadMesh had been paved. The construction technology and the treatment effect for reinforcing the joint section between new and existing asphalt pavement were researched to hope providing a scientific reference for similar projects.

2. Design scheme of RoadMesh reinforcement in joint section between new and existing asphalt pavement

In order to determine the reasonable design scheme for pavement reinforcement with the RoadMesh and effectively control the longitudinal reflection cracking and the transverse uncoordinated deformation in the joint section between new and existing pavement, and relied on the S415 Reconstruction and Extension Project from Xiapuzhen to Jietoutang in Tongshan County, Xianning City, Hubei Province, the test road section with a total length of 300 m with the RoadMesh reinforcement was determined to pave in the joint section between new and existing pavement. The highway is a connection line of ordinary dart and an important channel connecting Tongshan County and Chongyang County. The technical standard of the existing highway is a third class highway with a roadbed of 7 m, and it will be widened and reconstructed to a second class highway with a roadbed width of 10 m. It adopts the asphalt pavement structure. Among them, the newly repaving structure layers are “3 cm AC-13C upper surface layer + 5 cm AC-20C lower surface layer + 18 cm base layer with cement stabilized macadam”, and an additional 20 cm subbase layer with cement stabilized macadam is paved on the newly built roadbed in the widening parts to keep flush with the surface of existing pavement. The existing pavement structure layers are “5 cm AC surface layer + 20 cm base layer with cement stabilized macadam”.

Generally, because the total thicknesses of pavement structure layer on the RoadMesh reinforced layer must not be less than 6 cm, according to the fact that the upper surface layer is thinner (only 3 cm) and the field construction conditions, the reinforcement scheme between surface layers is not adopted. Meanwhile, due to the differential settlement in the embankment widening section is more obvious than that in the excavation widening section, the embankment widening section with a 3 m unilateral widening is selected as the test road section for reinforcement treatment. For this reason, two design schemes of pavement reinforcement with the RoadMesh are applied. One is to reinforce between the additional paving base layer with cement stabilized macadam and the subbase layer composed of the newly paved cement stabilized macadam and the existing asphalt pavement, i.e. reinforce between the base layer and the subbase layer (existing pavement) (Scheme 1: reinforcement between base layers), with the section of length 150 m. Another is to reinforce between the additional paving lower surface layer of AC-20C and the base layer with cement stabilized macadam, i.e. reinforce between the lower surface layer and the base layer (Scheme 2: reinforcement between base and surface layers), with the section of length 150 m. The pavement structure design drawings of the two reinforcement schemes are shown in figure 1. The specifications of RoadMesh are the length of

25 m and width of 3.6 m each roll, and the mesh size is 8 cm × 12 cm, as shown in figure 2. Its transverse direction is horizontally paved across the joint section between new and existing pavement with the width of 1.8 m on both sides of the new and existing pavement. In order to drive construction machinery conveniently and prevent the RoadMesh from being rolled up, a bonding coat with cement paste and crushed stone is added between the base layers in the Scheme 1, and a seal coat with emulsified asphalt and crushed stone is added between the base and surface layers in the Scheme 2. So as to compare the reinforcement effect with the RoadMesh, one unreinforcement section of original design for each scheme is selected as the comparison section near the test road section. The stack numbers of each section are shown in table 1.

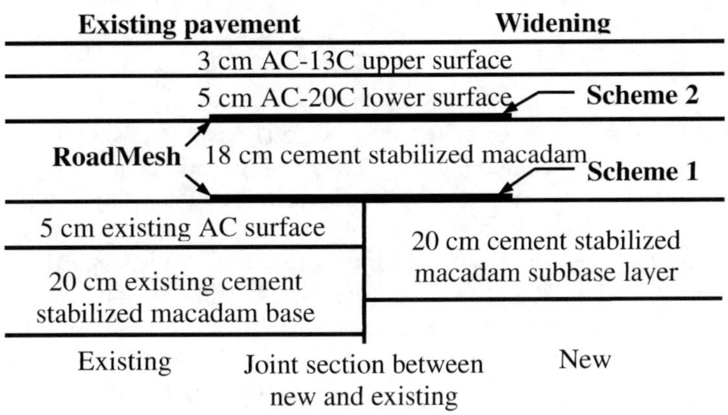

Figure 1. Design schemes of RoadMesh reinforcement in joint section between new and existing pavement

3. Construction technology of RoadMesh reinforcement in joint section between new and existing asphalt pavement

3.1. Reinforcement construction process between base layers

The construction processes of RoadMesh reinforcement for the Scheme 1 between base layers are mainly divided into the three major steps: laying RoadMesh, spreading bonding coat of cement paste and crushed stone and paving base layer of cement stabilized macadam. According to the actual situation of the test road, the specific construction techniques include the following five steps.

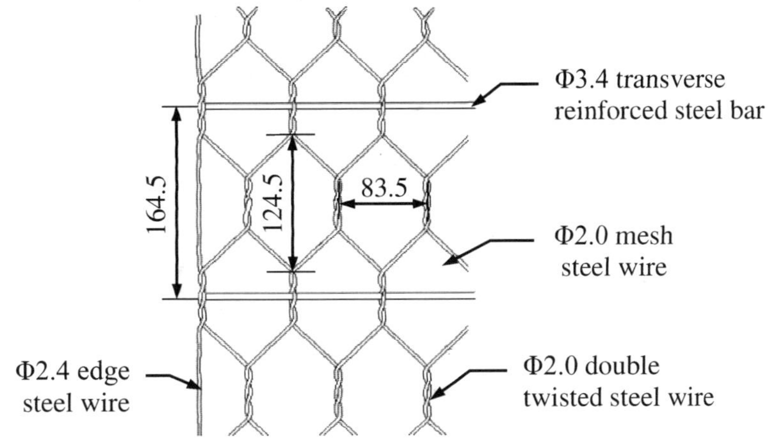

Figure 2. RoadMesh structure (the size unit: mm)

Table 1. Test section of RoadMesh reinforcement in joint section between new and existing asphalt pavement

No.	Origin and final stack number	Length (m)	Types of reinforced materials	Reinforced location
1	K3+250~K3+400 (right range)	150	RoadMesh	Scheme 1: between base layers
2	K2+980~K3+180 (right range)	200	Unreinforced	Comparison section 1
3	K2+725~K2+875 (left range)	150	RoadMesh	Scheme 2: between base and surface layers
4	K2+320~K2+550 (left range)	430	Unreinforced	Comparison section 2

1. Clean the underlayer. The distresses of existing asphalt pavement should be checked out. The structural diseases such as ravelling, pit slot etc. are dug and repaired in accordance with the technical requirements of maintenance. The newly paved subbase layer with cement stabilized macadam is cured at least 7 days. It should be ensured that the underlayer surface is smooth, dense without potholes, and the loose aggregates on the surface are cleaned up. Meanwhile, it should be ensured that the surface is flush and tidy between the existing asphalt pavement and the newly paved subbase layer with cement stabilized macadam.
2. Lay the RoadMesh. The artificial method is used to lay the RoadMesh along the longitudinal direction of the route, and it should be ensured that the longitudinal joints in the joint section between new and existing asphalt pavement is located in the middle of the RoadMesh. Since the finished product of RoadMesh is in the form of a roll, the RoadMesh surface will be some bending. When laying, the curved surface should be installed downwards so that the RoadMesh surface is close to the underlayer surface. After laying, the entire RoadMesh surface should be rolled and levelled twice round trip with the small medium-sized rubber roller, loader or truck at a uniform speed of 10 km/h, then manually adjusted.
3. Tension and fix the RoadMesh. Firstly, the beginning of RoadMesh about a length of 4 m is fixed with the high-density clamp nails consisting of steel nails and clips. The another end is tensioned by fixing on a tension rod and connecting to the small medium-sized vehicle, and also it is fixed with the high-density clamp nails. Lastly, the middle of RoadMesh is fixed with the clamp nails to eliminate the slack and fold of RoadMesh. Among them, the length of steel nail used for the asphalt surface layer and the semi-rigid base layer should be 6 cm and 10 cm respectively, and the density of clamp nail should be at least 1 per square meter. Then, the next roll of RoadMesh is laid, and the length of longitudinal lap joint is not less than 30 cm. Accordingly, the fixed RoadMesh is shown in Figure 3.

Figure 3. Fixed RoadMesh

4. Spread the bonding coat of cement paste and crushed stone. In order to improve the interlayer bonding capacity and prevent the construction machinery from pushing the RoadMesh, the bonding

coat of cement paste and crushed stone of "cement paste + 5~10 mm crushed stone" is spread on the fixed RoadMesh. So as to facilitate construction, the cement paste coat is taken the spreading method of "firstly sprinkling water, secondly sprinkling ash, and lastly sprinkling water" in site. Among them, the sprinkling amount of cement and crushed stone should be 0.5~1 kg/m^2 and 7~9 kg/m^2 respectively.

5. Pave the base layer of cement stabilized macadam. The base layer of cement stabilized macadam is paved and rolled in accordance with the technical requirements of *"Technical Guidelines for Construction of Highway Roadbases" (JTG/T F20-2015)*. In order to avoid the larger disturbance to the reinforced layer and the bonding coat during the construction process, the driving speed of dumper and paver should be strictly controlled to not exceed 10 km/h, and the rolling speed of roller should not exceed 5 km/h.

3.2. Reinforcement construction process between base and surface layers

The construction processes of RoadMesh reinforcement for the Scheme 2 between base and surface layers are mainly divided into the three major steps: laying RoadMesh, spreading prime coat of emulsified asphalt and seal coat of crushed stone and paving asphalt lower layer. Among them, the technical requirements of construction process for cleaning the underlayer and laying, tensioning and fixing the RoadMesh are basically the same with the above mentioned reinforced section between base layers. The other two construction processes are as follows:

1. Spread the prime coat of emulsified asphalt and the seal coat of crushed stone. The prime coat and seal coat of "emulsified asphalt + 5~10 mm crushed stone" are spread on the fixed RoadMesh, where the spraying amount of emulsified asphalt is 1~1.5 L/m^2 for covering the entire RoadMesh surface. The travelling speed of asphalt sprayer is strictly controlled within the range of 4~6 km/h to prevent the RoadMesh from being pushed or rolled up. The crushed stone is manually spread with the spreading amount of 7~9 kg/m^2, and it is evenly distributed and blackened while spreading. It should be ensured that the crushed stone is evenly embedded into the RoadMesh meshes. The finished prime coat and seal coat are shown in Figure 4.
2.

Figure 4. Finished prime coat and seal coat

3. Pave the asphalt lower layer. The lower layer of AC-20C is paved and rolled in accordance with the technical requirements of *"Technical Specifications for Construction of Highway Asphalt Pavements" (JTG F40-2004)*. The paving speed of paver shall be less than 3 m/min to prevent disturbing severely the RoadMesh. The rolling speed of roller should be under 2~3 km/h slowly and evenly. When semi-range constructing, it should be noted the roller to avoid rolling the exposed RoadMesh near the central line of road.

The following precautions should be noted during the construction of RoadMesh.

1. Strictly control the tensioning and fixing of RoadMesh. Tensioning and fixing are the important steps for reinforcing asphalt pavement with the RoadMesh. The tensioning quality and the fixing effect of the RoadMesh should be strictly guaranteed. If the tensioning and the fixing are poor, the RoadMesh reinforcement effect will be seriously influenced.
2. Strictly prohibit the arching and rolling up of RoadMesh. The RoadMesh arching would cause the upper layers on the reinforced layer, especially the asphalt surface layers, to be difficult to compact, additionally the thicknesses would become thinner. The rolling up can lead to the damage of RoadMesh. As a result, it is easy to cause early distresses such as rutting, swelling and loosening for the pavement at the places of arching and rolling up, and affect the intergral reinforcement performance of RoadMesh.

Accordingly, for the purpose of improving the construction quality of RoadMesh reinforcement, the following preventive measures should be taken.

1. The number of clamp nails should be increased at the both ends, edges and laps of RoadMesh, the bends and the up and down slopes.
2. The seal coat or bonding coat of crushed stone should be added on the reinforced layer of RoadMesh to protect the RoadMesh from being directly damaged by the construction machinery.
3. After fixing the RoadMesh, the construction machinery should be strictly forbidden to accelerate, emergency brake, turn and U-turn to prevent damage to the RoadMesh.
4. The crawler paver should be strictly prohibited for paving the upper structure layers to prevent the track shoe from damaging the RoadMesh because of hooking up.
5. The paver paving speed and the roller rolling speed should be strictly contolled during the construction of upper structure layers to prevent the moving and arching of RoadMesh.

4. Treatment Effect of RoadMesh reinforcement in joint section between new and existing asphalt pavement

In order to evaluate the treatment effect of RoadMesh reinforcement in joint section between new and existing asphalt pavement, the field detection, core drilling observation, tracking observation and economic comparison were carried out on the reinforced test section and the comparison road section after completion.

a. Field detection

The test results of pavement performance such as deflection, smoothness, pendulum value and water permeability coefficient in each section are shown in table 2. For convenience of comparison, the two comparison road sections were merged to analyze.

Table 2. Test results of pavement performance in each section

Road section	Representative deflection (0.01mm)	Maximum clearance (mm)	Pendulum value (BPN)	Water permeability coefficient (mL min^{-1})
Scheme 1	32.4	1.7	64	512
Scheme 2	17.4	1.9	64	1381
Comparison section	40.1	1.7	63	1872
Technical requirements	≤41.6	≤5	/	/

As can be seen from Table 2:

1. All the representative deflections in each section meet the design requirements of strength, and the representative deflection of Scheme 2 is the smallest, which is 46.3% and 56.6% lower than that of Scheme 1 and the comparison section respectively. The representative deflections in both reinforced sections are smaller than that of the unreinforced comparison section. The results show that the joint section between new and existing pavement reinforced by the RoadMesh can obviously improve the intergral bearing capacity and uniformity of asphalt pavement structure, and the reinforced effect between base and surface layers is better than that between base layers.

2. All the maximum clearances with the three-meter ruler in each section meet the design requirements of smoothness. Among them, Scheme 1 has the same test results as the comparison section, while the maximum clearance of Scheme 2 is increased by 11.8%. It indicates that the RoadMesh reinforcement between base layers has almost no influence on the pavement surface smoothness. The reinforcement between base and surface layers has a certain adverse effect on the pavement surface smoothness, but the driving comfort and safety would not be affected.
3. Although the pendulum value of RoadMesh reinforced section is slightly increased, it is basically consistent with the pendulum value of the comparison section, which indicates that the RoadMesh reinforcement will not affect the surface skid resistance of asphalt pavement.
4. Although the second class highway has no specific requirements for the impermeability of pavement surface, the test results of water permeability coefficient in each section are larger and obviously more than 300 mL/min. From the field detection situation, the main reason for the larger water permeability coefficient is the larger air voids of AC-13C in the upper surface layer to cause the reverse seepage phenomenon, but it is not directly related to the RoadMesh reinforcement.

In general, the joint section of new and existing asphalt pavement reinforced by the RoadMesh can obviously improve the bearing capacity of asphalt pavement structure, and has no obvious adverse effect on the pavement surface performances such as smoothness, skid resistance and impermeability.

b. Core drilling observation

For analyzing the interlayer reinforcement distribution of the RoadMesh, two core samples were drilled for each road section. Due to the height limitation of drill bit, the complete core sample reinforced between base layers could not be drilled. Wherein, the core samples in the comparison section were disconnected between the lower surface layer and the base layer, but the core samples in Scheme 2 show that there are completely bonded between the surface layer and the base layer, as shown in figure 5.

(a) (b)

Figure 5. Core drilling samples

The cross-sections of transverse reinforced steel bar and ordinary steel wire in the RoadMesh and the aggregate particles embedded in the RoadMesh can be clearly observed from figure 5. It shows that there are no gaps between the asphalt mixture and the RoadMesh, and the contacts are tight and evenly full. The results indicate that the RoadMesh hoops the asphalt mixture into its meshes in a three-dimensional way, and the asphalt mixture is evenly and compactly embedded in the RoadMesh meshes to form a whole structure which can effectively resist the reflective cracking and the differential settlement. Thus its excellent reinforcement effect is verified.

c. Tracking observation

To further verify the long-term pavement performances reinforced by the RoadMesh in joint section between new and existing asphalt pavement, the test road section has been tracked and observed in site after completion and opening to traffic for one year. The results are shown in figure 6 and figure 7.

IOP Conf. Series: Materials Science and Engineering 1075 (2021) 012029 doi:10.1088/1757-899X/1075/1/012029

It can be seen from figure 6 and figure 7 that after the comprehensive effect of vehicle loads and environmental factors for more than one year, the common distresses such as cracking and rutting etc. in the widening projects have been not found on the asphalt pavement of reinforced test section with the RoadMesh, but there are multiple longitudinal and transverse cracks and more obvious ruts on the unreinforced section. It shows that the unreinforced asphalt pavement is difficult to resist the stress concentration and the surface layer displacement caused by uncoordinated deformation, while the joint section between new and existing asphalt pavement reinforced by the RoadMesh can effectively disperse the stress concentration in the structure layers of asphalt pavement and improve the flexural tensile strength and intergral performance for the structure. The generation and development of typical longitudinal and transverse cracks for the asphalt pavement structure can be effectively delayed, prevented and treated in the widening projects. Therefore, the pavement reinforcement technology with RoadMesh has an excellent long-term performance for resisting the reflective cracking and differential settlement.

Figure 6. Reinforced test section

Figure 7. Unreinforced comparison section

d. Economic comparison

By calculation, for the reinforcement treatment in joint section between new and existing asphalt pavement, the unit cost of RoadMesh is about 5~15 RMB/m^2 higher than that of the traditional geosynthetic materials such as geotextile, geogrid and glass fiber grid etc. According to the calculation of laying reinforcement material of 3600 m^2 per kilometer in an unilateral widening project, the cost per kilometer has only increased by about 18~54 thousands RMB. However, the RoadMesh has excellent effect of pavement reinforcement which can effectively improve the long-term pavement performances and service ability of asphalt pavement structure, extend the service life and reduce the maintenance and repair costs. From the perspective of the total life cycle, the RoadMesh has higher performance price ratio to be worth popularizing and applying.

5. Conclusions

According to the actual conditions of test section in the entitative project, the two design schemes were proposed for reinforcing the joint section between new and existing asphalt pavement with the RoadMesh, including the reinforcement between base layer and subbase layer, and between lower surface layer and base layer. The field construction and detection were carried out. Thus the following conclusions can be obtained.

1. The construction processes of pavement reinforcement with the RoadMesh can be mainly divided into the three major steps: laying RoadMesh, spreading bonding coat or seal coat and paving upper structure layers. Among them, the tensioning and the fixing of RoadMesh are the key processes. During the process of construction, the RoadMesh should be strictly prohibited from arching and rolling up, and the paving and rolling speed of upper structure layers should be strictly controlled.

2. The treatment technology of RoadMesh reinforcement can effectively improve the intergral bearing capacity of asphalt pavement structure and the abilities of reflective cracking resistance and differential settlement resistance in the joint section between new and existing pavement, but there is almost no effect on the performances of pavement surface such as smoothness, skid resistance and impermeability etc. The reinforcement effect between lower surface layer and base layer is better than that between base layer and subbase layer.
3. The long-term pavement performances of asphalt pavement with the RoadMesh reinforcement is better than that without reinforcement. Compared to the reinforcement with other traditional geosynthetic materials, although its cost is slightly higher, its comprehensive performance price ratio is superior. Therefore, the RoadMesh has broad promotion and application prospects.

Acknowledgments
This work was supported by National Natural Science Fundation of China (No. 51878077).

References
[1] Tang YL and Wang YM 2014 Analysis of laboratory test on geogrid for widening subgrade based on the Guangsan Highway *Chinese Journal of Underground Space and Engineering*. 10(S2) 1904-09
[2] Zhang HW, Hao PW, Tang C and Pang Y 2017 Adherence of geotextile interlayer in asphalt overlay *Journal of Building Materials*. 20(5) 745-51
[3] Pan ZX 2011 A study of steel mesh reinforced asphalt overlays on existing concerte pavemrnts to combat reflective cracking *MD Thesis of Hunan University*
[4] Tang T, Zha XD, Xiao QM and Chen YQ 2018 Laboratory characterization and field validation of RoadMesh-reinforced asphalt pavement in China *Int J Civ Eng*. 16(3) 299-313
[5] Al-Qadi IL, Elseifi M and Leonard D 2003 Development of an overlay design model for reflective cracking with and without steel reinforcing nettings *J Assoc Asphalt Paving Technol*. 72 388-423.
[6] Xie G 2011 Application technology study of joint section between new and existing pavement reinforced with double twisted steel wire mesh *MD Thesis of Changsha University of Science & Technology*
[7] Huang X 2011 Mechanical analysis of antireflertive cracking for joint section of new and existing pavement reinforced with double twisted steel wire mesh *MD Thesis of Changsha University of Science & Technology*
[8] Zha XD, Zhu K and Tang T 2015 Mechanical analysis of AC+PCC composite pavement structure reinforced with RoadMesh steel mesh *Journal of Changsha University of Science and Technology (Natural Science)*. 12(1) 1-9
[9] Zhang ZC 2016 Research on RoadMesh applied in reinforcement of joint section between new and existing pavement in S415 of Hubei *MD Thesis of Changsha University of Science & Technology*
[10] Huang X, Zha XD and Wu H 2018 Numerical simulation on crack resistance of reinforced joint between new and existing pavement with RoadMesh steel mesh *Journal of Changsha University of Science and Technology (Natural Science)*. 15(1) 1-7

Prediction of Temperature Cooling Trend of Asphalt Mixtures

L. Chu[1], B. Zhu[1] and T. F. Fwa[1,2*]

[1]School of Highway, Chang'an University, South Erhuan Middle Section, Xi'an, 710064, China
[2]Dept of Civil and Environmental Engineering, National University of Singapore, Singapore
Corresponding author's e-mail: ceefwatf@nus.edu.sg

Abstract. Monitoring of the temperature changes of asphalt mixtures during the mix production and pavement construction phases is important to avoid over-heating, and to ensure that optimum mixing and compaction is achieved. Over-heating or prolonged heating of asphalt mixtures would cause ageing, while inadequate mixing or compaction would result in mixtures with a strength and durability less than design optimal. Careful temperature monitoring would help to avoid such undesirable problems. This study develops a finite-element asphalt mixture cooling temperature prediction software as an asphalt pavement mix design planning tool. For any given asphalt paving design mix, it is able to predict the temperature cooling characteristics of the design mix during the laying and compaction phase. The software can provide the temperature cooling characteristics during a multi-lift paving of asphalt pavement layer. The factors affecting the temperature cooling trend are determined and analysed. They include ambient temperature, laying temperature, solar influx intensity, wind velocity, number of lifts, and lift thickness. A sensitivity analysis is conducted to identify the most influencing factors under different paving conditions.

1. Introduction

Temperature monitoring and control of asphalt mixtures is vital in the production of asphalt mixtures and construction (including repair and maintenance) of asphalt pavements. During the mix production phase, too high a temperature would cause over-heating and ageing of the asphalt binder; too low a temperature could lead to inadequate or non-uniform mixing [1]. During the pavement construction phase, adequate temperature must be maintained for effective spreading and compaction [2]. Another phase in which temperature monitoring and control is important, but overlooked in many instances, is the time that a newly compacted pavement is open to traffic during the post compaction phase. When a newly compacted pavement section or a newly repaired pavement layer is opened to traffic too soon before the asphalt materials have cooled down sufficiently, premature failure of the pavement may occur due to excessive deformation or cracking in the asphalt layer under heavy wheel loads. Sufficient time must be allowed for the temperature to cool down and the asphalt material to gain strength before traffic is allowed. Close monitoring of the temperature in the asphalt layer is needed to prevent such premature failures of a newly compacted pavement section. Instead of tedious and manpower demanding field installation of sensors and temperature monitoring, an alternative means is to develop temperature prediction capability to estimate the cooling time required, and plan ahead for the time at which the pavement concerned can be opened to traffic. This is the topic of interest of this paper.

This paper describes the development of a 2-dimensional finite simulation model for predicting the cooling trend of a newly compacted asphalt layer. It provides a numerical solution based on heat transfer theories. The model developed can provide the time-temperature cooling history during a single- or multi-lift paving of an asphalt pavement layer. The solution presents the complete temperature regime within the asphalt layer, thus offering a complete quantitative information to site engineers to help plan for their paving operation to allow sufficient cooling time before the pavement is opened to traffic. It is highlighted in this paper the key factors that must be considered during paving operation in the monitoring of temperature changes in an asphalt layer. Such factors include ambient temperature, laying temperature, solar influx intensity, wind velocity, number of lifts, and lift thickness. A sensitivity analysis is conducted to identify the most influencing factors under different paving conditions.

2. Developing cooling model for multi-lift asphalt layer construction

The 3-dimensional finite element simulation model was developed in this study to obtain the time variation of temperatures within an entire asphalt layer under a given set of pavement thermal properties and environmental conditions. This section presents the formulation of the simulation model based on the theory of heat transfer.

2.1. Theoretical basis for heat transfer modelling

In the temperature cooling process of a newly compacted pavement layer, heat losses and heat gains are considered based on heat transfer theories. Heat losses take place to the ambient atmosphere and the existing underlying pavement layer. Heat gains are received mainly from solar radiation influx. Overall, there are the following three processes influencing temperature changes within a newly laid pavement layer:

(1) Heat exchange by means of conduction between newly paved layer and the existing surrounding materials.;
(2) Heat loss between newly paved layer and the ambient air;
(3) Heat gain by the newly paved asphalt mixture under the effect of solar influx.

Figure 1 presents these heat transfer processes schematically after the asphalt mixture is laid on the existing pavement structure.

Heat conduction occurs between the newly paved asphalt layer and the existing pavement structure and governs by Fourier's law [3]. The distribution of heat over time and space can be expressed by the following equation:

$$\frac{\partial^2 T}{\partial x^2}+\frac{\partial^2 T}{\partial y^2}+\frac{\partial^2 T}{\partial z^2}=\frac{1}{\alpha}\frac{\partial T}{\partial t} \quad (1)$$

$$\alpha=\frac{k}{\rho c} \quad (2)$$

where T = temperature, °C, α = thermal diffusivity, m^2/s, k = thermal conductivity, W/m·K, ρ = density, kg/m^3, c = specific heat, J/kg·K, and x, y, z are the three coordinate directions in a three-dimensional space.

Heat convection can be estimated by the following equation to describe heat transfer between a solid and the surrounding air [3]:

$$q_{cv}=h(T_a-T_s) \quad (3)$$

where h = convective heat transfer coefficient for air flow, W/m^2 ℃, Ta = temperature of surrounding air, ℃, and Ts = surface temperature of newly laid asphalt layer, ℃.

The radiation energy transmitted from the newly paved layer can be simply represented by the following equation [4]:

$$q_r = \varepsilon\sigma\left(T_s^4 - T_a^4\right) \tag{4}$$

where qr = radiation heat flux emitted, ε = emissivity coefficient of asphalt layer, σ = 5.669 x 10^{-8} W/m^{2}K^4 is the Stefan-Boltzman constant, Ts = surface temperature of pavement, and Ta = ambient air temperature.

Besides heat loss in different forms as described above, the newly paved asphalt layer will also absorb energy due to the effect of solar influx, which can be quantified by the following equation [3]:

$$q_s = aH_s \tag{5}$$

where qs = solar radiation energy absorbed, Hs is the net solar flux at a pavement surface, and a = solar radiation absorptivity coefficient, a dimensionless quantity.

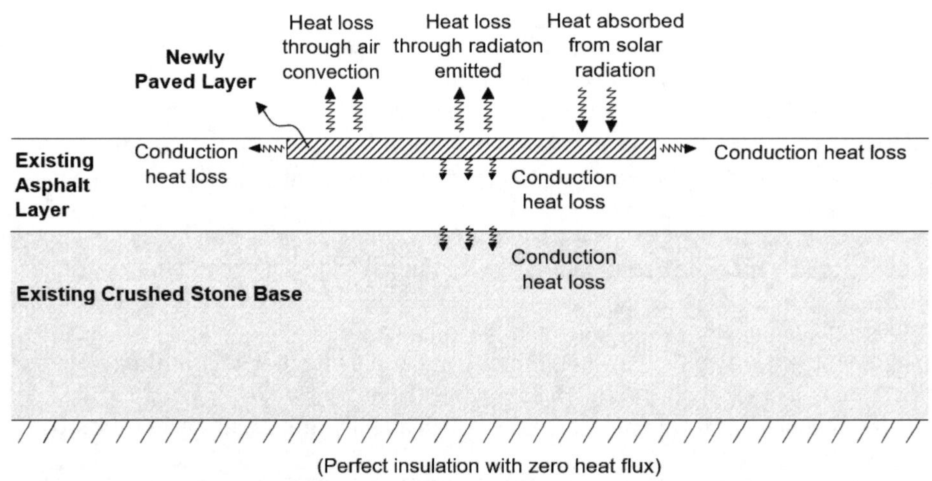

Figure 1. Heat transfer processes considered in thermal cooling simulation model

2.2. *Finite-Element Modelling of Cooling of Newly Paved Asphalt Layers*

Applying the heat transfer theory described in the preceding section, the temperature changes of newly paved asphalt layers are modeled taking into consideration the following parameters:

- Geometric dimensions of existing pavement structure;
- Geometric dimensions of newly paved asphalt layers;
- Thermal properties of existing pavement materials and newly paved asphalt layers;
- Ambient temperature, solar flux and wind speed;
- Time and thickness increments of newly paved asphalt layers;
- Initial temperature of existing pavement structure; and
- Initial temperature of each newly paved asphalt layers when laid.

The finite element simulation model was developed making use of the finite-element software ABAQUS [5]. The dimension for the simulated area to study the temperature trend is 12 m by 21 m. To minimize the computation time and also guarantee the accuracy of the simulation results, the layout of the meshes adopted in this study is found in figure 1. The smallest mesh size is 2 mm.

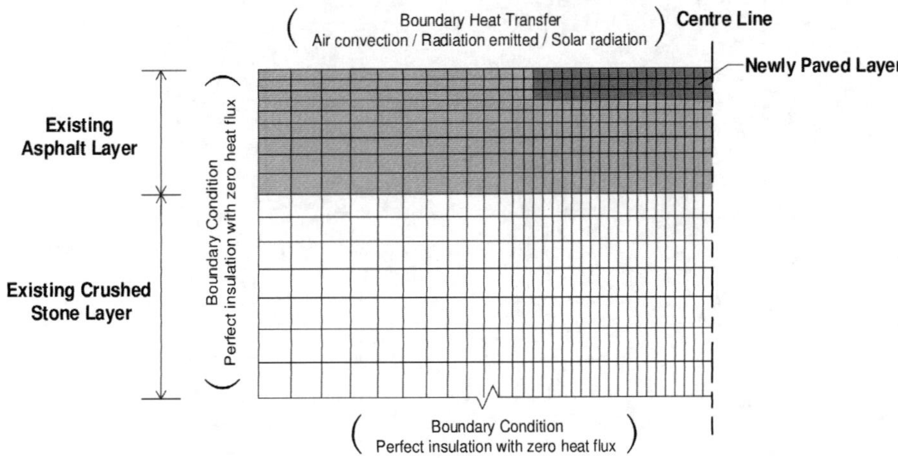

Figure 2. Schematic representation of finite-element mesh for proposed simulation model

3. Analysis of example problems

This section presents analyses of several example problems that consider the effects of different environmental conditions on the cooling trend of a newly paved asphalt layer. To study the variation in the length of cooling time needed, 60°C is set as the maximum temperature allowed before a newly paved pavement section can be opened to traffic. The effects of the following three factors on cooling time are examined: (i) Wind speed, (ii) ambient air temperature, and (iii) compaction temperature of asphalt layer. The effects of these factors on the length of pavement cooling time needed is considered for two overlay construction : (1) A 150 mm layer laid in two equal lifts of 75mm ; and (2) a 100mm layer laid in two equal lifts of 50 mm each.

3.1. Effects of wind speed

The results of wind effect study are shown in figure 3. The plot illustrates that as the wind speed increases, the time duration to reach to 60°C decreases for both the bottom and the surface layers,. Higher wind speed helps to reduce the length of cooling time needed. Compared with the case of 1 m/s wind speed, a wind speed of 8 m/s will cut down the required cooling time by more than 150 minutes. It is noted that the effect of wind speed tends to level off after wind speed reaches 9 m/s.

It can be seen that the controlling factor of the cooling time is the temperature of the bottom layer. For both cases of overlay construction, the bottom layer takes almost 150 minutes to reach 60°C, regardless of the magnitude of the wind speed. Figure 3 also shows that the case with thinner thickness takes shorter time to cool down to 60°C, even though the difference is not very substantial. The case of 100 mm thick layer compacted in two lifts of 50 mm needs about 10 minutes shorter to cool down to 60°C. This difference does not appear to be affected by the magnitude of wind speed.

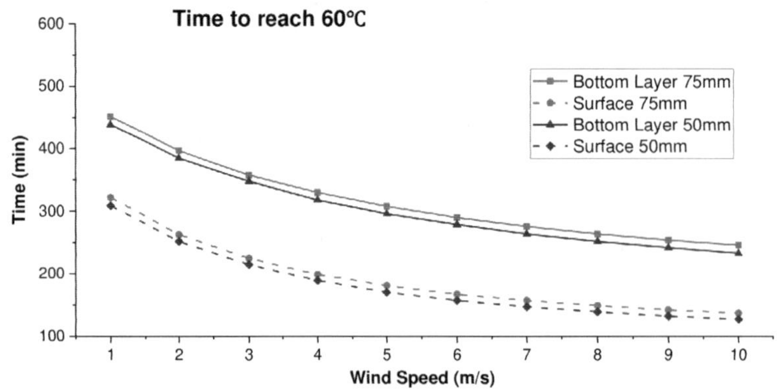

Figure 3. Time to reach 60℃ for different layers of two pavement structures

3.2. *Effects of ambient air temperature*

The range of ambient air temperature considered is from 24°C to 33°C. Figure 4 presents the bottom layer and surface layer temperature variation when the overlay constructions are carried out under different ambient air temperatures. As can be expected, the length of cooling time increases with the ambient air temperature. For both overlay construction cases at an ambient air temperature of 33°C, it takes about 50 minutes longer than when the ambient air temperature is 24°C. There is a slight tendency that the rate of increase of the cooling time picks up as the ambient air temperature rises.

The results show that the bottom lift pavement temperature governs the total length of cooling needed. Regardless of the magnitude of the ambient air temperature and the lift thickness, the bottom layer takes about 130 minutes longer to cool down to 60°C. As for the effect of lift thickness, when the ambient air temperature rises, the case of 75 mm lift thickness takes longer time to cool down to 60°C than the case of 50 mm lift thickness. The difference is slightly more than 10 minutes, and this difference appears to stay the same for the range of ambient air temperatures studied.

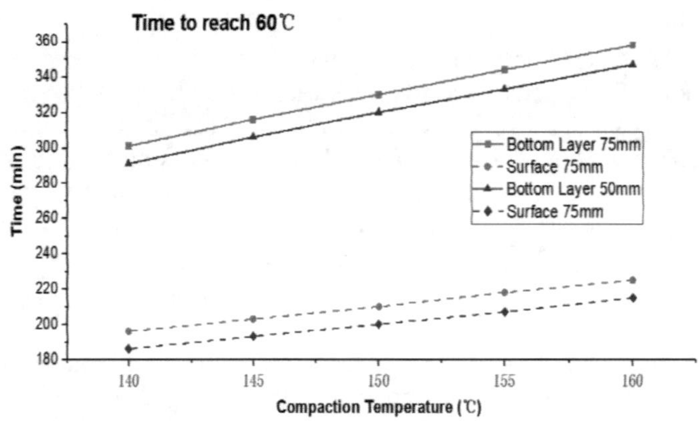

Figure 4. Effect of compaction temperature on time to reach 60°C after compaction

3.3. *Effects of compaction temperature*

The initial compaction temperatures is an important factor leading to different lengths of time needed for cooling down to 60°C. Figure 5 presents the effect of compaction temperature for the two overlay construction cases studied. In general, the length of cooling time needed increases with the magnitude of initial compaction temperature. Comparing the impacts of initial compaction temperatures of 140°C and 160°C on the bottom layer temperature, it is seen that the case of 160°C compaction temperature takes about 60 minutes longer to reach 60°C. It is observed again that the bottom lift takes much longer time to cool than the surface lift. At 140°C initial compaction temperature, the bottom lift takes slightly more than 100 minutes to cool down to 60°C than the surface lift ; while at 160°C initial compaction temperature, the difference is 120 minutes. The difference increases with the magnitude of initial compaction temperature.

Between the two overlay construction cases, the case with two 75 mm lifts takes about 10 minutes longer to cool down to 60°C than the case with two 50 mm lifts. This difference in the length of cooling time remains practically unchanged over the range of initial compaction temperature analyzed.

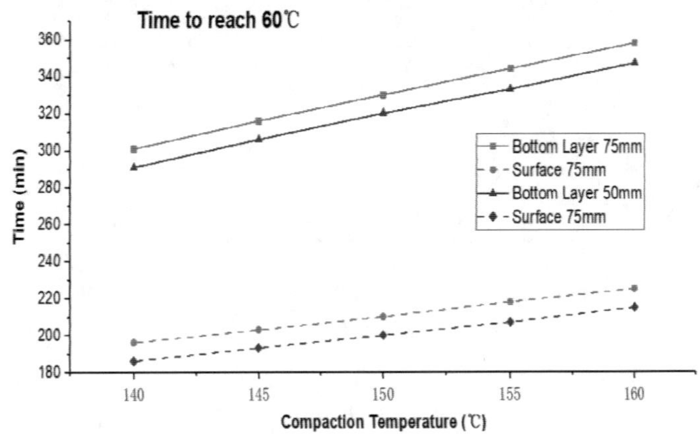

Figure 5. Effect of compaction temperature on time to reach 60°C after compaction

4. Conclusion

A 2-dimensional finite element temperature cooling simulation model for asphalt pavement has been developed and presented in this paper. It is applied in this study to analyse the cooling characteristics of two newly laid overlaid asphalt pavements under the influences of three factors: wind speed, ambient air temperature, and initial compaction temperature. It was found that the length of time needed for a newly laid overlay to cool down to 60°C decreases as wind speed increases, but increases as either ambient air temperature or initial compaction temperature increases. For the two cases of 2-lift overlay constructions analyzed, the bottom lift was found to take the longest time to cool down, and hence governed the length of cooling time needed before the newly laid overlay could be opened to traffic. All three factors considered had significant impacts on the length of cooling time needed, and have to be taken into consideration in determining the cooling time needed for opening a newly compacted asphalt pavement to traffic.

References

[1] B.A. Chadbourn, J.A. Luoma, D.E. Newcomb & V. R. Voller. Consideration of hot mix asphalt thermal properties during compaction. ASTM Special Technical Report STP 1299, Quality Management of Hot-Mix Asphalt, Dale S. Decker, Ed., American Society for Testing and Materials (1996).

[2] J.S. Corlew & P.F. Dickson. Methods for Calculating Temperature Profiles of Hot- Mix Asphalt Concrete as Related to the Construction of Asphalt Pavements. Journal of Association of Asphalt Paving Technologists, 37, pp. 101-140 (1968).

[3] M. N. Ozisik. Basic Heat Transfer, McGraw-Hill Book Company, New York (1979).

[4] M.R. Hall, P.K. Dehdezi, A.R. Dawson, J. Grenfell & R. Isola. Influence of the thermophysical properties of pavement materials on the evolution of temperature depth profiles in different

IOP Conf. Series: Materials Science and Engineering 1075 (2021) 012030 doi:10.1088/1757-899X/1075/1/012030

climatic regions. Journal of Materials in Civil Engineering. 24(1), 32-47 (2012).
[5] ABAQUS. ABAQUS User's Manual. Pawtucket, RI, USA (2010).

IOP Conf. Series: Materials Science and Engineering 1075 (2021) 012031 doi:10.1088/1757-899X/1075/1/012031

A review on the application of natural rubber as asphalt modifier

A H Ansari*, F M Jakarni, R Muniandy, and S Hassim

Department of Civil Engineering, Faculty of Engineering, Universiti Putra Malaysia, 43400 UPM Serdang, Selangor, Malaysia.

Corresponding author's e-mail: gs52876@student.upm.edu.my

Abstract. Natural rubber as bio-modifier of bitumen and asphalt mixtures has shown to have some beneficial effects. As initially produced, natural rubber consists of polymers of the organic compound isoprene, with minor impurities of other organic compounds, plus water. Natural rubber can consist of various forms including as latex, which is a sticky, milky colloid; cup lump which is a coagulated latex and ribbed smoke sheet. Natural rubber is used extensively in many products and applications, either alone or in combination with other materials as it has a large stretch ratio and high resilience, and extremely waterproof. Due to these properties together with a chemical composition which can be classified as elastomers, natural rubber modification can improve the performance of conventional asphalt mixtures that are currently faced limitation against increased traffic loading and higher axle loads together with the effect of ageing, moisture, and temperature. Also, the application of natural rubber as part of the construction of asphalt mixture pavement might support the national agenda to boost rubber industry. This paper will review the application of various forms of natural rubber as modifier either in bitumen or asphalt mixtures based on previous studies done by other researchers. Numerous studies have been conducted based on different types and composition of natural rubber to determine the performance especially in terms of rutting and fatigue cracking properties.

1. Introduction

Many studies have been reported over the past decades on polymer-modified bitumen and its benefits significantly increased [1,2]. The service life of highway pavement can be improved extensively by using polymer-modified bitumen [3]. Bitumen properties can be enhanced with the addition of different polymers such as stiffness, cohesion, and elasticity. Thus, it will become stiffer at a higher temperature and provides flexible properties at low temperature [4]. A Polymer can enhance the elasticity and flexibility of asphalt mixture and eventually improving the thermal cracking and fatigue characteristics of asphalt mixture [4–6]. Thermoplastic elastomer (TE) with elastomeric and thermoplastic properties is one of the frequently used polymers around the world. Styrene-butadiene–styrene (SBS) is the most universally recognized TE that can advance the rheological characteristics of bitumen. Styrene-butadiene– rubber (SBR) is an example of an elastomer polymer while using in pavement it enhances the resistant against cracks and more flexible at low temperature and also increase the ductile properties of asphalt [7]. Additionally, few other types of polymers have also been utilized as modifiers, i.e., as TE, such as styrene isoprene styrene (SIS), styrene-ethylene butadiene styrene (SEBS), ethylene-propylene–diene-terpolymer (EPDM), isobutene–isoprene-copolymer (IIP), crumb-rubber, polybutadiene (PBD), polyisoprene and natural rubber (NR) [8].

Researchers around the globe are concerned for the alternative binder and modifiers for the pavement industry due to the declination of a reserve of oil [9]. Therefore, research on the bio-binders has been conducted to replace conventional binder partially to plaid out their practicability [9]. Thus industries are becoming wider in utilizing natural rubber in numerous application for improving its practice on several kinds of products and Malaysia is leading country after Thailand and Indonesia [10]. The extracted natural rubber latex from the trees by tapping is used for the modification of bitumen. Latex form of rubber contains water and other impurities such as resin and protein in addition to raw rubber, and due to bacterial attack and of high water content it may not be used directly [11]. Utmost of the research on the modification of bitumen with NR were studied in the United Kingdom during the 1950–1960s and it was revealed that powder of NR had numerous benefits i.e. the enhancement of resistance against rutting and fatigue of asphalt pavements and becomes an alternative solution for the pavement advancement cost-effectively [12].

Since bitumen is a natural constituent and resulted upon the refinement of crude oil and has thermoplastic behaviour and waterproofing properties which is sensitive to a variety of temperature [13]. The main usage of bitumen is in road pavement construction which accompanied approximatively 85% to its total consumption. However, it may also use as waterproofing and sealing agent as well as insulating material [14]. This merely means that the usage of asphalt, without adding any modifier reduced the service life of road due to permanent deformation and upsurge the volatility keeping road performance and cost.

Natural rubber production is improved by about 6.9 million metric tons from 2001 to 2017. On the other way, the price has been dropped extensively since 2011 and it seems to continue to drop [15]. Hence the utilization of NR in the pavement industry can probably make a win-win situation for the farmers of rubber by releasing financial burdens [16]. On prior research, it is evaluated that the natural rubbers have been used mostly for the modification of asphalt in latex form and provide good mixing conditions as compared to another form in an equal amount and few other used Ribbed Smoked Sheet (RSS) and Liquid Natural Rubber (LNR) [17]. Despite using natural rubber as the modifier for asphalt pavement, some researchers around the world also used chemical addictive's an additional modifier to mitigate some deficiencies of using natural rubber as alone. Consequently, this paper summarizes the natural rubber introduction, production and manufacturing overall process, and suitable application for road pavement as a modifier with and without chemical addictive's, and its effectiveness on overall pavement performance. This review paper is an effort to summarize sufficient information for the researchers and people around the industry on the above-mentioned topic.

2. Origin of natural rubber

Natural rubber is a versatile agriculture product and found to be consumed in nearly 50000 products worldwide [18]. Rubber was first used during the Olmec civilization (*circa* 1300–300 BC), and its usage continued among the Mayans (Yucatan peninsula in Mexico), and Incas (Andes highlands, Peru) from about AD 1100 to 1500, however Aztecs used in 12th century in central Mexico up to the Spanish demolition of the Central and South American civilizations. The Olmeç had made rubber and well known to tap plants, most likely *Castilla elastica* [11,18]. A Spanish rector, P. Martyre d'Anghiera, committed to the attacking army, foremost wrote about rubber in his book '*De Orbo Novo*' in 1530, but the worthwhile use of rubber persisted unidentified among Europeans for approximately 200 years [19].

A revolution on rubber emanated from two French scientists F. Fresneau and C. M. de la Condamine while they remained in search of economically beneficial plant and travelling in Guiana and the Amazon they became intent in rubber-producing trees on which they prepared a first scientific report and presented a talk on rubber at the conference of the French Academy of Science in Paris [20]. Two Englishmen, Charles Macintosh (1766–1843), an industrialist and Thomas Hancock (1786–1865), an engineer in the 19th century, initiated the development of rubber goods. Macintosh found rubber as water-resistant and in support with Hancock they produce raincoats.

Charles Goodyear discovers the method of 'vulcanization was in 1839' then later natural rubber became a scientifically imperative reserve and a well indispensable material after the development of vulcanization [20,21]. After the discovery of vulcanization, the growth of the rubber industry was created promptly in the second half of the 19th century. The progression of rubber industry was obstructed due to complications in collecting natural rubber in enough quantities with bad and poor techniques as the only source was Brazil, and this uncaring collecting caused the speedy collapse of the sources [11].

Subsequently, on the initiative of Sir Clements Markhem the London botanical garden and the India office decided to cultivate Hevea brasiliensis and Henry Wickam was appointed to acquire the seeds of the rubber plant mainly found in Amazon valley, South America [11].In 1876, Sir Wickam brought about 70000 seeds to Royal Botanic Gardens at Kew, in England, nearly 2000 of the seeds were successfully cultivated into plants, some of which were sent to Peradeniya in Sri Lanka and to Malaysia via the Singapore Botanical Gardens and lastly consignments were sent to Indonesia. In 1880 Hevea seedlings were extensively disseminated in Southeast Asia [22]. In Malaysia, the seeds were rapidly grown, though the rubber plant was planted by cutting a large area of the jungle [22]. In Southeast Asia countries initially, by the end of 19th Century, there was about 2500 ha of land under cultivation and become the foremost traders of natural rubber globally at the beginning of the automotive period, and by 1910 the rubber plantation was increased by 500,000 ha.

Nearly 2000 plant species have rubber, and merely a limited of which have good end product application, the primary place is taken from a tree which is autochthonous to the tropical region name as *Hevea brasiliensis* [19]. Hevea is used mostly as it produced more latex upon proper and well management of tree wounding and taping. Species other than Hevea are also used as commercially low category natural rubber are (*Castilla elastica*), rubber fig (*Ficus elastica*), vine rubber (*Landlphia*), gutta-percha (*Palaquium gutta*) and *Funtumia* [21]. Some other plants which require extra processing to furnished rubber and some of these also very difficult to tap and not effective on large scale industry, such as *Taraxacum* species including *Taraxacum officinale* and *Taraxacum kok-saghyz*; refined by Russian scientist, and guayule (*Parthenium argentatum*); cultivated by the American scientist due to its hypoallergenic characteristics [21].

3. Natural rubber and its properties

Natural rubber (NR), particularly from Hevea brasiliensis, has gained interest due to its various beneficial properties, such as high tensile strength, high elasticity, flexibility, good crack growth resistance, and low heat build-up [23]. NR latex exists colloid, where rubber particles are suspended in the serum which contains many non-rubber components (Table 1) [24,25].

Table 1. Composition of fresh NR latex [26]

Constituent	Content%
Rubber	30-40
Protein	1.0-1.5
Resins	1.5-3.0
Minerals	0.7-0.9
Carbohydrates	0.8-0.1
Water	55-60

Natural rubber is essentially pure poly-cis-1.4 isoprene (consist of 99.9 % of cis 1.4 structural units) and particles vary in size from 0.15–3 um as of chemical view with a chemical formula of C_5H_8 (Figure 1) [19,23]. At the ends of its macromolecules, there may be bonded also non-isoprene structural units. The presence of non-rubber content which is varying in the range of 5 – 10% also has a significant effect on the properties of natural rubber [20,26].

The repeated units of isoprene have also double bond which easily crosslinked or vulcanized for the sake for the three-dimensional structure of the rubber to display its stable elastic properties. Vulcanization is the process of adding vulcanization agents (sulfur, peroxide, metal oxides, resins, quinone's and other) in latex to improve the rubber properties that from plastic sensitive material to technologically useful material by the formation of a cross-linked molecular network i.e. from thermoplastic to thermoset [21].

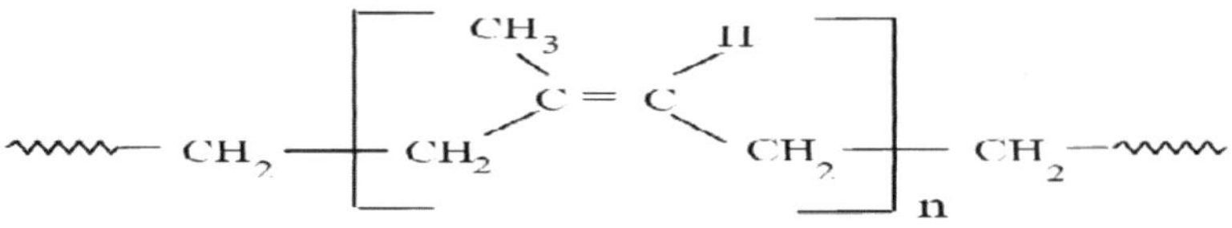

Figure 1. Chemical structure of cis-1, 4-polyisoprene in NR latex

In-depth reviews of NR physical properties, it has a flexible molecular chain due to its amorphous mass of coiled structures that make it too soft to be used for any useful purpose [27]. Natural rubber largely exhibits elasticity due to its high stretch ratio resilience characteristics and mainly happens due to a chain of polyisoprene and under loading, it performed like a linear chain while upon releasing load it wrinkled back [11]. Apart from the NR also shows some viscoelastic properties and an indication that its physical properties are partially liquid-like (viscous) and partially solid-like (elastic).

It is noted that the properties of Natural rubber also depend on modifiers or fillers such as carbon black, factice, whiting and other instead of the only polymer. NR processing mainly expands its application on other rubber raw products. After processing of tapping NR tree, it will be collected in the cup to avoid latex to be flowing on other parts of the tree. Mainly there are two main methods of collecting NR, the first is the preservation of latex by adding some quantity of ammonia also called ammunition and then proceed for a pattern in soft slabs rubber, and second is collecting latex in hard form by letting it coagulate in cups by itself for by adding commonly formic acid to turn in faster as cup lump. Coagulum sometimes forms a latex film on tapping cut called lace whereas latex which has dripped on the ground is called earth scrap which is also collected along with cup lump [22]. Asia is the main hub for the production of the natural rubber whereas Thailand on top after Indonesia and Malaysia, and about 13 million metric tons of NR produced every year worldwide [15]. In total production about 40% of rubber is in a natural state while remaining is consuming for synthetic rubber for the manufacturing for different rubber products [28]. Today with emergent science and with development different forms of rubbers modified while processing and utilized for commercialized aspects i.e. sheet rubber, crepe rubber, liquid rubber to see the more advanced in rubber industry instead of using simple raw rubber.

4. Compounding and vulcanization –Applications

High-grade rubber manufactured end products need a massive quantity of raw rubber. Since rubber applications for high-end products mainly productions of tires, belting, engine, rubber mounts, hot water bottles, and other mechanical goods are achievable by rubber compounding with some addictive's and it is justified method to achieve the properties in term of cost and the procedure involved for manufacturing [21]. Transformation of linear chains of NR into the three-dimensional matrix by insertion of some external chemical addictive's or cross-links such as Sulphur is called the process of vulcanization which is the main aspect of rubber compounding [26]. It holds high strength and resilience without reinforcing fillers and has quick repossession from deformation. Vulcanization may also be carried out by fillers or addictive's other than Sulphur i.e. urethanes, organic peroxides etc., but not as much frequent [19]. Compounding may also be carried out other than vulcanization by incorporating some fillers, addictive's and with oil to defend NR against heat, oxygen, and ozone to enhance resilient modulus, strength and reduce overall processing price of rubber production [21].

After a heavy demand for vulcanized rubber during the Second World War, it ultimately makes a shortage for vulcanized rubber [21]. Conversely, the polymer industry for synthetic rubber had boost up and emerged as a market force and changed the value for commercial NR. Chloroprene rubber, nitrile rubber, butyl rubber and most common styrene-butadiene rubber (SBR) are the examples for synthetic rubber. Alternative chemical method to advance applications of rubber is by modifying NR after it is collected from the tree as because NR has a structure of cis-polyisoprene which is fixed and cannot be polymerized by itself to provide superior pendant like synthetic rubber [29]. The modified shape will include epoxidized NR, cyclic NR, resin-modified NR, chlorinated NR, and hydrogenated NR [19].

5. Bitumen as the binder and its modification

Bitumen is one of the oldest known engineering materials. It has been used in many aspects i.e. as a sealant, binder for pavement, adhesive and waterproofing agent. Previously this material used directly and were taken from bitumen deposit at the earth surface. At the beginning of the 1900s, refined bitumen was primarily formed by refining crude oil in the USA, later then, the worldwide usage of bitumen has swiftly increased, and pavement construction was the major sector of consumption. Nearly 102 million tons annually bitumen used in various applications worldwide [30]. Truthfully, the chemical structure of generated bitumen is very intricate and variable; and the characteristics of bitumen is narrowly linked to the refinery process and crude oil [31].

Conversely, the production of good quality bitumen is on declination due to inadequate oil resources and due to an imposed force of earning peak economic benefits and due to deficiency of effective control actions during refinery force industries to focus further on bitumen modification. Conventional asphalt performs sound on most of the roads but demands made upon roads increase year by year i.e. growing traffic, high traffic load, rapid development and due to inadequate maintenance leads to pavement failure and decrease overall life. For fulfilment of this growing demands Highway Engineers there now exists a wide range of patented asphalts made with polymer-modified bitumen [32].

In countries with a tropical climate, asphalt mainly becomes soft with average temperature but inclines to viscous liquid or has flow like properties upon high temperature [33,34]. Whereas the binder poor rheological and viscoelastic properties may cause pavement deformation on road surface [16,35]. In many situations, conventional asphalt on severe weather and heavy traffic will cause serious failure on pavement i.e. cracking at low temperature, ravelling, fatigue and strapping failures [36]. Other researchers revealed that conventional asphalt also undergoes moisture susceptibility. These are the reason due to which researcher and Engineers around the world realized for improvement in a binder and to develop new methodologies to mitigate pavement problems.

Modification of bitumen is one of the ways to be considered to enhance the performance of asphalt concrete. Despite polymer modified asphalt or PMA, there are other approaches like modification with fillers like carbon black, hydrated lime, fly ash, sasobit wax and some others. However, the PMA modification is one of the premium ways which produces superior quality or superior form of road pavement. (Table 2) summarized the majority of modifier and addictive's which have been observed. The significant benefits of PMA comprise increasing elasticity and viscosity of bitumen and also enhance the ageing resistance, softening point and deduce the overall failures of pavement [37,38].

Since 1970 more and more research began on the achievement of enhancing bitumen properties for road pavement, and various kind of polymers has been used for the modification [38]. Hence these modifiers lead to enhancement in bitumen properties i.e. moisture resistance, extended fatigue life, greater stiffness on high temperature and resistance to cracking. In view of benefits of PMA, it also leads to some disadvantages upon using or before using such as in cost analysis some of the polymers are not suitable to use in broad quantities meanwhile few polymers have low property qualities as high-temperature sensitivity and low ageing resistance. Poor compatibility of bitumen and polymers during mixing and in-service life is also a chief issue.

IOP Conf. Series: Materials Science and Engineering 1075 (2021) 012031 doi:10.1088/1757-899X/1075/1/012031

Table 2. Some additives used to modify bitumen [32]

Modifier	Example
Thermoplastic elastomers	Styrene-butadiene-styrene (SBS) Styrene-butadiene-rubber (SBR) Styrene-isoprene-styrene (SIS) Styrene-ethylene-butadiene-styrene- (SEBS) Ethylene-propylene-diene terpolymer (EPDM) Isobutene-isoprene-copolymer (IIR) Natural Rubber (NR) Crumb Tyre rubber Polybutadiene (PBD) Polyisoprene
Thermoplastic polymers	Ethylene-vinyl acetate (EVA) Ethylene methyl acrylate (EMA) Ethylene butyl acrylate (EBA) Atactic polypropylene (APP) Polyethylene (PE) Polypropylene (PP) Polyvinyl chloride (PVC) Polystyrene (PS)
Thermosetting polymers	Epoxy resin Polyurethane resin Acrylic resin Phenolic resin
Chemical modifiers	Organo-metallic compounds Polyphosphoric acid (PPA) Sulphur Lignin
Fibres	Cellulose Alumino-magnesium Glass fiber Asbestos Polyester Polypropylene
Adhesion improvers	Organic amines Amides
Antioxidants	Amines Phenols Organo-zinc/oregano-lead compounds
Natural asphalt	Trinidad late asphalt (TLA) Gilsonite Rock asphalt
Fillers	Carbon black Hydrated lime Lime fly ash

Among all polymers advantages and disadvantages, NR is a way interesting modifier that can be added to bitumen efficiently [16]. Due to good tear strength, resistance against fatigue and of higher stability NR extending the stability of asphalt pavement [39]. Thus some studies also support the NR as modifier due to its excellent low-temperature properties and have also the minimum loss of hysteresis on dynamic loading [35]. Accordingly, the enhancement of road pavement made with cost-effective way and to increase the service life with a minimum cost of maintenance [40].

6. Natural rubber as modifier-blends of asphalt

The mixing of natural rubber in asphalt has been testing for many years and it is worthy to improve performance. Moreover, the improvement of asphalt characteristics by utilizing various raw latex i.e. low and high ammonium field latex, cream skim latex, revertex latex, and centrifuged latex had analyzed a long time ago in 1969 [41]. The results revealed that NR as field latex with high ammonium is effective in the reduction of penetration and increasing of softening point against all other forms of NR. The properties of liquid natural rubber (LNR) modified with bitumen was perceived in 1998 by Nair et al, [42] for soft and blown bitumen and results were founded increase in softening point by adding 5 to 10 % LNR and improvement in penetration values as it decreased in both cases, and also shear resistance increased for all cases upon adding of 5 to 10% LNR.

Recently, the improvement due to temperature variation has been analyzed mainly in cold conditions where cracks develop and summarized that latex in pavement mainly acts like an elastic band that helps to disintegrate stresses and at the same time also hold asphalt together. Conversely upon changing of rheological properties and flowing of asphalt latex acts as a membrane which boosts the shear strength and resists the flow [12]. Natural rubber built-in elastomers possessions proved that it has potential in enhancing pavement long term performance in asphalt concrete mix and to increase the cohesion and adhesion properties of binder [43]. Natural rubber or generally recognized as field latex comprises 30-40% rubber and rest is water including a small percentage of impurities. Though NR could not maintain its liquid form too long only last for 3 to 4 hours [21]. Considering this cause it is needed to preserve latex for extended time to get benefit from a wide range application of this raw material.

Presently, many studies have been conducted on modification of asphalt binder by using preserve NR latex as an enhancer. In additional research have been conducted to analyze the asphalt modification by utilizing field latex, concentrated latex and skim latex and results founded satisfactory upon improvement in softening, high penetration index (PI), high torsional recovery and high tenacity and toughness [16]. Next, a research conducted to summarized the effect of time and temperature on physical properties of modified NR bitumen and revealed that the temperature and blending time should not be more than 150°C and 10 minutes respectively and 9% NRL is maximum useful for improvement [44]. Moreover, the physical properties of bitumen by using epoxidized NR (ENR) have been analyzed and founded that based on softening and penetration values bitumen seems to involve stiffness however the ductile property at the start declined after adding ENR and then starts increasing which indicated the performance improvement. Additionally, the storage stability showed that 6 % ENR is stable for binder use in high temperatures [5].

Apart from rheological properties researchers have conducted studies on performance properties of modified NR bitumen and proved the enhancement of deformation and fatigue resistance of pavement. Furthermore, a study conducted by Khatijah Abu Bakar et al, [45] and reveal that natural latex is best to extend the service life of pavement upon improvement in its flexibility and stability properties. Therefore, it validates asphalt modified with rubbers have more resistance and strength in comparison with conventional asphalt. Next, a study on Hot Mix Asphalt (HMA) to estimate the stripping performance utilizing natural latex as modifiers reveal that on the enhancement of pavement performance and so concluded that NRL is a worthy enhancer for binder properties [43]. The ENR has been used to enhance the performance properties of asphalt concrete and from a study conducted it is revealed that 6% of ENR by weight of bitumen is optimally modified and improved that rutting resistance and also fatigue at a temperature of 30°C [5]. Additionally, Wen et al, [12] found that upon 6% of adding NR latex in asphalt improves the resistance against rutting, thermal cracking and fatigue resistance. Furthermore, krishnapriya et al, [34] revealed that modified rubberized asphalt mix provides a superior result and although not only improves resistance against rut but also upgrade resilient modulus to increase fatigue life. Next, NRL and carbon black utilizing in asphalt mix also found a satisfactory enhancement in mixture rutting resistance and stiffness [13]. Furthermore, Daniel et al, [46] studied the addition of fresh natural rubber latex on the rheological behaviour of the bitumen. It was evaluated from the results the addition of NRL in the bitumen decrease the penetration and simultaneously increasing the softening temperature of bitumen as (Figure 2).

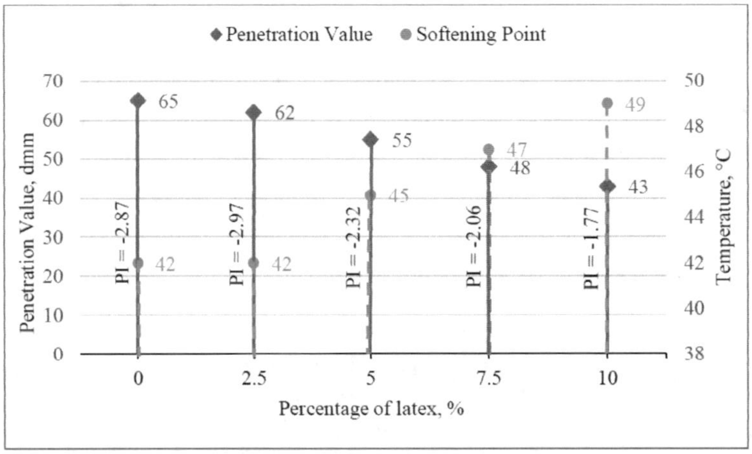

Figure 2. Penetration and softening point at a different percentage of NRL [46]

Presently, raw rubber in form of cup lump has been identified for the modification of bitumen as an addictive as it has similar properties as NRL and very much appropriate to mix with asphalt due to its less content of water for road improvement application. Furthermore, a study conducted by (Shaffie et al, [47] and summarized the performance properties of asphalt by using cup lump rubber (CLR) as a modifier. Thus, CLR significantly improves the rutting depth by 41% in contrast of unmodified bitumen mix Next, a study conducted by Othman et al, [48] for modified asphalt mixture incorporating cup lump rubber along Jalan Kualampur-Kuantan, Daerah Temerloh, Pahang, Malaysia. Thus it was concluded from results that 5% cup lump modified bitumen (CMB) utilized in cup lump asphalt (CMA) shows better physical properties along with performance in rutting and produced better in conditions which are 94% good in roughness conditions also proved improvement in skid resistance.

Rubber sheets with different form are also pertinent to utilize with asphalt and examples are a ribbed smoked sheet (RSS), creep rubber and block rubber. Additionally, the rubber sheet is reliable during blending with asphalt as it reduces the foaming and frothing. It was also concluded 6% by weight of natural rubber in asphalt enhanced the physical properties meanwhile develop high torsional recovery and toughness and tenacity [16].

Addition of NR in asphalt tends to physically disperse in, not chemically react with, asphalt. Hence, the gained asphalt properties are slightly enhanced and some properties particularly penetration or softening point needs a moderately large quantity of the NR, i.e. up to 10% by weight to attain required properties to encounter the specification [33]. Furthermore, chemical additives i.e. PPA and Sulphur have been utilized as a co- modifier to further enhance the blended mixture properties such as viscosity, softening, and storage stability and high-temperature properties and to also reduced the quantity of NR in asphalt and improves the blending conditions as well.

7. Natural rubber utilization and challenges

The constant utilization and expansion of the natural rubber all the impediments in the useful process must be eliminated. Natural rubber as biopolymer has high significance for many applications. The natural rubber producing countries worldwide and particularly Thailand, Malaysia and Indonesia always been concerned for the expansion of the natural rubber industry. The fluctuation in the price is the key factor which always been account for the buyer's due estimate their product gain profit and loss analysis. The source for the natural rubber is not in shortage however there is always been unpredictable demand for natural rubber. The production of natural rubber is 13 MMT in 2017 in comparison to synthetic rubber which was 15 MMT which is not as less. Alternatively, the demand and supply for the natural rubber in major producing countries is also struggling due to the parallel emerging growth of palm oil plantation. Presently Southeast Asian countries earning major economic growth and attention worldwide from palm oil production. Keeping all the facts and figure discussed and realizing its economic value initiative must be taken to mitigate the challenges and problems on the rubber industry before this valuable raw material lost its commercial value.

In recent years, many researchers around the world utilized natural rubber in a different form to achieve the desired properties in contrast to conventional road materials. RSS has been blended with asphalt to enhance the properties and seen as one of the ways to consume domestic consumption of NR [16]. Next, a new alternative form of NR i.e. cup lump which requires very minimum manufacture process has been utilized recently many researchers in rheological as well as performance testing and gained attention as an asphalt modifier [47,48]. Blending conditions has always been a key parameter upon asphalt modification as it impacts a lot on physical and mixture properties of asphalt. Furthermore, the modifier quantity in asphalt mixture also contributes a lot as for enhancement and up to 10% NR is used to improve properties which merely increase the viscosity and worse the mixing conditions. However, incorporating some chemical additives simultaneously reduce the quantity of rubber and may improve the properties of the mixture. The retardation of extensive usage of NR and its different form may also be enhanced by performing full-scale testing with proper techniques and blending conditions to produce an economical mix of rubberized asphalt.

8. Conclusion

Modification of road pavement with additives have been in practice since many decades ago to increase the performance and to protect the sources of asphalt. Concerning this and due to a declination of oil reserves researcher are interested around the globe in finding substitute pavement binder and modifier. Furthermore, NR recently expands its applications while attempting a combination with bitumen. The review enlightens the studies conducted for its usage, advancement and the challenges faced on the modification of natural rubber that influence the binder properties. Previously conducted studies on NR modified asphalt has evidenced that it is one of the bio-polymer that can be used as a modifier for pavement performance to mitigate the failures i.e. fatigue, rutting, ageing due to tire stresses or temperature variation. Extensive usage of NR due to its low cost can also enhance the economic growth of industry and simultaneously service life of roads increase with low maintenance cost and will also significantly improve the ride quality.

References

[1] Kök BV, Yılmaz M, Akpolat M. Evaluation of the conventional and rheological properties of SBS+Sasobit modified binder. *Constr. Build. Mater.*, **63**, 174–9 (2014).
[2] Yildirim Y. Polymer modified asphalt binders. *Constr. Build. Mater.*, **21**, 66–72 (2007).
[3] Padhan RK, Gupta AA, Badoni RP, Bhatnagar AK. Improved performance of a reactive polymer-based bituminous mixes – effect of cross-linking agent. *Road Mater. Pavement Des.*, **16**, 300–15 (2015).
[4] Biro S, Fazekas B. Asphalt rubber versus other modified bitumens. *Road Mater. Pavements Des.*, **X**, 1–9 (2005).
[5] Al-Mansob RA, Ismail A, Alduri AN, Azhari CH, Karim MR, Yusoff NIM. Physical and rheological properties of epoxidized natural rubber modified bitumens. *Constr. Build. Mater.*, **63**, 242–8 (2014).
[6] Pasandín AR, Pérez I. The influence of the mineral filler on the adhesion between aggregates and bitumen. *Int. J. Adhes. Adhes.*, **58**, 53–8 (2015).

[7] Lu X, Isacsson U. Rheological characterization of styrene-butadiene-styrene copolymer modified bitumens. *Constr. Build. Mater.*, **11**, 23–32 (1997).
[8] Mo L, Shu D, Li X, Huurman M, Wu S. Experimental investigation of bituminous plug expansion joint materials containing high content of crumb rubber powder and granules. *Mater. Des.*, **37**, 137–43 (2012).
[9] Fini EH, Kalberer EW, Shahbazi A, Basti M, You Z, Ozer H, Aurangzeb Q. Chemical characterization of biobinder from swine manure: sustainable modifier for asphalt binder. *J. Mater. Civ. Eng.*, **23**, 1506–13 (2011).
[10] Tansella M, De Salvia D, Williams P. The Italian psychiatric reform: some quantitative evidence. *Soc. Psychiatry.*, **22**, 37–48 (1987).
[11] Greve H-H. Rubber, 2. Natural. *Ullmann's Encyclopedia of Industrial Chemistry*. Wiley-VCH Verlag GmbH & Co. KGaA, Weinheim, Germany, p.14 (2000).
[12] Wen Y, Wang Y, Zhao K, Sumalee A. The use of natural rubber latex as a renewable and sustainable modifier of asphalt binder. *Int. J. Pavement Eng.*, **18**, 547–59 (2017).
[13] Ali SA, Yusof I, Hermadi M, B.S.Alfergani M, Sinusi AA. Pavement performance with carbon black and natural rubber (Latex). *Int. J. Eng. Adv. Technol.*, **2**, 124–9 (2013).
[14] Institute A. *The bitumen industry A global perspective production, chemistry, use,specification and occupatinal exposure*. 3rd ed. Vol. 89, Asphalt Institute Inc. and European Bitumen Association–Eurobitume, USA, (2015).
[15] Board MR. Natural rubber statistics, 2018.
[16] Zborowski A, Kaloush KE. Predictive Equations to Evaluate Thermal Fracture of Asphalt Rubber Mixtures. *Road Mater. Pavement Des.*, **8**, 819–33 (2007).
[17] L.M Smith. Some Viscous and Elastic Properties of Rubberised. *J. appliled Chem.*, 296–305 (1960).
[18] Nair KPP. Rubber (Hevea brasiliensis). *The Agronomy and Economy of Important Tree Crops of the Developing World*. Elsevier, pp.237–73 (2010).
[19] Shinzo Kohjiya, Ikeda Y. *Natural Rubber: Production, Properties and Applications*. Elsevier Science, (2014).
[20] Nakao T, Kohjiya S. Computer simulation of network formation in natural rubber (NR). *Chemistry, Manufacture and Applications of Natural Rubber*. Elsevier, pp.216–46 (2014).
[21] Mathew MRSN. *Natural Rubber,Biology, Cultivation and Technology*. Elsevier Science, (1992).
[22] Baker CSL, Fulton WS, Updated by Staff. Rubber, Natural. *Kirk-Othmer Encyclopedia of Chemical Technology*. John Wiley & Sons, Inc., Hoboken, NJ, USA, (2012).
[23] Thomas AG, Whittle JM. Tensile Rupture of Rubber. *Rubber Chem. Technol.*, **43**, 222–8 (1970).
[24] Nawamawat K, Sakdapipanich JT, Ho CC, Ma Y, Song J, Vancso JG. Surface nanostructure of Hevea brasiliensis natural rubber latex particles. *Colloids Surfaces A Physicochem. Eng. Asp.*, **390**, 157–66 (2011).
[25] Dararat Mekkriengkrai1, Jitladda T. Sakdapipanich YT. Structural Characterization of Terminal Groups in Natural Rubber : *Rubber Chem. Technol.*, **79**, 366–79 (2006).
[26] Matador Rubber. Rubber Chemistry, 2007.
[27] Tinavallie A/P Grumoorthy. University Tun Abdul Rahman, 2013.
[28] ResearchInChina. Global and China Natural Rubber Industry Report, 2017-2021, 2017.
[29] Bhowmick AK and HL stephens. *Handbook of Elastomers*. 2nd ed., CRC Press, New York, (2000).
[30] Institute A. The Asphalt Paving Industry. A Global Perspective, 2011.
[31] Zhu J, Birgisson B, Kringos N. Polymer modification of bitumen: Advances and challenges. *Eur. Polym. J.*, **54**, 18–38 (2014).
[32] Read J, Whiteoak D. *The Shell Bitumen Handbook*. fifth ed., thomas telford Ltd, USA, (2003).
[33] Saowapark W, Jubsilp C, Rimdusit S. Natural rubber latex-modified asphalts for pavement

application: effects of phosphoric acid and sulphur addition. *Road Mater. Pavement Des.*, **20**, 211–24 (2019).

[34] Krishnapriya. Performance evaluation of modified Bituminous mixes. *J. Civil, Struct. Environ. water Resour. Infrastruct. Eng. Res.*, **5**, 1–12 (2015).

[35] Vasavi Swetha, D, K DR. Effect of natural rubber on the properties of bitumen and bituminious mixes. *Int. J. Civ. Eng. Technol.*, **5**, 976–6308 (2014).

[36] Xiao F, Amirkhanian S, Wang H, Hao P. Rheological property investigations for polymer and polyphosphoric acid modified asphalt binders at high temperatures. *Constr. Build. Mater.*, **64**, 316–23 (2014).

[37] Sengoz B, Isikyakar G. Evaluation of the properties and microstructure of SBS and EVA polymer modified bitumen. *Constr. Build. Mater.*, **22**, 1897–905 (2008).

[38] Zhang F, Hu C. The research for SBS and SBR compound modified asphalts with polyphosphoric acid and sulfur. *Constr. Build. Mater.*, **43**, 461–8 (2013).

[39] Colin Ruggles. The use of natural rubber latex in modified asphalt road binders in the UK. (2005).

[40] Polacco G, Stastna J, Biondi D, Zanzotto L. Relation between polymer architecture and nonlinear viscoelastic behavior of modified asphalts. *Curr. Opin. Colloid Interface Sci.*, **11**, 230–45 (2006).

[41] Fernando MJ, Nadarajah M. Use of natural rubber latex in road construction. *Rubber Res. Inst. Malaya*, **22**, 430–40 (1969).

[42] Nair NR, Mathew NM, Thomas S, Chatterjee P, Siddiqui MA. Physical and rheological characteristics of liquid natural rubber modified bitumen. *J. Appl. Polym. Sci.*, **68**, 53–61 (1998).

[43] Shaffie E, Ahmad J, Arshad AK, Kamarun D, Kamaruddin F, Mara UT. Stripping performance and volumetric properties evaluation of hot mix asphalt (HMA) mix design using natural rubber latex polymer modified binder (NRMB), InCIEC, 2015, springer, singapore, pp.873–84.

[44] Shafii MA, Lai Yew Veng C, Mohamad Rais N, Ab Latif A. Effect of blending temperature and blending time on physical properties of NRL-modified bitumen. *Int. J. Appl. Eng. Res.*, **12**, 3844–9 (2017).

[45] Khatijah Abu Bakar S, Ezree Abdullah M, Mustafa Kamal M, Abd Rahman R, Putra Jaya R, Buhari R, Arifin Hadithon K. Influence of ageing process on rheological characteristic of waste natural rubber latex modified bitumen. *J. Phys. Conf. Ser.*, **1049**, 012100 (2018).

[46] Daniel NH, Hassan NA, Idham MK, Jaya RP, Hainin MR, Ismail CR, Puan OC, Azahar NM. Properties of bitumen modified with latex. *IOP Conf. Ser. Mater. Sci. Eng.*, **527**, 012063 (2019).

[47] Shaffie E, Wan Hanif WMM, Arshad AK, Hashim W. Rutting resistance of asphalt mixture with cup lumps modified binder. *IOP Conf. Ser. Mater. Sci. Eng.*, **271**, 012056 (2017).

[48] Othman Z, Hainin MR, Naqiuddin M, Warid M. Cup lump modified asphalt mixture along jalan Kuala Lumpur-Kuantan , daerah Temerloh , 2018, MATEC Web of Conferences 250 SEPKA-ISEED 2018, pp.1–8.

IOP Conf. Series: Materials Science and Engineering 1075 (2021) 012032 doi:10.1088/1757-899X/1075/1/012032

Use of Non-Metallic Powder Reclaimed from Waste Printed Circuit Boards in Rigid Concrete Pavement

K Kakria[1] and S Priya[1, *]

[1]School of Civil Engineering, Vellore Institute of Technology, Vellore, India
[*] Corresponding author: shanmugapriya.t@vit.ac.in

Abstract. Treating E-Waste is the need of the hour, about 2 million tons of E-Waste with a 5% share of PCB (Printed Circuit Boards) is generated by India out of which only few is treated, majority of which is disposed in water bodies and landfills. The disposal of e-waste is a worldwide concern and to deal with the same we propose to use the e-waste in rigid concrete pavement. India with 43,20,000 km of roads, gives a wide possibility of using e-waste in this industry. This research mainly focuses on the conversion of Non-Metallic Powder (NMP) waste recycled from PCBs into acceptable construction material. The objective of this study is to investigate the effect of e-waste as a filler replacement, on the strength parameters of concrete (M40 grade) and used as rigid pavement. NMP from waste PCB is mixed with 0%, 3%, 6%, 9% and 12% as an additive and based on Compressive Strength, Split Tensile Strength, Flexural Strength test and Impact Resistance test, the most suitable dosage of the additive is then to be calculated. It is observed that there is definite increase in compressive strength, split tensile strength and flexural strength with increase in percentage of NMP up to 9%. Here an attempt is made to use the PCB waste products in rigid road construction, which affect the environment and are difficult to process. The study concludes that 9% of e-waste can be used as filler material in bituminous mixes.

1. Introduction

The electronic and electrical waste (e-waste) is one of the fastest flourishing waste streams in the world. The increase of "market penetration" in developing countries, "replacement market" in developed countries and "high obsolescence rate" makes e-waste one of the fastest flourishing waste streams. Environmental issues and trade linked with e-waste at local, cross border and International level has led many countries to intervene [1]. In accordance with the National Environmental Policy (NEP) and to address sustainable development concerns, there is a need to facilitate the recovery, reuse and recycling of useful materials from waste generated by electronics by different processes thereby reducing the wastes for final disposal to safeguard the environment. The NEP also gives legal recognition and strengthening of the informal sector systems for the collection and recycling of unused or disposed materials. Considering the high recycling ability of e-waste such wastes should be subjected to recycling in an environmental friendly way [2].

India is ranked as the second largest mobile phone user and fifth largest e-waste producer in the world. Overall it produces 18.5 lakh tons of e-waste every year, which increases with the increment in the usage of electronic devices. The devices, which are sold in the beginning of the year, will end up in 25% of e-waste at the end [3]. Computer wastes account to virtually 70%, the contributions from telecom industry is around 12%, medical instruments being 8% and electrical instruments being 7% of the total annual e-waste accumulation in our country [3]. Several studies have been done on using the waste material in concrete as aggregate or filler. It is not only a solution to reduce the environmental and ecological problems but overall it improves the microstructure and hence increases its strength [4, 5, 6].

The printed circuit boards (PCB) are recycled by physical processing and hydrometallurgical treatment. The chemical analysis showed that PCB contains metals like Cu, Zn, Pb and Sn and some precious metals [7]. The boards are shredded at required size which can be used as a substitute for aggregate in construction. The chemical components are mostly toxic and it may cause adverse effects to the environment. Therefore the recycling should be done in such a way that it should not cause any harm to the surroundings and should reduce the demand of natural materials [8].

In this study PCB waste (NMP) is used as an additive in concrete used as rigid pavement and to determine the optimum percentage of PCB Board waste for M40 grade concrete based on compressive strength, three-point load test strength, split tensile strength and to test with different dosages (0%, 3%, 6%, 9% and 12%) of the additive and hence evaluate the potential of the PCB waste to be used in rigid pavement constructions.

2. Material and methods

2.1. Materials

2.1.1 Cement

Ordinary Portland cement (OPC 53) conforming to IS:12269-1987 [9] has been used in this research. The cement was free from lumps and had a grey with a light greenish shade. Table 1 gives the physical properties of the cement.

2.1.2 Non-metallic powder

Printed circuit boards were collected from discarded electrical appliances and gadgets, which were treated to remove the metallic component to obtain the non-metallic component. The non-metallic component was then shredded into fine particles of size less than 150 μm. The non-metallic processed powder (Figure 1) was obtained from Victory Recovery and Recycle Technologies India Pvt Ltd. The physical properties of NMP is discussed in Table 1.

Figure 1. Non-metallic powder (NMP)

IOP Conf. Series: Materials Science and Engineering 1075 (2021) 012032 doi:10.1088/1757-899X/1075/1/012032

Table 1. Physical properties

	Blaine fineness (cm^2/g)	**Soundness Autoclave (%)**	**Specific Gravity**	**Setting time in minutes**	
				Initial	**Final**
Cement	2280	0.85	3.15	94	232
E-waste	5210	0.72	2.3	115	243

Figure 2 shows the SEM images of the non-metallic powder. The non-metallic powder exists in small cylindrical solids with some content of irregular solids of thermo-setting resins.

Figure 2. SEM image of non-metallic powder

2.1.3 Fine aggregates

Clean river sand sieved through 4.75 mm IS sieve was used in this research. It had a specific gravity of 2.62. The sand was confirmed to Zone II and the tests were carried as per the specification laid by IS 383:1970 [10].

2.1.4 Coarse aggregates

These are the strongest and least porous component of the mix. Various aspects where taken care while selecting the aggregates like, low porosity, high crushing strength and interlocking textures. Coarse aggregates of 20 mm IS Sieve size was selected which had a specific gravity of 2.74.

2.2 Methods

2.2.1 Mix design

A concrete mix was prepared with help of the above-mentioned materials, following the guidelines given in IS 10262- 2009 [11]. For 0%, 3%, 6%, 9% and 12% addition of NMP the mix ratio chosen was 1:1.6:2.98 (Cement:Fine Aggregate:Coarse Aggregate) and 2.842 litres of water was added with 0.0392 kg of admixture. However, for 3% addition of NMP 0.1959 kg of additive was added, 0.3918 kg was used for 6% addition of NMP, 0.5877 kg for 9% addition of NMP and 0.7836 kg for 12% addition on NMP.

2.2.2 Preparation of cube specimens

To determine the compressive strength of the mix, five mixes M1-0% addition of NMP, M2-3% addition of NMP, M3-6% addition of NMP, M4-9% addition of NMP and M5-12% addition of NMP were prepared. 30 cube moulds of 100mm×100mm×100mm, 15 beam specimens of 500mm×100mm×100mm and 15 cylinders of 100mm diameter and 200mm height were used to cast the specimens.

3. Testing

3.1 Compressive strength test

The compressive strength test was performed on concrete with varying non-metallic powder contents as per IS 4031-6 (1988) [12]. The wet cubes were placed with the cast faces directly in contact with the testing machine plates as shown in Figure 3. The machine plates is to be cleaned first and the dimensions is to be taken to nearest 0.2 m. The cubes were then tested after a curing of 7 and 28 days in a Digital Compression testing machine having a capacity of 20,000 kN.

The load was applied uniformly starting from zero at the rate of 2.5 kN/sec till the specimen failed. Three specimens of each percentage of additive was tested and the average of compressive strength was found.

Figure 3. Compressive strength test of cube

3.2 Split tensile strength test

Concrete is very weak to direct shear stress and often develop cracks due to its brittle nature. Thus, it is necessary to determine the tensile strength of the concrete to determine the point of failure of concrete. The test cylinders of size 100 mm diameter were casted using different amounts of non-metallic powder and tested after 7 days curing. A uniform load of 2.5 kN/sec was applied starting from zero till the specimen failed (Figure 4).

Figure 4. Split tensile test of cylinder

3.3 Three-point loading test of beam

During flexure test the tensile stress is produced in the convex side of the specimen whereas the compression stress is produced in the concave side. This creates a shear stress area along the midline. The Flexural test is used to measure the force required to bend a beam by a Three-point loading system. This test data is used to select materials for components that will support loads without bending and flexing. Flexural modulus is an indicator of a material's stiffness when flexed. The flexural test was performed as per ASTM C348-14 [13] (Figure 5).

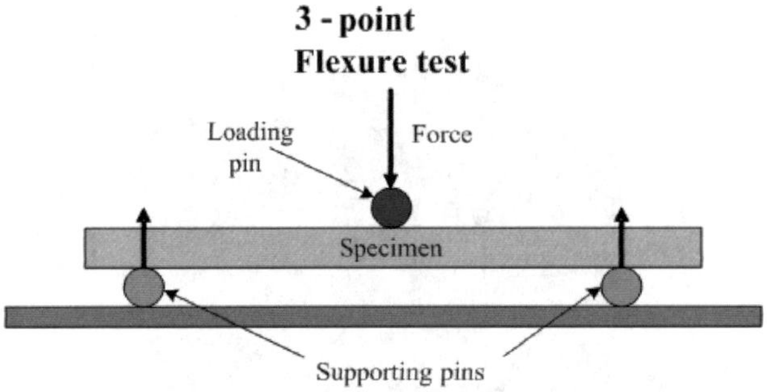

Figure 5. Three-point loading test of beam

3.4 Impact resistance test

Concrete structures are often prone to high force or shock applied for a short time period (impact loads) such as machine vibrations, vehicle impact, earthquake etc. Impact resistance energy for the specimen was measured by drop hammer test as recommended by ACI committee 544.2R-89 [14]. A concrete specimen of 150 mm and a height of 64 mm was subjected to drop hammer test. A hammer of weight 45 N was used and dropped from a height of 457mm to test the ultimate failure of the specimen. Ultimate failure is the number of blows to open the crack in the specimen, when the fractured piece touches three or four lugs on the base plate. The impact energy absorbed for complete failure was calculated by multiplying the number of blows required for ultimate failure and the impact energy (Figure 6).

Figure 6. Test specimens used for drop hammer test

4. Results and discussion

4.1. Compressive strength test

The compressive strength of concrete with different amount of non-metallic powder after 7 and 28 days of curing was found. Figure 7 shows the variation of compressive strength with additive percentage.

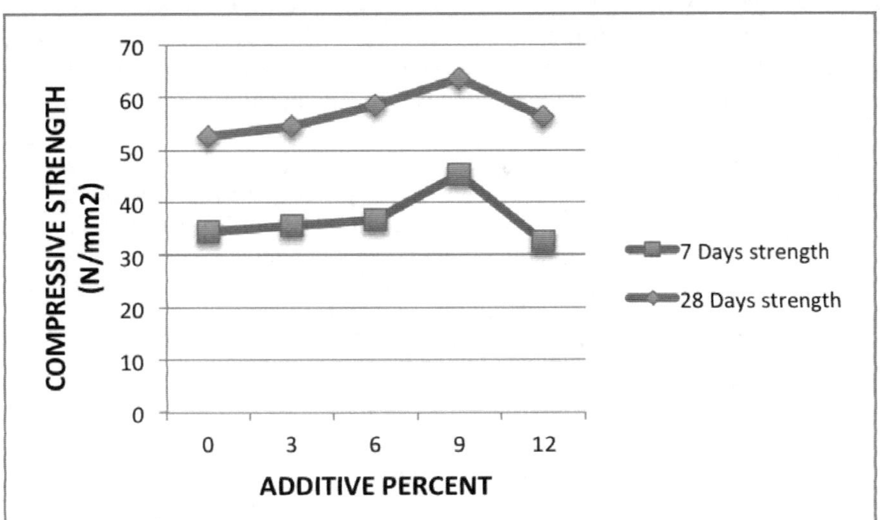

Figure 7. Compressive strength of cube for different percentage replacement

The compressive strength of cube with 0%, 3%, 6%, 9% and 12% additive after a curing of 7 as 34.51, 35.67, 36.63, 45.22 and 32.52 N/mm^2 and 28 days as 52.67, 54.52, 58.51, 63.57 and 56.32 N/mm^2 respectively. The compressive strength is seen to increase with addition of non-metallic powder, 9%(M4) being the largest. There is seen a 3.39% increase in compressive strength with the addition of 3% NMP when compared to 0% addition of NMP. The compressive strength is maximum on 9% addition of non-metallic powder but decreases on further addition of non-metallic powder. The compressive strength is seen to increase on addition of e-waste as due to the small size of particles they fill up the voids.

4.2 Tensile strength test

The tensile strength of specimen was found after 28 days of curing in water tank. Figure 8 shows the variation of tensile strength with additive percentage.

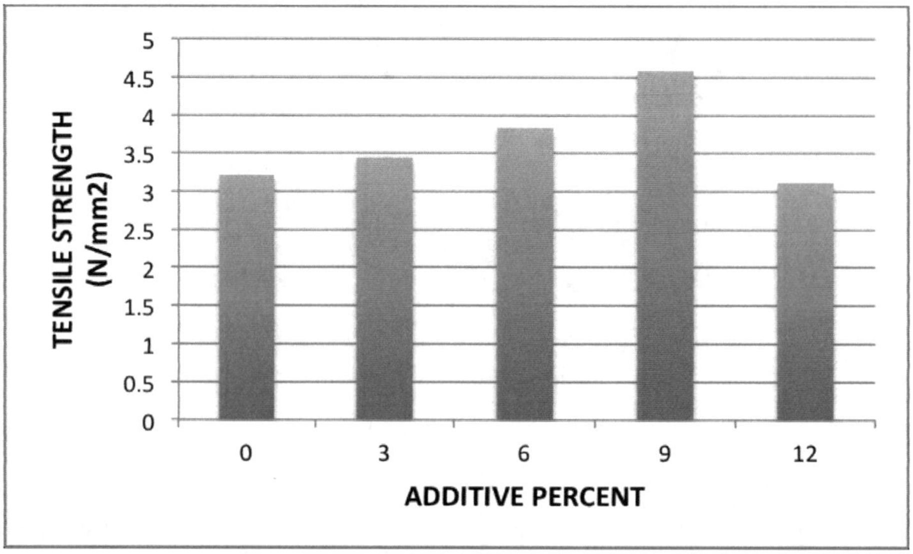

Figure 8. Tensile strength of cylinder for different percentage replacement

It is seen that the best result was shown on addition of 9% (M4) of non-metallic powder. The tensile strength of concrete increases on addition of non-metallic powder. Addition of non-metallic powder does not have major effect on compressive strength but increases the tensile strength of concrete. Non-metallic powder act as a fiber for concrete and decreases the permeability of pavement.

4.3 Flexural test

The flexural test was performed on 15 beams of size 500mm×100mm×100mm after 28 days of curing as per ASTM C348-14. Figure 9 shows the variation of flexural strength with additive percentage.

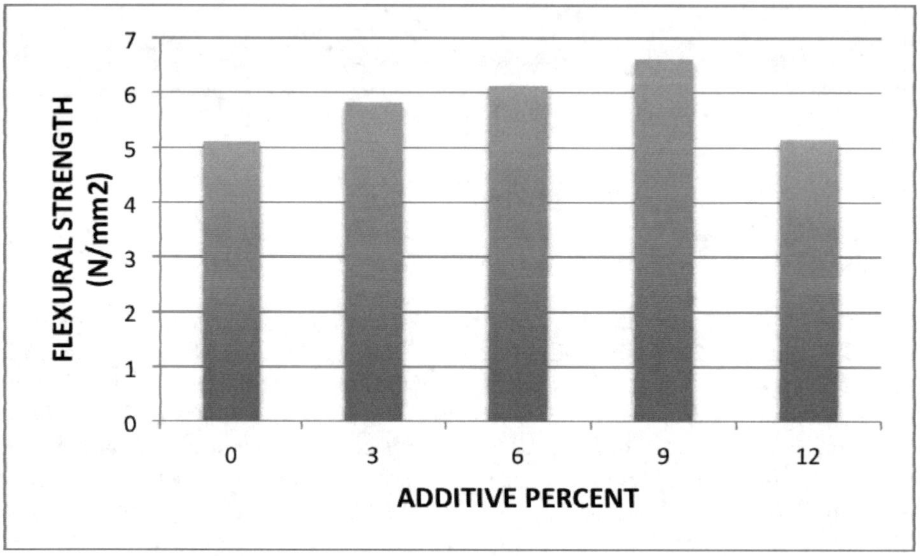

Figure 9. Flexural strength for different percentage replacement

The flexural strength of the mix for different percentage replacement (0%, 3%, 6%, 9% and 12%) was found to be in the range of 5.12 to 6.62, 6.62 N/mm^2 being the highest at 9% (M4) addition of NMP.

4.4 Impact resistance test
The impact resistance of the specimens were found with a hammer of weight 45kN. Fig 6 shows the test specimens after performing the drop hammer test. Table 2 shows the variation of impact energy with additive percentage.

Table 2. Impact resistance test

Mix	Impact resistance or energy absorbed for complete failure (Nm)	
	28 days	90 days
M1	110	128
M2	122	135
M3	135	142
M4	140	156
M5	132	120

The Impact resistance or energy absorbed for complete failure of the specimen with different mixes 0%, 3%, 6%, 9% and 12% was found to be in the range of 110 to 152, 156 Nm being the highest at 9%(M4) addition of NMP at the age of 90days.

This concrete can be used in rigid pavements for highway and airfield and for repair of existing pavement. NMP additive concrete helps in reducing the pavement thickness and increases the resistance to impact. It provides a smooth riding surface and increases the transverse and longitudinal joint spacing.

5. Conclusion
According to the test results the following conclusions can be drawn:

- There is no major effect on compressive strength of the concrete when the percentage of the additive increases and shows maximum compressive strength on 3% (M4) addition of non-metallic powder.
- The tensile strength increases as the percentage of PCB waste increases. The same characteristic is observed when the fibre is added to the concrete. So we can conclude that the PCB waste can be used as a fibre to increase the tensile strength of the concrete pavement.
- The flexural modulus increases with percentage addition of NMP and mix M4 (9% addition of NMP) shows the highest value of 6.62.
- The mix M4 shows best results for drop hammer test with a value of 156 Nm.
- The addition of fibre (in the form of PCB waste) decreases the permeability of the concrete pavement, not allowing water to seep inside and destroy the subgrade, hence increasing the life of concrete pavement.

Acknowledgments
The research work was carried in the Structural Research Laboratory, School of Civil Engineering, VIT, Vellore, India. The authors would like to thank the technical staff of the laboratory for their support and guidance.

References

[1] Duan H, Hu J, Yuan W, Wang Y, Yu D, Song Q and Li J 2016 Characterizing the environmental implications of the recycling of non-metallic fractions from waste printed circuit boards *Journal of Cleaner Production* 546-554

[2] Ministry of Environment and Forests 2008 Guidelines for Environmentally Sound Management of E-Waste, *Ministry of Environment and Forests Letter* No. 23-23/2007-HSMD Central Pollution Control Board Delhi.

[3] Muthukumar S, Bavithran OKC, Nandhagopal AR and Snehasree T 2017 Stability Study on Eco-Friendly Flexible Pavement Using E-Waste and Hips *International Journal of Civil Engineering and Technology (IJCIET)* 956–965

[4] Tavakoli D, Hashempour M and Heidari A 2018 Use of Waste Materials in Concrete: A review, *Pertanika J. Sci. & Technol.* 26 (2): 499 - 522

[5] Rui Silva JDB 2015 Use of waste materials in the production of concrete *Key Engineering Materials* 634:85-96

[6] Batayneh M, Marie I and Asi I 2006 Use of selected waste materials in concrete mixes *Waste Management* 27 (2007) 1870–1876

[7] Bizzo WA, Figueiredo RA and Andrade VFD 2014 Characterization of Printed Circuit Boards for Metal and Energy Recovery after Milling and Mechanical Separation *Materials* 7, 4555-4566

[8] Kadari R, Velchuri R, Malathi M, Vithal M and Munirathnam NR 2017 Degradation of organic pollutants by Ag, Cu and Sn doped waste non-metallic printed circuit boards *Waste Management* 60 629–635

[9] Ordinary Portland Cement 53 Grade-Specifications 2013 *Bureau of Indian Standards* 12269-1987

[10] Specification for Coarse and Fine Aggregates from Natural Sources for Concrete 2002 *Bureau of Indian Standards* 383:1970

[11] Concrete Mix Proportioning-Guidelines 2009 *Bureau of Indian Standards* 10262

[12] Methods of Physical Tests for Hydraulic Cement 2005 *Bureau of Indian Standards* 4031-6

[13] Standard Test Method for Flexural Strength of Hydraulic-Cement Mortars *ASTM* C348-14.

[14] Measurement of Properties of Fiber Reinforced Concrete (Reapproved 2009) 2009 *American Concrete Institute*

IOP Conf. Series: Materials Science and Engineering 1075 (2021) 012033 doi:10.1088/1757-899X/1075/1/012033

Improve Recycled Concrete Aggregate properties in Order to Use It in Paving Application as Aggregate

Gailan Ismat Safa Eldeen[1], Fauzan Mohd Jakarni[2], Alyaa A.Al-Attar[1], Ratnasamy Muniandy[2]and Salihudin Hassim[2]

[1]Northern Technical University- Iraq
[2]Department of Civil Engineering, Faculty of Engineering, University Putra Malaysia, Serdang, Malaysia
* Corresponding author: gailan2004@yahoo.com

Abstract.The amount of crushed cement concrete continues to increase daily as a result of the demolition of old structures, thereby increasing pollution. To cope with the pressure imposed by the rise in environmental awareness and the stringent disposal regulations set by environmental protection agencies, effective measures for handling and disposing of crushed concrete must be implemented. Instead of simply disposing of crushed concrete, alternative efforts should be considered to utilize it as a recyclable material. The objective of this study was to improve recycled concrete aggregate (RCA) to be able to using it in paving application. Three types of treatments applied to improve RCA properties. In this study three types of treatment were used to improve RCA properties. Theses treatment are soak RCA in HCL acid for 24 hours, grind RCA in Los Angeles Abrasion machine, re-grind RCA particle in jaw crush machine. In order to evaluate each treatment the aggregate properties are determined. All properties were determined for RCA passing and retained sieve size of 5 mm and 1.18 mm respectively. The calculated properties are bulk specific gravity of aggregate, absorption value, abrasion value of aggregate and angularity number of RCA.. The most important conclusions are Soaking RCA in HCL acid has side effect on the specific gravity of aggregate, Putting RCA in Los Angeles Abrasion machine make RCA particles more rounded. Finally, crush RCA in jaw crusher machine for second is best way to improve RCA properties

1. Introduction

According to the ASTM definition, aggregate is a granular material such as crushed stone, sand, gravel or iron blast furnace slag. Aggregate which is retained on a sieve size of 4.75 mm (No. 4) is called coarse aggregate, whereas aggregate which passes a sieve size of 4.75 mm (No. 4) and is retained by a sieve size of 75 μm (No. 200) is called fine aggregate [1]. Aggregate is usually used in Portland cement concrete, asphalt pavement, and as a filler to develop and maintain structures [2]. The basic skeleton of asphalt mixtures comes from the aggregate while the asphalt acts to hold them together in the matrix. The final performance of asphalt mixtures is greatly influenced by the mechanical and physical properties of the aggregate. The following figure 1 is based on the literature which addresses the relation of aggregate properties and the performance of asphalt mixtures [3].

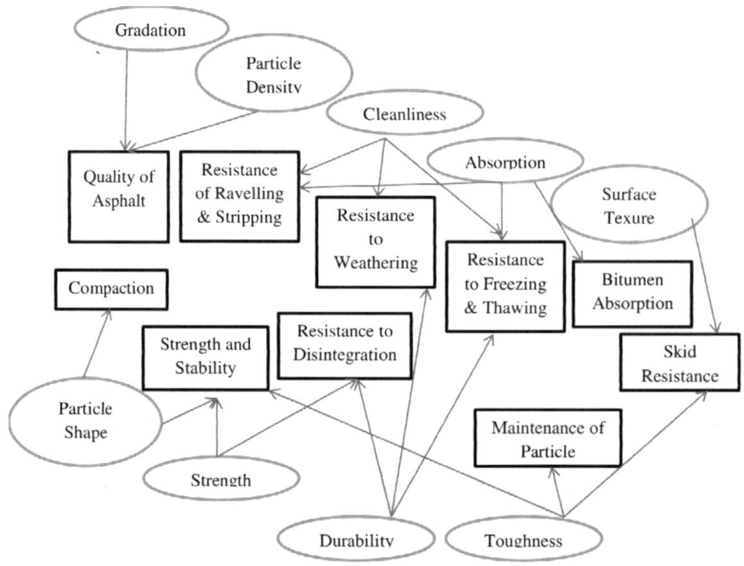

Figure 1. Effect of Aggregate Properties on Asphalt Mixture Performance [3]

The aggregate which is used in pavement construction is usually produced from crushing natural stone resources like basalt, granite and limestone. Meanwhile, natural resources are in great demand with the rapid development of human activities and especially the development of the highway industry. According to recent statistics, aggregate is the second most consumed material per capita on the Earth [4]. Hot mix asphalt contains 75–85% by volume and 90–95% by weight of aggregate [2–5]. For each kilometre of flexible pavement, 12500 tonnes of aggregate are used [5,6]. In the United Kingdom, 15 million tonnes per year of aggregate are used in road construction [7]. In Turkey, 290 million tonnes per year are created, around 35% of which is used in road construction [4]. In China, 5.6 ×108 tons of aggregate were used in asphalt mixes in 2010 [8]. Moreover, a shortage of natural aggregate and landfill sites is a common problem in Europe and other countries like Australia and Japan [9]. It is estimated that the concrete industry uses approximately 10 billion tons of aggregate worldwide [10,11]. This means that in the future the rate of production of natural aggregate will be too slow and unable to meet the demand for aggregate from the construction industry. The aggregate shortage problem can be solved by recycling concrete. in addition, Recycling waste concrete from the demolition of old structures as aggregate in hot mix asphalt is a good solution to decrease the amount of environment pollution.

The difference between the virgin aggregate and recycled concrete aggregate is the attached cement mortar. This attached cement mortar usually has higher porosity, lower strength and higher water absorption than natural aggregate. Therefore, it has a negative effect on the mechanical properties and performance of the mixture. Some studies suggest removing or decreasing the amount of attached mortar to improve the properties of RCA, whereas other studies suggest improving the properties of the attached mortar. It was concluded from reviewing the literature on improving RCA properties that there are two groups: the first group of treatments applies to the RCA in general without using it in the mixtures, while the second group of treatments applies to the RCA and the properties of the asphalt mixtures studied.

Treatments applied to RCA without evaluating asphalt mixes properties. These treatments are applied to the RCA before using it as aggregate in the asphalt or Portland cement mixtures. There are two common methods: First method is removing or decreasing the amount of attached mortar, and the other method is improving the properties of the attached mortar. The summary of these treatment is shown in figure 2.

IOP Conf. Series: Materials Science and Engineering 1075 (2021) 012033 doi:10.1088/1757-899X/1075/1/012033

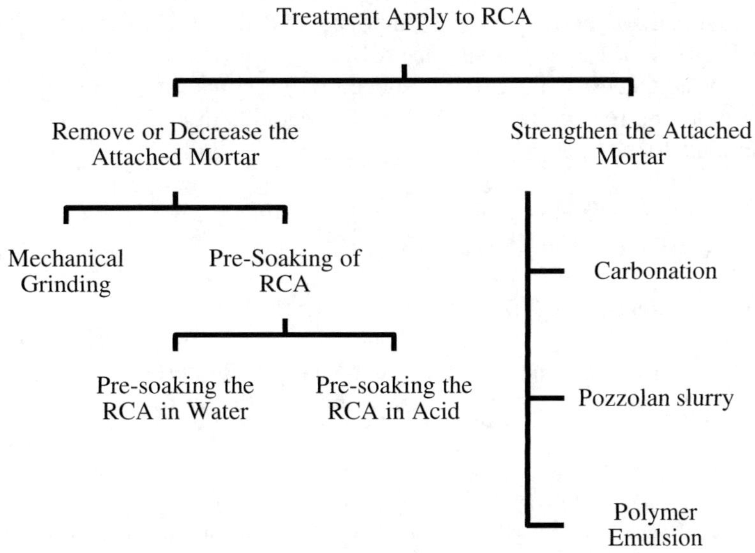

Figure 2. Summary of treatments apply to the RCA

Two common procedures are used to remove or decrease the amount of attached mortar before using the RCA as aggregate in the mixtures. These procedures are follows:

Mechanical Grinding: The principle of this method uses traditional grinding. Rolling vibration with a high-speed rotating eccentric gear in a grinding mill affects the attached mortar, which is ground by this action. It can be further modified by improving the eccentric gear to gain a higher speed, and thus to improve the attached mortars and the quality of the RCA [12].

Heating the RCA before grinding it makes removing the mortar easier. According to the study, recycled concrete aggregate was heated at around 300 °C to dehydrate the attached mortar and make it more brittle, before grinding in a grinder. Generally, the amount of removed mortar increases with increase of the temperature, whereas, when the temperature is higher than 500 °C, the properties of the RCA will be degraded [13].

Pre-soaking the RCA in water can remove some poor mortar from the RCA surface and separate out impurities [14]. In this method, stronger mortar cannot be removed.

Pre-soaking the RCA in acid can help to dissolve the cement. Therefore, acid solution can be used to remove the attached mortar and improve the properties of RCA [15]. The most effective acid is hydrochloric acid HCl, whereas sulphuric acid H_2SO_4 tends to crystallise and phosphoric acid H3PO4 is unstable. Soaking RCA in HCl (0.1 mol) for 24 hours improves the water absorption value [15]. However, this method increases the construction cost.

Strengthen the Attached Mortar there are many chemical materials that can improve the properties of attached mortar by filling the weak areas, but these increase the construction cost. These materials are as follows:

Polymer emulsion, which has adhesive properties and can solidify in a short period of time, can be used. Polyvinyl alcohol (PVA) emulsion and silane-based polymers can be used to decrease the sensitivity of RCA to water. When the RCA is soaked in polymer emulsion, the polymer molecules fill the pores of the attached mortar and seal the surface of the RCA [16]. The water absorption of the RCA decreases with increased PVA concentration and 10% of PVA is an adequate concentration. Polymer treatment can improve the water sensitivity for RCA, but it cannot improve its compressive strength [17].

Pozzolan slurry can be used to improve the porosity of the mortar attached on the surface of the RCA. When the RCA is immersed in or sprayed with pozzolanic slurry that contains silica fumes, the fumes fill the pores and voids in the attached mortar [18].
Moreover, the porosity of attached mortar can be reduced by carbonation. CO_2 can pass through the pores in the attached mortar and react with calcium hydroxide and hydrated calcium silicate which are the main hydration elements in the mortar [19].

$$Ca(OH)_2 + CO_2 \rightarrow CaCO_3 + H_2O \quad (1)$$
$$C\text{-}S\text{-}H + CO_2 \rightarrow CaCO_3 + SiO_2 + nH_2O \quad (2)$$

The carbonation increases the solid volume of calcium hydroxide by 11.5% and C-S-H by 23.1% [20]

Treatment Applied to the RCA and Evaluation of Asphalt Mixtures Properties

Several researchers have studied the use of treatments to improve the performance of asphalt mixtures containing RCA:

i. Pre-Coat Recycled Concrete Aggregate with Slag Cement

Slag cement is used for coating the recycled concrete aggregate with a coating thickness of 0.25, 0.45 and 0.65 mm to reinforce the ability of the RCA. The optimum thickness was 0.25 mm, which improves the Los Angeles abrasion value, increases the stiffness of the mix and improves the permanent deformation resistance of the mix. However, the treatment has a negative effect on the moisture sensitivity of the mixture [21].

ii.Coating Recycled Concrete Aggregate with a Liquid Silicone Resin

In order to improve the Recycled Concrete Aggregate, the coarse aggregate was soaked in liquid silicone resin for 1 hour and kept in an oven at 60 °C for 24 hours. This treatment improved the strength and absorption of the aggregate and increased the moisture damage resistance. However, the treatment has a negative effect on the permanent deformation resistance of the mixture [8].

iii.Keeping the Mixture in the Oven for Several Hours at Mixing Temperature Prior to Compaction

The mixture was kept in the oven for 0, 2 and 4 hours at mixing temperature prior to compaction. This treatment increases the resilient modulus value and makes the mixtures stiffer [22].

iv.Coating Recycled Concrete Aggregate with Bitumen Emulsion

In order to improve the properties of the RCA, it was coated with 5% bitumen emulsion prior to the mixing process. This treatment improved the moisture damage resistance and made the mixture more homogeneous [22].

methodology

Two types of aggregate, granite and recycled concrete aggregate were used in this study. The most important properties were determined for both aggregate types. The mechanical and chemical properties of recycled concrete aggregate were calculated for each fraction size separately. The mechanical properties of the aggregate are shown in table 1. The Flakiness Index, Elongation Index and Soundness Test met the specification requirements. The Soundness Test value for RCA was higher than for granite. The soundness value of RCA is higher than granite by 70 percent.

The results for the Angularity Number Test and Particle Shape Index showed that RCA was more angular than granite, and angularity for RCA grew with decreasing aggregate particle size.

Table 1. The Mechanical and Physical Properties of Aggregat

Aggregate size (mm) passing-retained	Test		Specification
	Los Angeles Abrasion value (%)		ASTM C131
	Granite	RCA	
19-9.5	20	35	<30 %
9.5-4.75	20	36	
4.75-2.36	19	28	
Test	Bulk specific gravity		ASTM C127
20-14	2.5	2.22	
14-10	2.55	2.32	
10-5	2.601	2.38	
5-3.35	2.618	2.36	
3.35-1.18	2.621	2.285	
1.18-0.075	2.66	2.37	
Test	Absorption (%)		ASTM C127
20-14	0.6	4	< 2%
14-10	0.5	6	
10-5	0.8	7	
5-3.35	0.9	10	
3.35-1.18	0.7	12	
1.18-0.075	1.2	18	
Test	Flakiness Index and Elongation Index		BS 812: Part 3
Flakiness Index	18.1	16	20 max
Elongation Index	7	5	20 max
	Soundness Test (%)		ASTM C88
19-9.5	2.14	3.7	< 20 %
9.5-4.75	2.3	3.9	
4.75-2.36	2.43	4.1	
2.36-1.18	2.55	4.36	
	9.42	16.06	
	Impact Test (%)		BS 812: Part III
Impact test	11	26	15% Max
	Angularity number		MS 7.5
19-12.5	5.3	7.1	
12.5-9.5	6.4	11.6	
9.5-4.75	9.3	15.8	
	Particle Shape Index		ASTM D3398
19-12.5	11	15	
12.5-9.5	11	19	
9.5-4.75	14	18	

IOP Conf. Series: Materials Science and Engineering 1075 (2021) 012033 doi:10.1088/1757-899X/1075/1/012033

2. Methodology

In this study three types of treatment were used to improve RCA properties. Theses treatment are soak RCA in HCL acid for 24 hours, grind RCA in Los Angeles Abrasion machine, re-grind RCA particle in jaw crush machine. In order to evaluate each treatment the aggregate properties are determined. all properties were determined for RCA passing and retained sieve size of 5 mm and 1.18 mm respectively.

RCA was soaked in HCL acid at 0.1 M concentration for 24 hours. RCA acid was used for two reasons. The first reason was to remove or decrease the attached mortar amount to improve RCA particles properties. The second reason was this acid is one of the least harmful sturdy acids to handle regardless of its acidity and it has the non-volatile and harmless chloride ion.

One of the methods which used to decrease amount of attached mortar is putting RCA in Los Angeles Abrasion machine In this study 5000 g of RCA was poured in the machine with 11 standard steel balls and operated 150 revolution. 150 revolutions were used to not effect on the aggregate particles and just decrease the attached mortar.

The last treatment which used in this study was using jaw crusher machine. After primary crushing of RCA in jaw crusher, the RCA which retained sieve size of 5 mm is crushed again in jaw crusher. In this method the attracted mortar where removed

Result and Discussion

In order to evaluate each treatment, the most critical properties after treatment are determined. The first property which determined was aggregate specific gravity the result is shown in figure 3. It can be noted the specific gravity improved for treatment2 and 3 that is because of attached mortar was removed whereas the specific gravity decreased after soaking RCA in HCL acid for 24 hours. That is because of interaction between aggregate and HCl acid.

The second property which determined was aggregate absorption value. Figure 4 shows absorption value of each treatment. The absorption was decreased by applying each treatment. The best treatment is re-crushing the RCA in jaw crusher. The absorption value decreased by 31%, 74% and 75% by applying treatment 1, 2 and 3 respectively.

The third property which determined was aggregate abrasion value. Figure 5 shows abrasion value of each treatment. The abrasion was decreased by applying each treatment. The best treatment is soaking RCA HCL acid. The abrasion value decreased by 11%, 7% and 3% by applying treatment 1, 2 and 3 respectively.

The last property which determined was aggregate Angularity number. Figure 6 shows Angularity number of each treatment. The Angularity number was increased by applying treatment 1 and 3 whereas the Angularity number was decreased by applying treatment 2. that RCA particles becomes more rounded . The best treatment is soaking RCA HCL acid. The abrasion value decreased by 11%, 7% and 3% by applying treatment 1, 2 and 3 respectively.

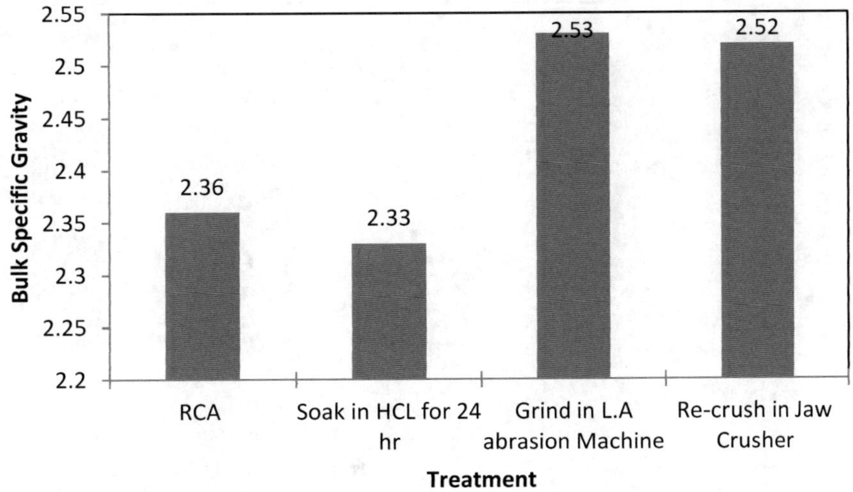

Figure. 3. Bulk Specific Gravity after Each Treatment

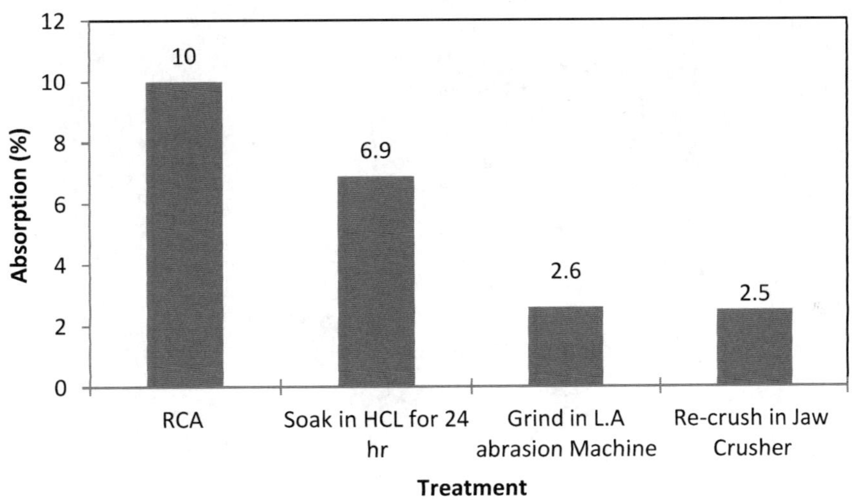

Figure. 4. Absorption value after Each Treatment

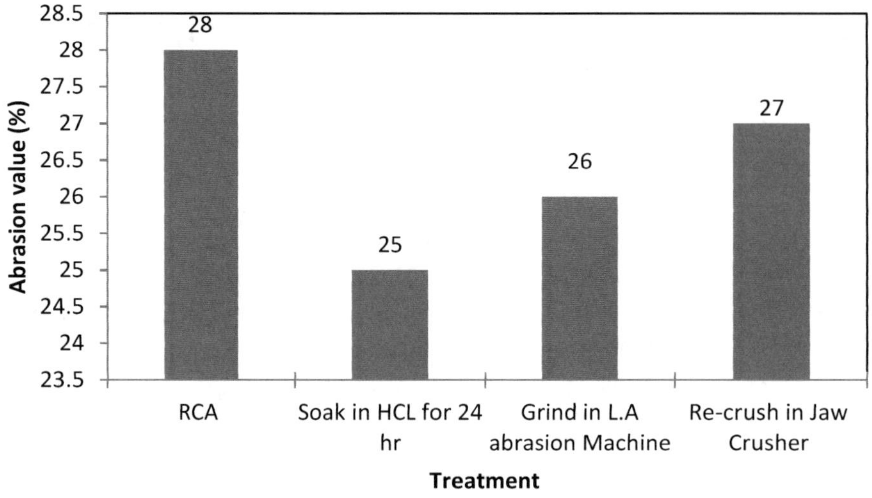

Figure. 5. Abrasion value after Each Treatment

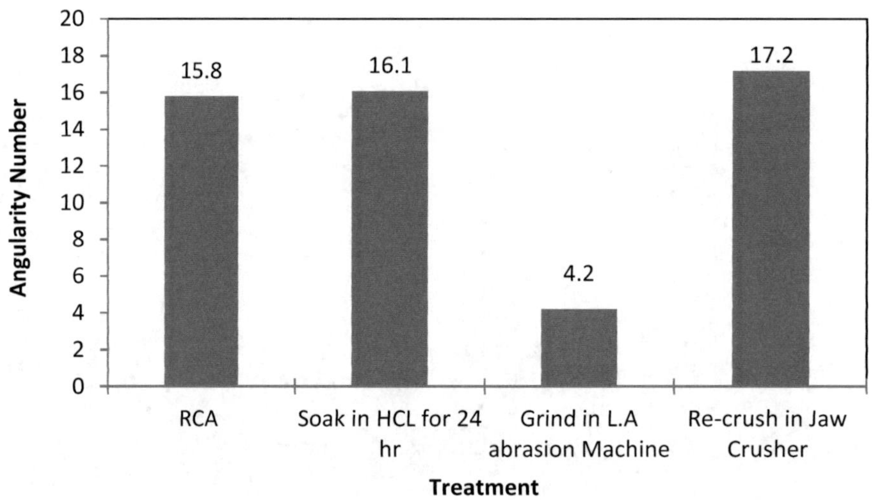

Figure 6. Angularity Number after Each Treatment

3 Conclusion

i. All treatments have significant effect to decrease Absorption value of RCA by decreasing attached mortar amount.
ii. All treatments have significant effect to decrease Abrasion value of RCA by decreasing attached mortar amount.
iii. Soaking RCA in HCL acid has side effect on the specific gravity of aggregate
iv. Putting RCA in Los Angeles Abrasion machine make RCA particles more rounded
v. crush RCA in jaw crusher machine for second is best way to improve RCA properties.

References

[1] Sang YM, Gu QB, Sun TC, Li FS. Color and organic compounds removal from secondary effluent of landfill leachate with a novel inorganic polymer coagulant. Water Sci Technol 2008;58:1423–32.

[2] Robinson GR, Menzie WD, Hyun H. Recycling of construction debris as aggregate in the Mid-Atlantic Region, USA. Resour. Conserv. Recycl., vol. 42, 2004, p. 275–94.

[3] Tahmoorian F, Samali B, Tam VWY, Yeaman J. Evaluation of mechanical properties of recycled material for utilization in asphalt mixtures. Appl Sci 2017;7.

[4] Gul WA, Guler M. Rutting susceptibility of asphalt concrete with recycled concrete aggregate using revised Marshall procedure. Constr Build Mater 2014;55:341–9.

[5] Pourtahmasb MS, Karim MR. Utilization of Recycled Concrete Aggregates in Stone Mastic Asphalt Mixtures. Adv Mater Sci Eng 2014;2014.

[6] Abdulkareem-Alsultan G, Asikin-Mijan N, Lee H V., Rashid U, Islam A, Taufiq-Yap YH. A Review on Thermal Conversion of Plant Oil (Edible and Inedible) into Green Fuel Using Carbon-Based Nanocatalyst. Catalysts 2019;9:350.

[7] Huang Y, Bird RN, Heidrich O. A review of the use of recycled solid waste materials in asphalt pavements. Resour Conserv Recycl 2007;52:58–73.

[8] Zhu J, Wu S, Zhong J, Wang D. Investigation of asphalt mixture containing demolition waste obtained from earthquake-damaged buildings. Constr Build Mater 2012;29:466–75.

[9] Arabani M, Azarhoosh AR. The effect of recycled concrete aggregate and steel slag on the dynamic properties of asphalt mixtures. Constr Build Mater 2012;35:1–7.

[10] Mehta PK. Greening of the Concrete Industry for Sustainable Development. Concr Int 2002;24:23–8.

[11] Abdulkareem-Alsultan G, Asikin-Mijan N, Mustafa-Alsultan G, Lee H V., Wilson K, Taufiq-Yap YH. Efficient deoxygenation of waste cooking oil over Co_3O_4–La_2O_3-doped activated carbon for the production of diesel-like fuel. RSC Adv 2020;6:17423–9.

[12] Montgomery D. Workability and compressive strength properties of concrete containing recycled concrete aggregate. Sustain Constr Use Recycl Concr Aggreg 1998:287–96.

[13] Tateyashiki. H., Shima H, Matsumoto Y, Koga Y. Properties of concrete with high. Proc JCI 2001;23:61–6.

[14] Katz A. Treatments for the Improvement of Recycled Aggregate. J Mater Civ Eng 2004;16:597–603.

[15] Tam VWY, Tam CM, Le KN. Removal of cement mortar remains from recycled aggregate using pre-soaking approaches. Resour Conserv Recycl 2007;50:82–101.

[16] Kou SC, Poon CS. Properties of concrete prepared with PVA-impregnated recycled concrete aggregates. Cem Concr Compos 2010;32:649–54.

[17] Shi C, Li Y, Zhang J, Li W, Chong L, Xie Z. Performance enhancement of recycled concrete aggregate - A review. J Clean Prod 2016;112:466–72.

[18] Kou SC, Poon CS, Wan HW. Properties of concrete prepared with low-grade recycled aggregates. Constr Build Mater 2012;36:881–9.

[19] Groves GW, Brough A, Richardson IG, Dobson CM. Progressive Changes in the Structure of Hardened C3S Cement Pastes due to Carbonation. J Am Ceram Soc 1991;74:2891–6.

[20] Thiery M, Villain G, Dangla P, Platret G. Investigation of the carbonation front shape on cementitious materials: Effects of the chemical kinetics. Cem Concr Res 2007;37:1047–58.

[21] Lee CH, Du JC, Shen DH. Evaluation of pre-coated recycled concrete aggregate for hot mix asphalt. Constr Build Mater 2012;28:66–71.

[22] Pasandín AR, Pérez I. Effect of ageing time on properties of hot-mix asphalt containing recycled concrete aggregates. Constr Build Mater 2014;52:284–93.

AUTHOR INDEX